Stochastic Hybrid Systems

edited by
Christos G. Cassandras
Boston University

John Lygeros
ETH Zürich, Switzerland

CRC is an imprint of the Taylor & Francis Group,
an informa business

CRC Press
Taylor & Francis Group
6000 Broken Sound Parkway NW, Suite 300
Boca Raton, FL 33487-2742

Printed in the United States of America on acid-free paper
10 9 8 7 6 5 4 3 2 1

International Standard Book Number-10: 0-8493-9083-4 (Hardcover)
International Standard Book Number-13: 978-0-8493-9083-8 (Hardcover)

Library of Congress Cataloging-in-Publication Data

Cassandras, Christos G.
Stochastic hybrid systems / Christos G. Cassandras and John Lygeros.
p. cm.
Includes bibliographical references and index.
ISBN-13: 978-0-8493-9083-8 (alk. paper)
ISBN-10: 0-8493-9083-4 (alk. paper)
1. Stochastic systems. 2. Control theory. I. Lygeros, John. II. Title.

QA274.2.C37 2007
003'.76--dc22 2006025156

Visit the Taylor & Francis Web site at
http://www.taylorandfrancis.com

and the CRC Press Web site at
http://www.crcpress.com

Preface

This book gives a tempting glimpse at what is arguably the most ambitious type of dynamic systems that have been studied to date: stochastic hybrid systems. Stochastic hybrid systems combine time-driven and event-driven dynamics and incorporate the ubiquitous uncertainty elements within which a system must operate.

Problems in mathematical finance such as pricing of options and insurance were among the first and very successful areas of application of stochastic hybrid methods. More recently, however, the great importance of stochastic hybrid systems in engineering has also been widely recognized. Stochastic hybrid systems are among the most common technological creations of modern society: every time a physical process interacts with computerized equipment in an uncertain environment, a stochastic hybrid system is present. In modern automobiles, the physical processes involved in the engine, brakes, or functions such as climate control and remote access are subject to computerized controllers responsible for overall proper coordination and response to unexpected occurrences. The same can be said of everyday devices such as photocopiers, printers, and even computers themselves. Communication networks, manufacturing systems, and air traffic management are other examples of technological environments where the same combination of physical (time-driven) processes coordinated by computerized (event-driven) equipment in the presence of uncertainty arises. More recently, many biological processes have also been cast into the stochastic hybrid setting, opening up an exciting new way of viewing these processes and, potentially, controlling them.

Following the usual scientific path in the study of dynamic systems, the book begins with models for stochastic hybrid systems. Based on these models, one can then develop specific analysis and synthesis techniques. In layman's terms, one can describe in plain English how a system ought to behave in order to meet some specifications. For example, a manufacturing process should produce x items of a given type, each meeting a quality criterion y and each being delivered within t minutes from being ordered with probability q. To implement this design specification, one needs to develop a precise model of the manufacturing process as a stochastic hybrid system, to incorporate discrete events such as the completion of an item, time-driven dynamics, and the uncertainty inherent in the manufacturing process. One then needs to apply a particular set of design techniques geared toward stochastic hybrid systems, to synthesize controllers achieving the desired goals.

The material contained in the first part of the book provides the underlying principles behind this process and a rigorous understanding of the basic design limitations within which one must operate. Building on this fundamental exposition, the second part of the book presents methods for implementing on a computer the calculations necessary for applying stochastic hybrid systems analysis and synthesis techniques in practice. Finally, the book casts examples of systems encountered in a wide range of application areas into the stochastic hybrid systems framework and explains how one can resolve practical problems associated with these systems.

The universal nature of stochastic hybrid systems makes the target audience of this book unusually broad. The use of stochastic hybrid systems has gained particular appeal for designers and managers of communication networks and automated manufacturing systems. It has also been associated with air traffic management and, very recently, an interest in better understanding biological processes. However, given the systems and control flavor of the material, the target audience is more likely to be concentrated among academic researchers and R&D personnel in industry with a background and interest in systems and control engineering and, to a lesser extent, computer science. Specific examples include those working in settings that rely on emerging embedded system technologies including automotive engineering, aerospace, digital signal processing, and automated manufacturing.

Acknowledgments. The editors of this volume are grateful to all the contributing authors for their exciting and timely contributions. They would also like to thank Badis Djeridane for his comments on early drafts, Nora Konopka and Helena Redshaw of Taylor & Francis for their editorial guidance, and the CRC LaTex help desk for their prompt assistance with the numerous formatting problems encountered during the production of the volume. The editorial work was supported by the European Commission, under the project HYGEIA, FP6-NEST-04995, by the U.S. National Science Foundation under grant DMI-0330171, and by the U.S. Air Force Office of Scientific Research under grants FA9550-04-1-0133 and FA9550-04-1-0208.

Christos G. Cassandras
Boston

John Lygeros
Zürich

September 19, 2006

About the Editors

Christos G. Cassandras is Professor of Manufacturing Engineering and Professor of Electrical and Computer Engineering at Boston University. He is also co-founder of Boston University's Center for Information and Systems Engineering (CISE). He received degrees from Yale University (B.S., 1977), Stanford University (M.S.E.E., 1978), and Harvard University (S.M., 1979; Ph.D., 1982). In 1982–1984 he was with ITP Boston, Inc. where he worked on the design of automated manufacturing systems. In 1984–1996 he was a faculty member at the Department of Electrical and Computer Engineering, University of Massachusetts/Amherst. He specializes in the areas of discrete event and hybrid systems, stochastic optimization, and computer simulation, with applications to computer and sensor networks, manufacturing systems, and transportation systems. He has published over 200 refereed papers in these areas, and two textbooks. He has guest-edited several technical journal issues and serves on several journal editorial boards. Dr. Cassandras is currently Editor-in-Chief of the *IEEE Transactions on Automatic Control* and has served as Editor for Technical Notes and Correspondence and Associate Editor. He is a member of the IEEE CSS Board of Governors, chaired the CSS Technical Committee on Control Theory, and served as Chair of several conferences. He has been a plenary speaker at various international conferences, including the *American Control Conference* in 2001 and the *IEEE Conference on Decision and Control* in 2002. He is the recipient of several awards, including the Distinguished Member Award of the IEEE Control Systems Society (2006), the 1999 Harold Chestnut Prize (IFAC Best Control Engineering Textbook) for *Discrete Event Systems: Modeling and Performance Analysis*, and a 1991 Lilly Fellowship. He is a member of Phi Beta Kappa and Tau Beta Pi. He is also a Fellow of the IEEE.

John Lygeros completed a B.Eng. degree in electrical engineering in 1990 and an M.Sc. degree in control in 1991, both at Imperial College of Science Technology and Medicine, London. He obtained a Ph.D. in 1996 from the Electrical Engineering and Computer Sciences Department, University of California, Berkeley. In 1996–2000 he held a series of postdoctoral research appointments at the National Automated Highway Systems Consortium, M.I.T., and U.C. Berkeley. In parallel, he also worked as a part-time research engineer at SRI International, Menlo Park, California, and as a Visiting Professor at the Mathematics Department of the Université de Bretagne Occidentale, Brest, France. Between July 2000 and March 2003 he was a University Lecturer at the Department of Engineering, University of Cambridge, Cambridge, U.K., and a Fellow of Churchill College, Cambridge. Between March 2003 and July 2006 he was an Assistant Professor at the Department of Electrical and Computer Engineering, University of Patras, Patras, Greece. Since July 2006 he has been an Associate Professor at the Automatic Control Laboratory, ETH, Zürich, Switzerland. His research interests include modeling, analysis, and control of hierarchical hybrid systems, with applications to biochemical networks and large-scale systems such as automated highways and air traffic management. He is a senior member of the IEEE, and a member of the IEE and the Technical Chamber of Greece.

CONTRIBUTORS

Richelle Adams Department of Electrical & Computer Engineering, Georgia Institute of Technology, Atlanta, GA, U.S.A.

Arunabha Bagchi Department of Applied Mathematics, University of Twente, Enschede, The Netherlands

G.J. (Bert) Bakker National Aerospace Laboratory NLR, Amsterdam, The Netherlands

Henk A.P. Blom National Aerospace Laboratory NLR, Amsterdam, The Netherlands

Christos G. Cassandras Department of Manufacturing Engineering, Boston University, Boston, MA, U.S.A.

João Hespanha Department of Electrical & Computer Engineering, University of California, Santa Barbara, CA, U.S.A.

Jianghai Hu Department of Electrical Engineering & Computer Science, Purdue University, West Lafayette, IN, U.S.A.

Joost-Pieter Katoen Department of Computer Science, RWTH Aachen, Germany

Margriet B. Klompstra National Aerospace Laboratory NLR, Amsterdam, The Netherlands

Panagiotis Kouretas Department of Electrical & Computer Engineering, University of Patras, Rio, Patras, Greece

Konstantinos Koutroumpas Automatic Control Laboratory, ETH Zürich, Switzerland

Jaroslav Krystul Department of Applied Mathematics, University of Twente, Enschede, The Netherlands

John Lygeros Automatic Control Laboratory, ETH Zürich, Switzerland

Zoi Lygerou School of Medicine, University of Patras, Rio, Patras, Greece

Bart Klein Obbink National Aerospace Laboratory NLR, Amsterdam, The Netherlands

Maria Prandini Dipartimento di Elettronica e Informazione, Politecnico di Milano, Milano, Italy

George Riley Department of Electrical & Computer Engineering, Georgia Institute of Technology, Atlanta, GA, U.S.A.

Arjan van der Schaft Institute for Mathematics and Computing Science, University of Groningen, Groningen, The Netherlands

Stefan Strubbe Department of Applied Mathematics, University of Twente, Enschede, The Netherlands

Yorai Wardi Department of Electrical & Computer Engineering, Georgia Institute of Technology, Atlanta, GA, U.S.A.

Contents

1 Stochastic Hybrid Systems: Research Issues and Areas **1**
Christos G. Cassandras and John Lygeros
1.1 Introduction 1
1.1.1 The Origin of Hybrid Systems 1
1.1.2 Deterministic and Non-deterministic Hybrid Systems 3
1.1.3 Stochastic Hybrid Systems 3
1.2 Modeling of Non-deterministic Hybrid Systems 4
1.3 Modeling of Stochastic Hybrid Systems 7
1.4 Overview of this Volume 9

2 Stochastic Differential Equations on Hybrid State Spaces **15**
Jaroslav Krystul, Henk A.P. Blom, and Arunabha Bagchi
2.1 Introduction 15
2.2 Semimartingales and Characteristics 18
2.3 Semimartingale Strong Solution of SDE 22
2.4 Stochastic Hybrid Processes as Solutions of SDE 27
2.5 Instantaneous Hybrid Jumps at a Boundary 31
2.6 Related SDE Models on Hybrid State Spaces 33
2.6.1 Stochastic Hybrid Model GB1 of Ghosh and Bagchi 34
2.6.2 Stochastic Hybrid Model GB2 of Ghosh and Bagchi 36
2.6.3 Hierarchy Between Stochastic Hybrid Models 38
2.7 Markov and Strong Markov Properties 40
2.8 Concluding Remarks 43

3 Compositional Modelling of Stochastic Hybrid Systems **47**
Stefan Strubbe and Arjan van der Schaft
3.1 Introduction 47
3.2 Semantical Models 48
3.2.1 Transition Mechanism Structure 49
3.2.2 Continuous Flow Spontaneous Jump System (CFSJS) 50
3.2.3 Forced Transition Structure (FTS) 53
3.2.4 CFSJS Combined with FTS 53
3.2.5 Non-deterministic Transition System (NTS) 54
3.3 Communicating PDPs 55
3.3.1 Definition of the CPDP Model 56
3.3.2 Semantics of CPDPs 60

3.3.3 Composition of CPDPs 61
3.3.4 Value Passing CPDPs 68
3.4 Conclusions 75

4 Stochastic Model Checking **79**
Joost-Pieter Katoen
4.1 Introduction 79
4.1.1 Stochastic Model Checking 80
4.1.2 Topic of this Survey 81
4.2 The Discrete-time Setting 81
4.2.1 Discrete-time Markov Chains 82
4.2.2 Rewards 84
4.3 The Continuous-time Setting 87
4.3.1 Continuous-time Markov Chains 87
4.3.2 Rewards 89
4.3.3 Time-inhomogenity 92
4.4 Bisimulation and Simulation Relations 94
4.4.1 Strong Bisimulation 94
4.4.2 Weak Bisimulation 95
4.4.3 Strong Simulation 97
4.4.4 Logical Characterization 98
4.5 Epilogue 100
4.5.1 Summary of Results 100
4.5.2 Further Research Topics 102

5 Stochastic Reachability: Theory and Numerical Approximation **107**
Maria Prandini and Jianghai Hu
5.1 Introduction 107
5.2 Stochastic Hybrid System Model 109
5.3 Reachability Problem Formulation 114
5.4 Numerical Approximation Scheme 116
5.4.1 Markov Chain Approximation 116
5.4.2 Locally Consistent Transition Probability Functions 123
5.5 Reachability Computations 124
5.6 Possible Extensions 128
5.6.1 Probabilistic Safety 128
5.6.2 Regulation 129
5.7 Some Examples 130
5.7.1 Manufacturing System 131
5.7.2 Temperature Regulation 133
5.8 Conclusions 134

6 Stochastic Flow Systems: Modeling and Sensitivity Analysis **139**
Christos G. Cassandras
6.1 Introduction 139

6.2 Modeling Stochastic Flow Systems 142
6.3 Sample Paths of Stochastic Flow Systems 146
6.4 Optimization Problems in Stochastic Flow Systems 148
6.5 Infinitesimal Perturbation Analysis (IPA) 150
6.5.1 Single-Class Single-Node System 150
6.5.2 Multi-node Tandem System 155
6.6 Conclusions 164

7 Perturbation Analysis for Stochastic Flow Systems with Feedback 169
Yorai Wardi, George Riley, and Richelle Adams
7.1 Introduction 169
7.2 SFM with Flow Control 171
7.3 Retransmission-based Model 178
7.4 Simulation Experiments 186
7.5 Conclusions 188

8 Stochastic Hybrid Modeling of On-Off TCP Flows 191
João Hespanha
8.1 Related Work 193
8.1.1 Models for Long-lived Flows 193
8.1.2 Models for On-Off Flows 197
8.2 A Stochastic Model for TCP 199
8.3 Analysis of the TCP SHS Models 203
8.4 Reduced-order Models 204
8.4.1 Long-lived Flows 205
8.4.2 Mixed-exponential Transfer-sizes 206
8.5 Conclusions 211

9 Stochastic Hybrid Modeling of Biochemical Processes 221
Panagiotis Kouretas, Konstantinos Koutroumpas, John Lygeros, and Zoi Lygerou
9.1 Introduction 221
9.2 Overview of PDMP 223
9.2.1 Modeling Framework 223
9.2.2 Simulation 226
9.3 Subtilin Production by *B. subtilis* 228
9.3.1 Qualitative Description 228
9.3.2 An Initial Model 228
9.3.3 A Formal PDMP Model 229
9.3.4 Analysis and Simulation 234
9.4 DNA Replication in the Cell Cycle 235
9.4.1 Qualitative Description 235
9.4.2 Stochastic Hybrid Features 237
9.4.3 A PDMP Model 239
9.4.4 Implementation in Simulation and Results 243

9.5 Concluding Remarks 244

10 Free Flight Collision Risk Estimation by Sequential MC Simulation 249
Henk A.P. Blom, Jaroslav Krystul, G.J. (Bert) Bakker, Margriet B. Klompstra, and Bart Klein Obbink
10.1 Introduction 249
10.1.1 Safety Verification of Free Flight Air Traffic 249
10.1.2 Probabilistic Reachability Analysis 250
10.1.3 Sequential Monte Carlo Simulation 251
10.1.4 Development of MC Simulation Model 252
10.2 Sequential MC Estimation of Collision Risk 253
10.2.1 Stochastic Hybrid Process Considered 253
10.2.2 Risk Factorisation Using Multiple Conflict Levels 255
10.2.3 Characterisation of the Risk Factors 256
10.2.4 Interacting Particle System Based Risk Estimation 257
10.2.5 Modification of IPS Resampling Step 4 258
10.3 Development of a Petri Net Model of Free Flight 259
10.3.1 Specification of Petri Net Model 259
10.3.2 High Level Interconnection Arcs 260
10.3.3 Agents and LPNs to Represent AMFF Operations 261
10.3.4 Interconnected LPNs of ASAS 264
10.3.5 Interconnected LPNs of "Pilot Flying" 265
10.3.6 Model Verification, Parameterisation, and Validation 268
10.3.7 Dimensions of MC Simulation Model 269
10.4 Simulated Scenarios and Collision Risk Estimates 271
10.4.1 Parameterisation of the IPS Simulations 271
10.4.2 Eight Aircraft on Collision Course 271
10.4.3 Free Flight Through an Artificially Constructed Airspace . . 273
10.4.4 Reduction of the Aircraft Density by a Factor Four 274
10.4.5 Discussion of IPS Simulation Results 275
10.5 Concluding Remarks 276

Index 283

Chapter 1

Stochastic Hybrid Systems: Research Issues and Areas

Christos G. Cassandras
Boston University

John Lygeros
ETH Zürich

1.1 Introduction 1
1.2 Modeling of Non-deterministic Hybrid Systems 4
1.3 Modeling of Stochastic Hybrid Systems 7
1.4 Overview of this Volume 9
References 12

1.1 Introduction

1.1.1 The Origin of Hybrid Systems

Historically, scientists and engineers have concentrated on studying and harnessing natural phenomena which are modeled by the laws of gravity, classical and non-classical mechanics, physical chemistry, etc. In so doing, one typically deals with quantities such as the displacement, velocity, and acceleration of particles and rigid bodies, or the pressure, temperature, and flow rates of fluids and gases. These are continuous variables in the sense that they can take on any value as time itself continuously evolves. Based on this fact, a vast body of mathematical tools and techniques has been developed to model, analyze, and control these time-driven systems around us. But in the day-to-day life of our technological and increasingly computer-dependent world, we notice two things: First, many of the quantities we deal with are discrete; and second, what drives many of the processes we use and depend on are instantaneous "events" such as pushing a button, or hitting a keyboard key. In fact, much of the technology we have invented and rely on is event-driven: Communication networks, manufacturing facilities, or the execution of a computer program are typical examples. This has motivated the development of a theory for Discrete Event Systems (DES) [8], mostly during the 1980s, leading to new modeling frameworks, analysis techniques, design tools, testing methods, and systematic control and optimization procedures for this new generation of event-driven systems.

By the mid 1990s a natural merging of the classical time-driven with the new event-driven systems took place, giving rise to the so-called Hybrid Systems (HS). By its very nature, the study of hybrid systems has evolved as the merging of these two complementary points of view of dynamic systems.

On one hand, a hybrid system may be viewed as an extension of a classical *time-driven* system, typically modeled through differential or difference equations, with occasional discrete events causing a change in its dynamic behavior. This change may preserve the continuity of the state variables characterizing the system, or it can cause discontinuities. When such an event takes place, the system is thought of as switching from one operating *mode* to another. The precise nature and timing of the events can dramatically affect the behavior of the system, especially when events are controllable and may be chosen from a given set, adding a combinatorial dimension to the analysis.

On the other hand, our starting point may be a purely *event-driven* system, typically modeled through a state automaton or Petri net with states belonging to a discrete (finite or countable) set. In such a system, each state is simply labeled by a symbol (e.g., a non-negative integer). One, however, can replace this simple labeling by associating to a discrete state a set of differential (or difference) equations describing the evolution of a time-driven system. In this case, a state automaton, for example, is enriched by the inclusion of a time-driven system component behavior incorporated within each of its states. This naturally leads to a modeling framework based on *hybrid automata*.

As an illustration of these two points of view, consider an automobile engine where the power train and air dynamics are time-driven processes described through differential equations: As time evolves, variables such as pressure or temperature continuously vary. However, events such as changing gears or sensing wheel slippage that engages an anti-lock braking action cause switches in the operating mode of this system. Thus, the various modes along with the associated transition mechanisms define an event-driven system co-existing with the continuous-time engine model. The presence of multiple sensors and actuators networked in an automobile and controlled by "embedded" microprocessors responding to or triggering various events naturally leads to the image of a modern car as a hybrid system. On the other hand, a manufacturing workstation is often viewed as a queuing system where parts arrive, wait to be processed, and then depart. In this viewpoint, processing in the workstation simply causes parts to be delayed by some amount of time. If, however, one chooses to explicitly account for the physical process that each part undergoes, then a hybrid system arises where events such as starting and completing processing of a part cause switches from one mode of operation to another, depending on the specific part and corresponding equipment settings.

1.1.2 Deterministic and Non-deterministic Hybrid Systems

Much of the work on hybrid systems has focused on *deterministic* models that completely characterize the future of the system without allowing any uncertainty. In practice, it is often desirable to introduce some levels of uncertainty in the models, to allow, for example, under-modeling of certain parts of the system. To address this need, researchers in discrete event and hybrid systems have introduced what are known as *non-deterministic* models. Here the evolution is defined in a *declarative* way (the system specifies what solutions are allowed) as opposed to the *imperative* way more common in time-driven dynamical systems (the system specifies what the solution must be). Non-deterministic hybrid systems allow uncertainty to enter in a number of places: choice of continuous evolution (modeled, for example, by a differential inclusion), choice of discrete transition destination, or choice between continuous evolution and a discrete transition. "Choice" in this setting may reflect disturbances that add uncertainty about the system evolution, but also control inputs that can be used to steer the system evolution.

Deterministic and non-deterministic hybrid systems have been a topic of intense research in recent years. Embedded systems were a key motivation for the study of these systems, since they involve, by their very nature, the interaction of digital devices with an analog environment. Deterministic and non-deterministic hybrid models are very versatile, can capture a wide range of behaviors encountered in practice and have proved invaluable in a number of applications (among them automated highways, automotive control, air traffic management, manufacturing, and robotics; see [2] for an overview). They do, however, have their limitations that make them too "coarse" for certain applications. In particular, non-deterministic systems provide no way of distinguishing between solutions, such as whether one is more likely than another. This implies that only worst case analysis is possible with non-deterministic hybrid systems; one can only pose *qualitative*, yes-no type questions. For example, a safety question for non-deterministic hybrid systems admits only one of two answers: "The system is safe" (if none of the solutions of the system ever reaches an unsafe state), or "the system is not safe" (if some solution reaches some unsafe state).

In some applications this type of analysis may be too coarse. For example, in Air Traffic Management (ATM) the question "Is it possible for a fatal accident to happen in the ATM system?" may be interesting, but the answer (which is most likely "yes") does not convey nearly as much information as the answers to the questions "What is the probability that a fatal accident happens in the ATM system?" and "How can the probability of a fatal accident be reduced?" This provides the motivation for developing explicit means that can provide *quantitative*, hence more useful, answers to questions such as these through stochastic models of hybrid systems.

1.1.3 Stochastic Hybrid Systems

The need for finer, probabilistic analysis of uncertain systems has led to the study of an even wider class of hybrid systems, that allow things such as random failures causing unexpected transitions from one discrete state to another, or random task execution times which affect how long the system spends in different modes. For example, the events in a hybrid system may be controllable (e.g., deciding to switch gears when driving a car) or uncontrollable (e.g., some equipment failure). Uncontrollable events may occur at random points in time, in which case the hybrid system becomes *stochastic*. Randomness may also enter the picture through noise in one or more time-driven components of the system, in which case we must resort to stochastic differential equations. In this case, if a mode switch is the result of a continuous variable reaching a certain level (e.g., a tank containing fluid whose level exceeds a specific value), then the random fashion in which this variable evolves in time affects the associated switching event.

To allow the realistic modeling of such phenomena, researchers extended their study of hybrid systems beyond continuous and discrete dynamics, to include probabilistic phenomena. This has led to the more general class of Stochastic Hybrid Systems (SHS), which have found applications to, among others, insurance pricing [12], capacity expansion models for the power industry [11], flexible manufacturing, and fault tolerant control [14]. A number of chapters on some more recent applications of SHS are included in the present volume.

In this opening chapter, we give an introduction to non-deterministic and stochastic hybrid systems to set the stage for the subsequent chapters that present key advances in the theory, computational methods, and application areas of stochastic hybrid systems. We start with a high level view of non-deterministic hybrid system modeling (Section 1.2) to highlight the different areas where uncertainty can enter hybrid system evolution. We then discuss how different classes of stochastic hybrid systems replace these sources of uncertainty by appropriate probabilistic phenomena (Section 1.3). We conclude with an overview of the remaining chapters of the book, highlighting the key points of each (Section 1.4).

1.2 Modeling of Non-deterministic Hybrid Systems

We start by giving a high level overview of a general class of non-deterministic hybrid systems. Even though many of the technical details are omitted, the discussion is given in sufficient detail to allow us to highlight the different types of uncertainty that can arise in hybrid systems. We restrict our attention to continuous time hybrid systems; for discrete time hybrid systems the reader is referred to (among others) [4, 16].

The dynamical systems we consider involve the interaction of a continuous state (denoted by x) and a discrete state (commonly referred to as the "mode" and denoted by q). To allow us to capture different types of uncertainty, we assume that the evolution of the state is influenced by two different kinds of inputs: control inputs and disturbance inputs. Similarly to the state, we partition the inputs of each kind into discrete and continuous, and use υ to denote discrete controls, u to denote continuous controls, δ to denote discrete disturbances, and d to denote continuous disturbances.

Four functions determine the evolution of a state: a vector field f that determines the continuous time-driven evolution, a reset map r that determines the outcome of discrete transitions, a function G giving the "guard" sets that determine when discrete transitions can take place, and a function *Dom* giving the "domain" sets that determine when continuous evolution is possible. The following definition formalizes the above description.

DEFINITION 1.1 *A hybrid automaton characterizes the evolution of*

- *discrete state variables* $q \in Q$ *and continuous state variables* $x \in X$*,*
- *control inputs* $\upsilon \in \Upsilon$ *and* $u \in U$*, and*
- *disturbance inputs* $\delta \in \Delta$ *and* $d \in D$

by means of four functions

- *a vector field* $f : Q \times X \times U \times D \to X$*,*
- *a domain set* $\mathrm{Dom} : Q \times \Upsilon \times \Delta \to 2^X$*,*
- *guard sets* $G : Q \times Q \times \Upsilon \times \Delta \to 2^X$*, and*
- *a reset function* $r : Q \times Q \times X \times \Upsilon \times \Delta \to X$*.*

Here 2^X denotes the set of all subsets (power set) of X; in other words, *Dom* and G are set valued maps. We assume that the sets Q, Υ, and Δ are countable and that $X = \mathbb{R}^n$, $U \subseteq \mathbb{R}^m$, and $D \subseteq \mathbb{R}^p$ for integers n, m and p. To avoid pathological situations (for example, lack of solutions, chattering, etc.) one needs to introduce assumptions on the functions f, r, G and *Dom*. We will not go into the details here, we refer the reader to [13] for examples of such well-posedness conditions.

Roughly speaking, the solution of a hybrid automaton (often called a "run" or an "execution") is defined as a sequence of intervals of continuous evolution followed by a discrete transition. Starting at some initial state (q_0, x_0) the continuous state moves along the solution of the differential equation

$$\dot{x} = f(q_0, x, u, d)$$

as long as it does not leave the set $Dom(q_0, \upsilon, \delta)$. The discrete state remains constant throughout this time. If at some point x reaches a set $G(q_0, q', \upsilon, \delta)$ for some $q' \in Q$, a discrete transition can take place. If this transition does take place, the state instantaneously resets to (q', x') where x' is determined by the reset map $r(q, q', x, \upsilon, \delta)$. The process is then repeated. Notice that one can think of changes in υ and δ as discrete events that enable discrete transitions (e.g., a transition from q to q' by making sure $x \in G(q, q', \upsilon, \delta)$) or force discrete transitions (e.g., a transition out of q by making sure $x \notin Dom(q, \upsilon, \delta)$).

Considerable freedom is allowed when defining the solution in this "declarative" way. In particular:

- The direction of the continuous motion at any point in time may not be unique, since it depends on the continuous inputs u and d. One can think of the continuous motion as being described by a differential inclusion

$$\dot{x} \in F(x) = \{f(q,x,u,d) \mid u \in U, d \in D\},$$

  that admits different solutions, depending on the choice of $u(\cdot)$ and $d(\cdot)$.

- The mode after a discrete transition may not be uniquely defined. If, for example,

$$x \in G(q,\hat{q},\upsilon,\delta) \cap G(q,\hat{q}',\upsilon',\delta'),$$

  then from state (q,x) it is possible to transition to either mode $\hat{q}$ (if discrete inputs (υ,δ) are applied) or mode $\hat{q}'$ (if discrete inputs (υ',δ') are applied). It may even be possible to have

$$x \in G(q,\hat{q},\upsilon,\delta) \cap G(q,\hat{q}',\upsilon,\delta),$$

  in which case a choice to transition to either $\hat{q}$ or $\hat{q}'$ exists even for the same discrete inputs (υ,δ).

- The continuous state after a discrete transition may not be uniquely defined, since it depends on the continuous inputs u and d. If from state (q,x) a discrete transition to mode $\hat{q}$ takes place the continuous set can simultaneously change to any value in the set

$$\{r(q,\hat{q},x,u,d) \mid u \in U, d \in D\}.$$

- There may be a choice between continuous evolution and a discrete transition. For example, if we have

$$x \in Dom(q,\upsilon,\delta) \cap G(q,q',\upsilon',\delta'),$$

  then from state (q,x) it may be possible to either evolve continuously in mode q (under discrete inputs (υ,δ)) or to take a discrete transition to mode q' (under discrete inputs (υ',δ')). It may even be possible to have

$$x \in Dom(q,\upsilon,\delta) \cap G(q,q',\upsilon,\delta),$$

  in which case this choice between continuous evolution and discrete mode transition is possible even for the same inputs (υ,δ).

Being able to capture all these aspects of "choice" in the system is generally desirable, since it allows one to model a wide variety of phenomena and include different types of uncertainty. Moreover, this flexibility also allows one to formulate a range of interesting analysis and control problems, among others:

- Stability, stabilization, and robust stabilization problems (respectively, if no inputs, only control inputs (u, υ), or both control inputs (u, υ) and disturbance inputs (d, δ) are present).
- Optimal control and robust control problems (respectively, if control inputs, or both control and disturbance inputs are present).
- Safety, controlled invariance, and hybrid pursuit-evasion problems (respectively, if no inputs, only control inputs, or both control and disturbance inputs are present).

What the non-deterministic framework outlined in this section cannot accommodate is randomness. All the choices listed above are possible under the non-deterministic framework, but there is nothing to distinguish between them. Selecting among these choices leads to different system solutions, but there is no implication that some of these solutions are more likely than others. As a consequence, all analysis and control problems formulated for non-deterministic systems are of the "yes-or-no," "worst case" type. As we have seen, this all or nothing approach may be too coarse for some applications. To show how this difficulty can be alleviated we introduce next the class of *stochastic* hybrid systems.

1.3 Modeling of Stochastic Hybrid Systems

The observation that the inclusion of stochastic terms in the hybrid systems framework may be crucial in some applications has led to a flurry of research activity since the turn of the century into the area that has come to be known as Stochastic Hybrid Systems (SHS). The great interest of the research community in SHS has produced a number of different types of stochastic hybrid models. The main difference between these classes of stochastic hybrid models lies in the way the stochasticity enters the process [21]. Roughly speaking, stochasticity can manifest itself in any of the places where "choice" is possible in non-deterministic hybrid systems: Continuous evolution may be governed by stochastic differential equations, transitions may occur spontaneously at random times (at a given, possibly state-dependent, "rate"), the destinations of discrete transitions may be given by probability kernels on the state space, etc.

It is easy to see that a stochastic hybrid system can acquire an arbitrary level of complexity in terms of the physical processes it encompasses, the operating rules that guide event occurrences, and the stochastic elements involved. Thus, it is natural to try and categorize SHS into classes that describe a sufficiently rich number of processes or applications while preserving some structural properties that facilitate analysis. Two examples of SHS models that are featured in this book are *Piecewise Deterministic Markov Processes* (PDMP) and *Stochastic Fluid Models* (SFM). In the former, the probabilistic nature of mode transitions is assumed to be Marko-

Table 1.1: Overview of stochastic hybrid models.

Characteristics	[9, 10]	[14, 15]	[5, 20]	[19, 22]	[17]	[18]	[6, 7]
Stochastic diff. equation		√	√	√		√	√
Probabilistic resets	√		√		√	√	√
Spontaneous transitions	√	√		√	√		√
Forced transitions	√		√			√	√
Continuous control	√	√	√	√			
Transition rate control	√	√					
Forcing transition control	√		√				
Continuous reset control	√		√				

vian (memoryless), while the time-driven behavior of the system in each mode can be arbitrarily complex, though entirely deterministic. In the latter class, it is the time-driven behavior which is limited to flow dynamics (describing the contents of tanks, reservoirs, buffers, and the like), whereas the probabilistic nature of the flow processes is allowed to be virtually arbitrary.

The situation gets even more complicated if one considers inputs in addition to the stochastic terms. Input variables are necessary in many applications to allow for the introduction of control, or non-deterministic (as opposed to stochastic) disturbances, such as an adversary in a stochastic game. Input variables essentially introduce an extra element of choice into the system and require a modeling formalism that can accommodate both stochastic and non-deterministic features. Similar to the stochastic terms, input variables can enter the continuous motion and the timing and destination of discrete transitions.

Clearly, all these alternatives allow for the formulation of countless variants of modeling, analysis and control problems. Consequently, the literature on SHS is very diverse. Table 1.1 attempts to summarize the modeling choices made in some of the key references in the literature. Notice that the models in the last three columns are autonomous (i.e., do not accommodate any inputs). The table only contains SHS models that evolve in continuous time. Modeling frameworks for discrete time SHS can be found in [3, 1].

1.4 Overview of this Volume

This book comprises a number of cutting edge studies in the rapidly evolving area of SHS and integrates them in a comprehensive manner. In addition to this introduction, the book contains nine chapters, that can roughly be divided into three categories:

1. Theoretical foundations for SHS, including such fundamental issues as deriving unifying modeling frameworks for SHS (Chapter 2) and composition and abstraction frameworks of complex SHS (Chapter 3).

2. Analysis and computational methods for SHS, including model checking techniques (Chapter 4) and the numerical solution of reachability problems (Chapter 5).

3. Applications to areas where SHS are most prominent, including communication networks (Chapters 6–8), biological processes (Chapter 9), and air traffic management (Chapter 10).

It is fair to say, however, that a number of chapters cut across these categories. For example, Chapter 5 contains an extensive discussion of the theoretical foundations of reachability problems for stochastic hybrid systems, while Chapter 10 provides an introduction into computational sequential Monte Carlo methods, that are then used to study the safety of air traffic situations.

In terms of the theoretical foundations of SHS, the discussion in Section 1.3 outlines the wide range of possibilities and alternatives one has to consider when modeling SHS. It also lists some of the different attempts that have been made in the literature to bring these alternatives together. Chapter 2 provides a formal discussion of these points. A general framework for modeling autonomous SHS (i.e., SHS without inputs) is proposed. The framework combines stochastic differential equations, spontaneous and forced transitions and probabilistic rests of the discrete and continuous states. To highlight the generality of the proposed framework, a formal comparison of the descriptive power of different SHS modeling frameworks found in the literature is given. The chapter also presents a series of results to show that, under certain technical assumptions, the stochastic processes defined in this framework are well posed and have desirable properties, such as the strong Markov property. These results are exploited in later chapters of this volume, for example, to ensure the well-posedness of reachability problems (Chapter 5), or to enable the use of powerful sequential Monte Carlo methods (Chapter 10).

Autonomous SHS are not the end of the story, however. To build models of large systems, one typically needs to combine models of simpler components, that can be developed from first principles. To be able to do this, one needs a compositional modeling framework, that allows one to formally compose subsystem models and argue about the properties that the resulting model inherits from its components. Such a modeling framework is presented in Chapter 3. The chapter concentrates on the class of SHS known as PDMP [10] and presents a method by which such processes can be composed. A related approach, using concepts from the area of Petri nets is outlined in Chapter 10.

The price one has to pay for the enhanced modeling capabilities of SHS is that the analysis of SHS is in general much more difficult that than of deterministic or non-deterministic hybrid systems. Chapters 4 and 5 highlight two of the most powerful and general purpose methods that can be used for the analysis of SHS.

Chapter 4 presents model checking, an analysis approach motivated by research in computer science. Roughly speaking, the idea is to establish classes of SHS for which it is possible to "code" problems in such a way that the analysis can be carried out automatically by a computer. A number of such classes are identified in Chapter 4. For these classes the chapter provides termination guarantees (guarantees that the computation will terminate in a finite amount of time) and complexity estimates (bounds on how long this time can be).

As will become apparent from Chapter 4, the class of systems for which such automated analysis is possible is not nearly as general as the class of SHS considered in Chapter 2. For more general classes of SHS other analysis methods, such as numerical methods, have to be found. Chapter 5 concentrates on reachability analysis for SHS, in particular the problem of estimating the probability that the trajectories of a given stochastic hybrid system will enter a certain subset of the state space during a possibly infinite look-ahead time horizon. The chapter looks into the theoretical foundations of the reachability problems and provides conditions under which the problem is well posed. It then develops a numerical algorithm to compute an estimate of the desired probability for a class of hybrid systems known as *switching diffusion processes* [14, 15]. This algorithm can be applied to system verification and safety analysis, and also to solve related problems such as probabilistic invariance and regulation. These uses are illustrated through examples.

Chapters 6–10 concentrate on some specific classes of SHS and application areas that include communication networks, molecular biology, and air traffic management.

Chapter 6 deals with the class of stochastic flow systems where the time-driven dynamics are of the form $\dot{x}(t) = \alpha(t) - \beta(t)$ with $\alpha(t)$ and $\beta(t)$ representing incoming and outgoing stochastic flow rate processes generally varying with time. Thus, $x(t)$ may be thought of as the time-varying state of a container of fluid. Of particular interest are the buffer contents of communication networks, computer systems, transportation networks, and manufacturing systems which define stochastic processes with such flow dynamics. The buffer contents in these settings are in fact discrete entities (packets, tasks, vehicles, or parts), but a Stochastic Fluid Model (SFM) is a powerful abstraction that facilitates their analysis and allows simulation that would otherwise be prohibitively slow. A particularly attractive feature of SFMs is the very efficient means by which one can perform sensitivity analysis through stochastic gradient estimation. Infinitesimal Perturbation Analysis (IPA) is a gradient estimation technique originally developed for discrete event systems in order to evaluate sensitivities of performance metrics with respect to controllable parameters based solely on observable sample path data. The strength of IPA is not only its implementation simplicity, but also the fact that it applies to virtually arbitrary characterizations of the stochastic processes involved. However, IPA gradient estimates cease to be unbiased (therefore, they may no longer be reliable) when the system of interest includes complexities such as multiclass networks and feedback control. In SFMs on the other hand, IPA may be used to provide unbiased estimators for many interesting types of environments that include models of the Internet and of complex manufacturing processes. Chapter 6 introduces the fundamentals of IPA and describes its use in a single node with a single class of fluid and then extends the analysis to multiple nodes in tandem.

Chapter 7 discusses the use of IPA in stochastic flow systems that incorporate feedback mechanisms. Such mechanisms are crucial in dealing with congestion in networks and part of this chapter studies the effect of buffer sizes on packet loss due to the inherent delay in acknowledging successfully received packets and the need for retransmitting them if no acknowledgment is received within a certain timeout interval. Among other uses, sensitivity analysis provides valuable insights about the dynamic behavior of large networks. As an example, the analysis of a serial multi-node network in Chapter 6 reveals that congestion in a network can generally not be regulated through control exercised several hops away, thus motivating the study of more "localized" schemes for congestion control.

Chapter 8 considers a more elaborate stochastic hybrid model for the analysis of the Transmission Control Protocol (TCP) widely used for congestion control in the Internet. Based on this model, an infinite-dimensional system of ordinary differential equations is derived; these equations describe the dynamics of the moments of the sending rate process induced by the TCP. By appropriate truncations, approximations are obtained which allow numerical solutions that provide new insights to the behavior of TCP-controlled flows. For instance, one of the conclusions in this chapter is that high-order moments appear to dominate the dynamics of TCP flows in many situations of practical interest and the standard deviation of the sending rate can be much larger than its mean. The stochastic hybrid models and techniques presented in Chapters 6–8 open up interesting possibilities for gaining further insight on the dynamics of flows in the Internet and for developing novel mechanisms for managing congestion.

Chapter 9 presents the use of stochastic hybrid systems on a different type of networks, the so called *biochemical networks*. The sequencing of the entire genome of organisms, the determination of the expression level of genes in a cell by means of DNA micro-arrays, and the identification of proteins and their interactions by high-throughput proteomic methods have produced enormous amounts of data on different aspects of the development and functioning of cells. A consensus is now emerging among biologists that to exploit these data to its full potential one needs to complement experimental results with formal models of biochemical networks. Chapter 9 presents formal stochastic hybrid models for two biochemical processes: the production of subtilin by the bacterium *Bachillus subtilis* and the process controlling DNA replication in the cell cycle of eukaryotic cells. Both models fall under the class of PDMP. Some basic analysis of these models, both theoretical and by means of Monte Carlo simulation is also presented.

Finally, Chapter 10 presents an analysis of the safety of free flight operations in air traffic based on sequential Monte Carlo simulation. Under free flight, air-crews have the freedom to select their trajectory and also the responsibility of resolving conflicts with other aircraft. There is general agreement that free flight can be made safe under low traffic conditions. Increasing traffic, however, raises safety verification issues. This problem is formulated as one of estimating the probability that the state of a large scale stochastic hybrid system reaches a small collision set. The size of the state space prohibits the use of existing numerical approaches (such as the ones presented in Chapter 5) to address this problem. The alternative is to study randomization methods. The simplest such method would be to run many Monte Carlo simulations of a stochastic hybrid system model of free flight operations, and count the number of runs during which a collision between two or more aircraft occurs. Such a straightforward approach, however, would require an impractically large number of Monte Carlo runs. Chapter 10 develops a sequential Monte Carlo simulation method for a much more efficient estimation of collision risk in free flight. The approach is demonstrated on an initial application of these novel Monte Carlo methods to a free flight air traffic concept of operations.

References

[1] S. Amin, A. Abate, M. Prandini, J. Lygeros, and S.S. Sastry. Reachability analysis for discrete time stochastic hybrid systems. In J. Hespanha and A. Tiwari, editors, *Hybrid Systems: Computation and Control*, number 3927 in LNCS, pages 49–63. Springer-Verlag, Berlin, 2006.

[2] P.J. Antsaklis, Editor. Special issue on hybrid systems: Theory and applications. *Proceedings of the IEEE*, 88(7), July 2000.

[3] A. Bemporad and S. Di Cairano. Optimal control of discrete hybrid stochastic automata. In L. Thiele and M. Morari, editors, *Hybrid Systems: Computation and Control*, number 3414 in LNCS, pages 151–167. Springer-Verlag, Berlin, 2005.

[4] A. Bemporad and M. Morari. Control of systems integrating logic dynamics and constraints. *Automatica*, 35(3):407–427, March 1999.

[5] A. Bensoussan and J.L. Menaldi. Stochastic hybrid control. *Journal of Mathematical Analysis and Applications*, 249:261–288, 2000.

[6] M.L. Bujorianu. Extended stochastic hybrid systems and their reachability problem. In R. Alur and G.J. Pappas, editors, *Hybrid Systems: Computation and Control*, number 2993 in LNCS, pages 234–249. Springer-Verlag, Berlin, 2004.

[7] M.L. Bujorianu and J. Lygeros. Toward a general theory of stochastic hybrid systems. In H.A.P. Blom and J. Lygeros, editors, *Stochastic hybrid systems: theory and safety applications*, volume 337 of *Lecture Notes in Control and Informations Sciences*, pages 3–30. Springer, Berlin, 2006.

[8] C.G. Cassandras and S. Lafortune. *Introduction to Discrete Event Systems*. Kluwer Academic Publishers, Norwell, MA, 1999.

[9] M.H.A. Davis. Piecewise-deterministic Markov processes: A general class of non-diffusion stochastic models. *Journal of the Royal Statistical Society, B*, 46(3):353–388, 1984.

[10] M.H.A. Davis. *Markov Processes and Optimization*. Chapman & Hall, London, 1993.

[11] M.H.A. Davis, M.A.H. Dempster, and S.P. Sethi D. Vermes. Optimal capacity expansion under uncertainty. *Adv. Appl. Prob.*, 19:156–176, 1987.

[12] M.H.A. Davis and M.H. Vellekoop. Permanent health insurance: a case study in piecewise-deterministic Markov modelling. *Mitteilungen der Schweiz. Vereinigung der Versicherungsmathematiker*, 2:177–212, 1995.

[13] Y. Gao, J. Lygeros, and M. Quincapoix. The reachability problem for uncertain hybrid systems revisited: A viability theory perspective. In J. Hespanha and A. Tiwari, editors, *Hybrid Systems: Computation and Control*, number 3927 in LNCS, pages 242–256. Springer-Verlag, Berlin, 2006.

[14] M.K. Ghosh, A. Arapostathis, and S.I. Marcus. Optimal control of switching diffusions with application to flexible manufacturing systems. *SIAM Journal on Control Optimization*, 31(5):1183–1204, September 1993.

[15] M.K. Ghosh, A. Arapostathis, and S.I. Marcus. Ergodic control of switching diffusions. *SIAM Journal on Control Optimization*, 35(6):1952–1988, November 1997.

[16] W. P. M. Heemels, B. De Schutter, and A. Bemporad. Equivalence of hybrid dynamical models. *Automatica*, 37(7):1085–1091, 2001.

[17] J. Hespanha. Stochastic hybrid systems: Application to communication networks. In R. Alur and G.J. Pappas, editors, *Hybrid Systems: Computation and Control*, number 2993 in LNCS, pages 387–401. Springer-Verlag, Berlin, 2004.

[18] J. Hu, J. Lygeros, and S.S. Sastry. Towards a theory of stochastic hybrid systems. In N. Lynch and B.H. Krogh, editors, *Hybrid Systems: Computation and Control*, number 1790 in LNCS, pages 160–173. Springer-Verlag, Berlin, 2000.

[19] X. Mao. Stability of stochastic differential equations with Markovian switching. *Stochastic Processes and Applications*, 79:45–67, 1999.

[20] J.L. Menaldi. Stochastic hybrid optimal control models. *Aportaciones Matematicas (Sociedad Matematica Mexicana)*, 16:205–250, 2001.

[21] G. Pola, M.L. Bujorianu, J. Lygeros, and M. di Benedetto. Stochastic hybrid models: An overview with applications to air traffic management. In *IFAC Conference on Analysis and Design of Hybrid Systems (ADHS03)*, Saint Malo, France, June 16-18 2003.

[22] C. Yuan and X. Mao. Asymptotic stability in distribution of stochastic differential equations with Markovian switching. *Stochastic Processes and Applications*, 103:277–291, 2003.

Chapter 2

Stochastic Differential Equations on Hybrid State Spaces

Jaroslav Krystul
University of Twente

Henk A.P. Blom
National Aerospace Laboratory NLR

Arunabha Bagchi
University of Twente

2.1	Introduction	15
2.2	Semimartingales and Characteristics	18
2.3	Semimartingale Strong Solution of SDE	22
2.4	Stochastic Hybrid Processes as Solutions of SDE	27
2.5	Instantaneous Hybrid Jumps at a Boundary	31
2.6	Related SDE Models on Hybrid State Spaces	33
2.7	Markov and Strong Markov Properties	40
2.8	Concluding Remarks	43
	References	44

2.1 Introduction

In studying a wide variety of real-world phenomena we usually encounter processes the course of which cannot be predicted beforehand. For example: sudden deviation of the altitude of an aircraft from a prescribed flight level; reproduction of bacteria in a favorable environment; movement of a stock price on a stock exchange. Such processes can be represented by stochastic movement of a point in a particular space specially selected for each problem. The proper choice of the phase space turns physical, mechanical, or any other real-world system into a dynamical system (it means that the current state of the system determines its future evolution). Similarly, by a proper choice of the phase space (or state space) an arbitrary stochastic process can be turned into a Markov process, i.e., a process the future evolution of which depends on the past only through its present state. This property is called the Markov property. From a whole set of stochastic processes this Markov property singles out a class of Markov processes for which powerful mathematical tools are available.

Continuous time Markov processes have been successfully used for years in stochastic modelling of various continuous time real-world dynamical systems with either Euclidean or discrete valued phase spaces. Recently, there is a great interest in more complex continuous time stochastic processes with components being hybrid, i.e., containing both Euclidean and discrete valued components. Such processes are called stochastic hybrid processes. Euclidean and discrete valued components may interact, i.e., Euclidean valued components may influence the dynamics of discrete valued component and vice versa. This makes the modelling and the analysis of stochastic hybrid processes quite involved and challenging. Several classes of stochastic hybrid processes have been studied in the literature, e.g., counting processes with diffusion intensity [21, 17], diffusion processes with Markovian switching parameters [22, 18], Markov decision drift processes [20], piecewise deterministic Markov processes [5, 6, 14], controlled switching diffusions [7, 8, 1], and more recent stochastic hybrid systems of [12, 19]. All these stochastic hybrid processes arise in various applications, have different degrees of modelling power, and have different properties inherent to the problems that they have been developed for.

There exist two directions in the development of theory of Markov processes: an analytical and a stochastic direction. Transition densities or transition probabilities are the starting point of the analytical Markov process theory. It studies various classes of transition densities and transition probabilities, which are described by equations (for example, by partial differential equations). When proving the existence of the corresponding Markov processes, any obtained conditions and properties on transition densities and probabilities are simply interpreted as certain properties of these processes. Broadly speaking, the approach taken by analytical Markov process theory could be compared with the analysis of the properties of random variables on the basis of their distribution functions or densities. In the stochastic theory a Markov process is constructed directly as a solution to a stochastic differential equation (SDE). The main advantage is that it is easier to study a Markov process as a solution of a particular equation than a Markov process that is implicitly defined through its transition density or probability.

Moreover, the theory of SDE became a powerful tool for constructive description of various classes of stochastic processes including the processes which are semimartingales. Semimartingales form one of the most important and general class of stochastic processes which includes diffusion-type processes, point processes, and diffusion-type processes with jumps that are widely used for stochastic modelling. Considering SDE with semimartingale solutions gives an advantage. It allows the use of the powerful stochastic calculus available for the semimartingale processes when performing complex stochastic analysis. This has motivated many studies in the past to consider Markov processes that are solutions of SDE. However, most of the studies consider only Euclidean valued Markov processes and only a few of them treat SDE, the solutions of which are Markov processes with a hybrid state space. This chapter aims to give an overview of stochastic approaches of modelling hybrid state Markov processes as solutions to stochastic differential equations. In a series of recent studies, Blom [2], Ghosh and Bagchi [9], and Krystul and Blom [15] developed distinct classes of stochastic hybrid processes as solutions of SDE on a hybrid state space. These classes have different modelling power and cover a wide range of interesting phenomena (see the first column of Table 2.1), though, all they contain, as a subclass, the switching diffusion processes of Ghosh *et al.* [8], described in detail in Chapter 5 of this volume.

Table 2.1: Combinations of features for various stochastic hybrid processes.

Features	[2], [9]	[3], [15]	[9]	[15]
Switching diffusion	✓	✓	✓	✓
Random hybrid jumps	✓	✓	-	✓
Boundary hybrid jumps	-	✓	✓	✓
Martingale inducing jumps	-	-	-	✓
Mode dependent dimension	-	-	✓	-

The features of stochastic hybrid processes in Table 2.1 are:

- Switching diffusion: between the random switches of the discrete valued component, the Euclidean valued component evolves as diffusion.
- Random hybrid jumps: simultaneous and dependent jumps and switches of discrete and Euclidean valued components are driven by a Poisson random measure.
- Boundary hybrid jumps: simultaneous and dependent jumps and switches of discrete and Euclidean valued components are initiated by boundary hittings.
- Martingale inducing jumps: the Euclidean valued components driven by a compensated Poisson random measure may jump so frequently that it is no longer a process of finite variation.
- Mode dependent dimension: the dimension of the Euclidean state space depends on the discrete valued component (i.e., the mode).

In the first part of the chapter we pay special attention to the modelling approach taken by Krystul and Blom [15]. Then we relate this to the models of Blom [2], Blom *et al.* [3], and Ghosh and Bagchi [9] and provide a comparison of these classes of stochastic hybrid systems.

This chapter is organized as follows. Section 2.2 provides a brief introduction to semimartingales. Section 2.3 presents the existence and uniqueness results for $\mathbb{R}^n$-valued jump-diffusions. Section 2.4 extends these results to hybrid state processes with Poisson and hybrid Poisson jumps [15]. In Section 2.5 we characterize a general stochastic hybrid process which includes jumps at the boundaries [15]. Section 2.6 briefly describes stochastic hybrid models of Blom [2] and Ghosh and Bagchi [9] and compares various stochastic hybrid models. Finally, the Markov and the strong Markov properties for a general stochastic hybrid process [2], [15] are shown in Section 2.7.

2.2 Semimartingales and Characteristics

In this section, following [13], we provide basic results concerning semimartingales, their canonical representation, and their relation with the large class of SDE to be studied in this chapter.

Throughout this chapter we assume that a probability space $(\Omega, \mathscr{F}, P)$ is equipped with a right-continuous filtration $(\mathscr{F}_t)_{t\geq 0}$. The stochastic basis $(\Omega, \mathscr{F}, (\mathscr{F}_t)_{t\geq 0}, P)$ is called complete if the σ-algebra $\mathscr{F}$ is P-complete and if every $\mathscr{F}_t$ contains all P-null sets of $\mathscr{F}$. Note that it is always possible to "complete" a given stochastic basis, if it is not complete, by adding all subsets of P-null sets to $\mathscr{F}$ and $\mathscr{F}_t$. We will therefore assume throughout this chapter that the stochastic basis $(\Omega, \mathscr{F}, (\mathscr{F}_t)_{t\geq 0}, P)$ is complete.

The *predictable σ-algebra* is the σ-algebra $\mathscr{P}$ on $\Omega \times \mathbb{R}_+$ that is generated by all left-continuous adapted processes (considered as mappings on $\Omega \times \mathbb{R}_+$). A process or random set that is $\mathscr{P}$-measurable is called *predictable*.

DEFINITION 2.1 **The canonical setting.** *Ω is the "canonical space" (also denoted by $\mathbb{D}(\mathbb{R}^n)$) of all càdlàg (right-continuous and admit left hand limits) functions $\omega : \mathbb{R}_+ \to \mathbb{R}^n$; X is the "canonical process" defined by $X_t(\omega) = \omega(t)$; $\mathscr{H} = \sigma(X_0)$; finally $(\mathscr{F}_t)_{t\geq 0}$ is generated by X and $\mathscr{H}$, by which we mean:*

(i) $\mathscr{F}_t = \bigcap_{s>t} \mathscr{F}_s^0$ and $\mathscr{F}_s^0 = \mathscr{H} \vee \sigma(X_r : r \leq s)$ (in other words, $(\mathscr{F}_t)_{t\geq 0}$ is the smallest filtration such that X is adapted and $\mathscr{H} \subset \mathscr{F}_0$);

(ii) $\mathscr{F} = \mathscr{F}_{\infty-} (= \bigvee_t \mathscr{F}_t)$.

Throughout this chapter we assume that canonical setting of Definition 2.1 is in force. The $\mathbb{R}^n$-valued càdlàg stochastic process $\{X_t\}$ defined on a probability space $(\Omega, \mathscr{F}, (\mathscr{F}_t)_{t\geq 0}, P)$ is a *semimartingale* if X_t admits a decomposition of the form

$$X_t = X_0 + A_t + M_t,\ t \geq 0, \tag{2.1}$$

where X_0 is a finite-valued and $\mathscr{F}_0$-measurable, $\{A_t\} \in \mathscr{V}^n$ is a process of *bounded variation*, $\{M_t\} \in \mathscr{M}_{loc}^n$ is an n-dimensional *local martingale* starting at 0, and for each $t \geq 0$, A_t and M_t are $\mathscr{F}_t$-measurable. Recall that $\{M_t\} \in \mathscr{M}_{loc}^n$ if and only if there exists a sequence of $(\mathscr{F}_t)_{t\geq 0}$-stopping times $(\tau_k)_{k\geq 1}$ such that $\tau_k \uparrow \infty$ (P-a.s.)

for $k \longrightarrow \infty$ and for each $k \geq 1$, the *stopped process*

$$\{M_t^{\tau_k}\} \text{ with } M_t^{\tau_k} = M_{t \wedge \tau_k},\ k \geq 1, \tag{2.2}$$

is a *martingale*:

$$\mathbb{E}|M_t^{\tau_k}| < \infty,\ \mathbb{E}[M_t^{\tau_k} \mid \mathscr{F}_s] = M_s^{\tau_k}\ (P-\text{a.s.}),\ s \leq t. \tag{2.3}$$

Denote by $\mu = \mu(\omega; ds, dx)$ the measure describing the jump structure of $\{X_t\}$:

$$\mu(\omega; (0,t] \times B) = \sum_{0<s\leq t} I_{\{\omega:\Delta X_s(\omega)\in B\}}(\omega),\ t > 0, \tag{2.4}$$

where $B \in \mathscr{B}(\mathbb{R}^n \setminus \{0\})$, i.e., the σ-algebra of Borel sets on $\mathbb{R}^n$, $\Delta X_s = X_s - X_{s-}$, and $I_{\{\omega:\Delta X_s(\omega)\in B\}}(\omega)$ is the indicator function of set $\{\omega : \Delta X_s(\omega) \in B\}$. By $\nu = \nu(\omega; ds, dx)$ we denote a compensator of μ, i.e., a predictable measure with the property that $\mu - \nu$ is a local martingale measure. This means that for each $B \in \mathscr{B}(\mathbb{R}^n \setminus \{0\})$:

$$(\mu(\omega; (0,t] \times B) - \nu(\omega; (0,t] \times B))_{t>0} \tag{2.5}$$

is a local martingale with value 0 for $t = 0$.

A semimartingale $\{X_t\}$ is called *special* if there exists a decomposition (2.1) with a *predictable* process $\{A_t\}$. Every semimartingale with *bounded jumps* ($|\Delta X_t(\omega)| \leq b < \infty, \omega \in \Omega, t > 0$) is special [see 13, Chapter I, 4.24].

Let h be a truncation function, i.e., $\Delta X_s - h(\Delta X_s) \neq 0$ if and only if $|\Delta X_s| > b$ for some $b > 0$. Hence

$$\widetilde{X}_t = \sum_{0<s\leq t} (\Delta X_s - h(\Delta X_s)) \tag{2.6}$$

denotes the jump part of $\{X_t\}$ corresponding to *large jumps*. The number of the large jumps still is finite on $[0,t]$, for all $t > 0$, because for all semimartingales [13, Chapter I, 4.47]

$$\sum_{0<s\leq t} (\Delta X_s)^2 < \infty,\ P-a.s. \tag{2.7}$$

The process $\{X_t - \widetilde{X}_t\}$ is a semimartingale with *bounded jumps* and hence it is special:

$$X_t - \widetilde{X}_t = X_0 + \widetilde{B}_t + \widetilde{M}_t \tag{2.8}$$

where $\{\widetilde{B}_t\}$ is a predictable process and $\{\widetilde{M}_t\}$ is a local martingale. The "tilde" above the process denotes the dependence on the truncation function h.

Every local martingale $\widetilde{M}_t$ can be decomposed as:

$$\widetilde{M}_t = M_t^c + \widetilde{M}_t^d \tag{2.9}$$

where M_t^c is a *continuous* (martingale) part and $\widetilde{M}_t^d$ is a *purely discontinuous* (martingale) part which satisfies:

$$\widetilde{M}_t^d = \int_0^t \int h(x)(\mu(ds, dx) - \nu(ds, dx)). \tag{2.10}$$

Note that the continuous martingale part M_t^c does not depend on h. By definition of μ and $\{\widetilde{X}_t\}$ we have

$$\widetilde{X}_t = \int_0^t \int (x - h(x))\mu(ds,dx). \tag{2.11}$$

Consequently, substitution of (2.9)–(2.11) into (2.8) yields the following canonical representation of semimartingale $\{X_t\}$:

$$X_t = X_0 + \widetilde{B}_t + M_t^c + \int_0^t \int h(x)(\mu(ds,dx) - \nu(ds,dx)) + \int_0^t \int (x - h(x))\mu(ds,dx). \tag{2.12}$$

Next we may assume $h(x) = x \cdot I_{\{x:|x|<1\}}(x)$ and replace $\widetilde{B}_t$ by B_t. Then (2.12) takes on the form:

$$X_t = X_0 + B_t + M_t^c + \int_0^t \int_{|x|<1} x(\mu(ds,dx) - \nu(ds,dx)) + \int_0^t \int_{|x|\geq 1} x\mu(ds,dx). \tag{2.13}$$

We denote by $\langle M_t^c \rangle$ the predictable quadratic variation of $\{M_t^c\}$, hence $(M_t^c)^2 - \langle M_t^c \rangle$ is a local martingale.

We call the *characteristics* associated with h of the semimartingale $\{X_t\}$ (if there may be an ambiguity on h) the triplet (B_t, C_t, ν) consisting of:

(i) A predictable process $B_t = (B_t^i)_{i \leq n}$ in $\mathscr{V}^n$, namely the process $B_t = \widetilde{B}_t$ appearing in (2.8);

(ii) A continuous process $C_t = (C_t^{ij})_{i,j \leq n}$ in $\mathscr{V}^{n \times n}$, namely $C_t = \langle M_t^c \rangle$;

(iii) A predictable random measure ν on $\mathbb{R}_+ \times \mathbb{R}^n$, namely the compensator of random measure μ associated to the jumps of X by (2.4).

DEFINITION 2.2 **Jump diffusion.** *Let P be a probability measure on $(\Omega, \mathscr{F})$. Then $\{X_t\}$ is called a jump diffusion on $(\Omega, \mathscr{F}, (\mathscr{F})_{t\geq 0}, P)$ if it is a semimartingale with the following characteristics:*

$$\begin{cases} B_t^i(\omega) = \int_0^t \alpha^i(s, X_s(\omega))ds & (= +\infty \text{ if the integral diverges}) \\ C_t^{ij}(\omega) = \int_0^t \beta^{ij}(s, X_s(\omega))ds & (= +\infty \text{ if the integral diverges}) \\ \nu(\omega; dt \times dx) = dt \times K_t(\omega, X_t(\omega), dx) & \end{cases} \tag{2.14}$$

where:

$$\begin{cases} \alpha : \mathbb{R}_+ \times \mathbb{R}^n \longrightarrow \mathbb{R}^n & \text{is Borel} \\ \beta : \mathbb{R}_+ \times \mathbb{R}^n \longrightarrow \mathbb{R}^n \times \mathbb{R}^n & \text{is Borel, } c(s,x) \text{ is symmetric nonnegative} \\ K_t(\omega, x, dy) & \text{is a Borel transition kernel from } \Omega \times \mathbb{R}^n \times \mathbb{R}^n \\ & \text{into } \mathbb{R}^n, \end{cases}$$

with $K_t(\omega, x, \{0\}) = 0$.

Next, we relate the above with stochastic differential equations, partially following [13].

Let $(\Omega, \mathscr{F}, (\mathscr{F}_t)_{t\geq 0}, P)$ be a stochastic basis endowed with:

(i) $W = (W^i)_{i\leq m}$, an m-dimensional standard Wiener process (i.e., each W^i is a standard Wiener process, and the W^i's are independent);

(ii) p_i are Poisson random measures on $\mathbb{R}_+ \times U$ with intensity measure $dt \cdot m_i(du)$, $i = 1,2$; here, $(U, \mathscr{U})$ is an arbitrary Blackwell space (one may take $U = \mathbb{R}^d$ for practical applications), and m_i, $i = 1,2$, is a positive σ-finite measure on $U, \mathscr{U}$; We denote the compensated Poisson random measure by $q_i(dt,du) = p_i(dt,du) - dt \cdot m_i(du)$, $i = 1,2$.

Let us also be given the coefficients:

$$\begin{cases} a = (a^i)_{i\leq n}, & \text{a Borel function: } \mathbb{R}_+ \times \mathbb{R}^n \longrightarrow \mathbb{R}^n \\ b = (b^{ij})_{i\leq n, j\leq m}, & \text{a Borel function: } \mathbb{R}_+ \times \mathbb{R}^n \longrightarrow \mathbb{R}^n \times \mathbb{R}^m \\ f_1 = (f_1^i)_{i\leq n} & \text{a Borel function: } \mathbb{R}_+ \times \mathbb{R}^n \times U \longrightarrow \mathbb{R}^n \\ f_2 = (f_2^i)_{i\leq n} & \text{a Borel function: } \mathbb{R}_+ \times \mathbb{R}^n \times U \longrightarrow \mathbb{R}^n. \end{cases} \tag{2.15}$$

Let the initial variable be an $\mathscr{F}_0$-measurable $\mathbb{R}^n$-valued random variable X_0. The stochastic differential equation is as follows:

$$dX_t = a(t,X_t)dt + b(t,X_t)dW_t + \int_U f_1(t,X_{t-},u)q_1(dt,du) + \int_U f_2(t,X_{t-},u)p_2(dt,du). \tag{2.16}$$

Define two stochastic sets:

$$D_1 = \{(\omega,t) : p_1(\omega;\{t\} \times U) = 1\},$$
$$D_2 = \{(\omega,t) : p_2(\omega;\{t\} \times U) = 1\}.$$

If at least one of the Poisson random measures, p_1 or p_2, has a "jump" at point (t,u), then

$$\Delta X_t(\omega) = I_{D_1}(\omega,t) \cdot f_1(t,X_{t-}(\omega),u) + I_{D_2}(\omega,t) \cdot f_2(t,X_{t-}(\omega),u).$$

Next, let us assume that the following integrals make sense.

$$\int_0^t |a(s,X_s)|ds < \infty, \ P\text{-a.s.} \tag{2.17}$$

$$\int_0^t \int_U |f_1(s,X_{s-},u)|^2 ds\, m_1(du) < \infty, \ P\text{-a.s.} \tag{2.18}$$

$$\int_0^t \int_U |f_2(s,X_{s-},u)| p_2(ds,du) < \infty, \ P\text{-a.s.} \tag{2.19}$$

$$\int_0^t |b^{ij}(s,X_s)|^2 ds < \infty, \ P\text{-a.s. for any } i,j \in \{1,\ldots,n\} \tag{2.20}$$

for every $t \in \mathbb{R}_+$. By a solution to the SDE (2.16) we mean a càdlàg $\mathscr{F}_t$-adapted process $\{X_t\}$ such that the following equation is satisfied with probability one for every $t \in \mathbb{R}_+$

$$X_t = X_0 + \int_0^t a(s,X_s)ds + \int_0^t b(s,X_s)dW_s + \int_0^t \int_U f_1(s,X_{s-},u)q_1(ds,du) + \int_0^t \int_U f_2(s,X_{s-},u)p_2(ds,du). \tag{2.21}$$

If such process $\{X_t\}$ exists and conditions (2.17)–(2.20) are satisfied then it is a semimartingale with the characteristics, associated with truncation function $h = x \cdot I_{\{x\,:\,|x|<1\}}(x)$, given by (2.14), where

$$\begin{aligned}
\alpha(t,X_t(\omega)) &= \Big[a(t,X_t(\omega)) - \int_{|f_1|\geq 1} f_1(t,X_{t-}(\omega),u)m_1(du) \\
&\qquad + \int_{|f_2|<1} f_2(t,X_{t-}(\omega),u)m_2(du)\Big], \\
\beta(t,X_t(\omega)) &= b(t,X_t(\omega))b^T(t,X_t(\omega)), \\
K_t(\omega,X_t(\omega),A) &= I_{D_1}(\omega,t)\cdot \int_U I_{A\setminus\{0\}}\big(f_1(t,X_{t-}(\omega),u)\big)m_1(du) \\
&\qquad + I_{D_2}(\omega,t)\cdot \int_U I_{A\setminus\{0\}}\big(f_2(t,X_{t-}(\omega),u)\big)m_2(du).
\end{aligned}$$

2.3 Semimartingale Strong Solution of SDE

There are two important notions of the sense in which a solution to stochastic differential equation can be said to *exist* and also two senses in which *uniqueness* is said to hold.

DEFINITION 2.3 **Strong Existence.** *We say that strong existence holds if given a probability space $(\Omega,\mathscr{F},P)$, a filtration $\mathscr{F}_t$, an $\mathscr{F}_t$-Wiener process W, two $\mathscr{F}_t$-Poisson random measures p_1, p_2, and an $\mathscr{F}_0$-measurable initial condition X_0, then an $\mathscr{F}_t$-adapted process $\{X_t\}$ exists satisfying (2.21) for all $t \geq 0$.*

DEFINITION 2.4 **Weak Existence.** *We say that weak existence holds if given any probability measure η on $\mathbb{R}^n$ there exists a probability space*

$(\Omega, \mathscr{F}, P)$, a filtration $\mathscr{F}_t$, an $\mathscr{F}_t$-Wiener process W, two $\mathscr{F}_t$-Poisson random measures p_1, p_2, and an $\mathscr{F}_t$-adapted process $\{X_t\}$ satisfying (2.21) for all $t \geq 0$ as well as $P(X_0 \in B) = \eta(B)$.

Strong existence of a solution requires that the probability space, filtration, and driving terms (W, p_1, p_2) be given first and that the solution $\{X_t\}$ then be found for the given data. Weak sense existence allows these objects to be constructed together with the process $\{X_t\}$. Clearly, strong existence implies weak existence.

DEFINITION 2.5 **Strong Uniqueness**. *Suppose that a fixed probability $(\Omega, \mathscr{F}, P)$, a filtration $(\mathscr{F}_t)_{t \geq 0}$, an $\mathscr{F}_t$-Wiener process W, and two $\mathscr{F}_t$-Poisson random measures p_1 and p_2 are given. Let $\{X_t\}$ and $\{X'_t\}$ be two solutions of (2.16) for the given driving terms (W, p_1, p_2). We say that strong uniqueness holds if*

$$P(X_0 = X'_0) = 1 \Longrightarrow P(X_t = X'_t \text{ for all } t \geq 0) = 1, \tag{2.22}$$

i.e., $\{X_t\}$ and $\{X'_t\}$ are indistinguishable.

REMARK 2.1 Since solutions of (2.16) are càdlàg processes the requirement (2.22) can be relaxed to:

$$P(X_0 = X'_0) = 1 \Longrightarrow P(X_t = X'_t) = 1, \text{ for every } t \geq 0. \tag{2.23}$$

■

DEFINITION 2.6 **Weak Uniqueness.** *Suppose we are given weak sense solutions*

$$\{(\Omega_i, \mathscr{F}_i, P_i), (\mathscr{F}_{i,t})_{t \geq 0}, \{X_{i,t}\}\}, \; i = 1, 2,$$

to (2.16). We say that weak uniqueness holds if equality of the distributions induced on $\mathbb{R}^n$ by $X_{i,0}$ under P_i, $i = 1, 2$, implies the equality of the distributions induced on $\mathbb{D}(\mathbb{R}^n)$ by $\{X_{i,t}\}$ under P_i, $i = 1, 2$.

Strong uniqueness is also referred to as *pathwise uniqueness*, whereas weak uniqueness is often called *uniqueness in (the sense of probability) law*. Strong uniqueness implies weak uniqueness.

Next we present strong existence and strong uniqueness theorems for SDE (2.16). We assume that Wiener process W and Poisson random measures p_1 and p_2 are mutually independent. Suppose $\{W_t\}$, p_1 and p_2 are adapted to the given filtration $(\mathscr{F}_t)_{t \geq 0}$. If τ is a stopping time relative to $\mathscr{F}_t$ and X_τ is an $\mathscr{F}_\tau$ measurable random variable, then we will be looking for an $\{\mathscr{F}_t\}$-adapted process $\{X_t\}$, defined for

$t > \tau$, for which the following equation holds with probability 1

$$X_t = X_\tau + \int_\tau^t a(s,X_s)ds + \int_\tau^t b(s,X_s)dW_s + \int_\tau^t \int_U f_1(s,X_{s-},u)q_1(ds,du) + \int_\tau^t \int_U f_2(s,X_{s-},u)p_2(ds,du). \quad (2.24)$$

If equality (2.24) holds for all $t \in (\tau,\zeta)$, with ζ another stopping time, $\zeta > \tau$, then we will say that $\{X_t\}$ is the solution of SDE (2.16) on interval (τ,ζ), if started at X_τ.

THEOREM 2.1 *A solution of Equation (2.16) for any given X_0 is strongly unique if the coefficients of Equation (2.16) satisfy the following conditions:*

(i) for each $r > 0$ there exist a constant l_r, for which

$$|a(s,x)-a(s,y)|^2 + |b(s,x)-b(s,y)|^2 + \int_U |f_1(s,x,u)-f_1(s,y,u)|^2 m_1(du) \leq l_r|x-y|^2,$$

for all $|x| \leq r$, $|y| \leq r$, $s \leq r$.

(ii) $\int_0^t \int_U |f_2(s,X_{s-},u)|p_2(ds,du) < \infty$, *P-a.s.,*

(iii) $m_2(S_u) < \infty$, where S_u is the projection on U of the support of $f_2(\cdot,\cdot,\cdot)$.

PROOF See Theorem 3.8 in [15]. ∎

Related to Theorem 2.1 is that two solutions of two different equations with equal initial conditions coincide as long as their coefficients coincide. We formulate this statement precisely, known as the theorem of local uniqueness.

THEOREM 2.2 *Suppose $\{X_t\}$ is a solution of Equation (2.21), and $\{\tilde{X}_t\}$ is a solution of Equation*

$$\tilde{X}_t = \tilde{X}_0 + \int_0^t \tilde{a}(s,\tilde{X}_s)ds + \int_0^t \tilde{b}(s,\tilde{X}_s)dW_s + \int_0^t \int_U \tilde{f}_1(s,\tilde{X}_s,u)q_1(ds,du) + \int_0^t \int_U \tilde{f}_2(s,\tilde{X}_s,u)p_2(ds,du).$$

If the conditions of Theorem 2.1 are satisfied and $a(s,x) = \tilde{a}(s,x)$, $b(s,x) = \tilde{b}(s,x)$, $f_k(s,x,u) = \tilde{f}_k(s,x,u)$ given $|x| \leq N$, then $X_s = \tilde{X}_s$ for $s \leq \tau$, where $\tau = \inf\{s : |X_s| \geq N\}$.

Next, we state the classical existence results for the following equation [11]:

$$X_t = X_0 + \int_0^t a(s,X_s)ds + \int_0^t b(s,X_s)dW_s + \int_0^t \int_U f_1(s,X_s,u)q_1(ds,du). \quad (2.25)$$

THEOREM 2.3 *Assume that the coefficients of Equation (2.25) satisfy the following conditions:*

(i) $a(s,0)$, $b(s,0)$, $\int |f_1(s,0,u)|^2 m_1(du)$ *are locally bounded with respect to* s,

(ii) there exists increasing function $l(s)$ *such that*

$$|a(s,x)-a(s,y)|^2+|b(s,x)-b(s,y)|^2 + \int_U |f_1(s,x,u)-f_1(s,y,u)|^2 m_1(du) \le l(s)|x-y|^2.$$

Let us denote by $\mathscr{F}_t$ *the* σ*-algebra generated by* X_0, $q_1(ds,du)$, W_s *with* $s \le t$. *If* X_0 *is independent of* W_s, $q_1(ds,du)$ *and* $\mathbb{E}|X_0|^2 < \infty$, *then Equation (2.25) has* $\mathscr{F}_t$*-measurable solution, moreover* $\mathbb{E}|X_s|^2 < \infty$.

THEOREM 2.4 *Assume that for the coefficients of Equation (2.25) the following conditions hold:*

$$|a(t,x)|^2+|b(t,x)|^2+\int_U |f_1(t,x,u)|^2 m_1(du) \le l(1+|x|^2),$$

and for any $r > 0$ *one can specify constant* l_r *such that*

$$|a(s,x)-a(s,y)|^2+|b(s,x)-b(s,y)|^2 + \int_U |f_1(s,x,u)-f_1(s,y,u)|^2 m_1(du) \le l_r|x-y|^2$$

for $s \le r$, $|x| \le r$, $|y| \le r$. *If* X_0 *is independent of* $\{W_s, q_1(ds,du)\}$, *and* σ*-algebras* $\mathscr{F}_t$ *are constructed as in Theorem 2.3, then there exists an* $\mathscr{F}_t$*-measurable solution of (2.25) for every* $t \in \mathbb{R}_+$.

REMARK 2.2 Suppose $\{\hat{\mathscr{F}}_t\}$ is some admissible filtration, and τ is a stopping time relative to this filtration. Let us consider the SDE for $t > \tau$:

$$X_t = X_\tau + \int_\tau^t a(s,X_s)ds + \int_\tau^t b(s,X_s)dW_s + \int_\tau^t \int_U f_1(s,X_s,u)q_1(ds,du). \quad (2.26)$$

Under conditions of Theorem 2.4, Equation (2.26) has $\hat{\mathscr{F}}_t$-measurable solution, no matter what the $\hat{\mathscr{F}}_\tau$-measurable variable X_τ is. To prove this, it suffices to consider the process $\hat{X}_t$ which is a solution of the following equation.

$$\hat{X}_t = \hat{X}_0 + \int_0^t a(s+\tau,\hat{X}_s)ds + \int_0^t b(s+\tau,\hat{X}_s)d\hat{W}_s + \int_0^t \int_U f_1(s+\tau,\hat{X}_s,u)\hat{q}_1(ds,du), \quad (2.27)$$

where

$$\hat{W}_s = W(s+\tau) - W_\tau; \quad \hat{q}_1([s_1,s_2] \times du) = q_1([s_1+\tau, s_2+\tau] \times du). \qquad (2.28)$$

Obviously, $\hat{W}$ and $\hat{q}_1$ possess the same properties as W and q_1, and are independent of $\mathscr{F}_\tau$. Thus, for Equation (2.27), all derivations which were verified for Equation (2.25), hold as well, if expectations and conditional expectations with given X_0 are substituted by conditional expectation with respect to σ-algebra $\hat{\mathscr{F}}_\tau$. Obviously, then $X_t = \hat{X}_{t-\tau}$ will be the solution of Equation (2.26).
■

Now we state the existence theorem for general SDE (2.16).

THEOREM 2.5 *Assume that for Equation (2.16) the following conditions are satisfied:*

(i) The coefficients a, b, f_1 satisfy the conditions of Theorem 2.4.

(ii) X_0 is independent of $\{W_s, q_1(ds,du), p_2(ds,du)\}$.

(iii) Conditions (ii) and (iii) of Theorem 2.1 are satisfied.

Let $\mathscr{F}_t$ denote the σ-algebra generated by $\{W_s,\ q_1([0,s],du),\ p_2([0,s],du),\ s \le t\}$ and X_0. Then there exists an $\mathscr{F}_t$-measurable solution of Equation (2.16).

PROOF See Theorem 3.13 in [15]. ■

REMARK 2.3 The solution, whose existence was established in Theorem 2.5, is unique. Indeed, by Theorem 2.1 we have that for any enlargement of the initial probability space, any admissible filtration of σ-algebras $\tilde{\mathscr{F}}_t$, and any $\mathscr{F}_0$-measurable initial variable X_0, $\tilde{\mathscr{F}}_t$-measurable solution of Equation (2.16) is unique. Since $\mathscr{F}_t \subset \tilde{\mathscr{F}}_t$, the solution X_t constructed in Theorem 2.5 will be also $\tilde{\mathscr{F}}_t$-measurable, and therefore, there will be no other $\tilde{\mathscr{F}}_t$-measurable solutions of Equation (2.16). ■

REMARK 2.4 The solution constructed in Theorem 2.5 is fully determined by the initial condition, Wiener process W and Poisson random measures p_1 and p_2, i.e., it is a strong solution (solution-process). Thus, Theorem 2.5 states that there exists a strong solution of Equation (2.16) (strong existence), and from Remark 2.3 it follows that under conditions of Theorem 2.5 any solution of (2.16) is unique (strong uniqueness). ■

REMARK 2.5 Under the conditions of Theorem 2.5 the solution of SDE

(2.16) admits the decomposition (2.1) with

$$A_t = \int_0^t a(s,X_s)ds + \int_0^t \int_U f_2(s,X_{s-},u)p_2(ds,du) \in \mathscr{V}^n,$$
$$M_t = \int_0^t b(s,X_s)dW_s + \int_0^t \int_U f_1(s,X_{s-},u)q_1(ds,du) \in \mathscr{M}^n_{loc},$$

hence it is a semimartingale. ∎

2.4 Stochastic Hybrid Processes as Solutions of SDE

In this section we construct a switching jump diffusion $\{X_t, \theta_t\}$ taking values in $\mathbb{R}^n \times \mathbb{M}$, where $\mathbb{M} = \{e_1, e_2, \ldots, e_N\}$ is a finite set. We assume that for each $i = 1, \ldots, N$, e_i is the i-th unit vector, $e_i \in \mathbb{R}^N$. Note that the hybrid state space $\mathbb{R}^n \times \mathbb{M} \subset \mathbb{R}^{n+N}$ can be seen as a special subset of $(n+N)$-dimensional Euclidean space. Let $\{X_t, \theta_t\}$ be an $\mathbb{R}^n \times \mathbb{M}$-valued process given by the following stochastic differential equation of Ito-Skorohod type.

$$dX_t = a(X_t, \theta_t)dt + b(X_t, \theta_t)dW_t + \int_{\mathbb{R}^d} g_1(X_{t-}, \theta_{t-}, u)q_1(dt,du) \qquad (2.29)$$
$$+ \int_{\mathbb{R}^d} g_2(X_{t-}, \theta_{t-}, u)p_2(dt,du),$$
$$d\theta_t = \int_{\mathbb{R}^d} c(X_{t-}, \theta_{t-}, u)p_2(dt,du). \qquad (2.30)$$

Here:

(i) for $t = 0$, X_0 is a prescribed $\mathbb{R}^n$-valued random variable.

(ii) for $t = 0$, θ_0 is a prescribed $\mathbb{M}$-valued random variable.

(iii) W is an m-dimensional standard Wiener process.

(iv) $q_1(dt,du)$ is a martingale random measure associated to a Poisson random measure p_1 with intensity $dt \times m_1(du)$.

(v) $p_2(dt,du)$ is a Poisson random measure with intensity $dt \times m_2(du) = dt \times du_1 \times \bar{\mu}(d\underline{u})$, where $\bar{\mu}$ is a probability measure on $\mathbb{R}^{d-1}$, $u_1 \in \mathbb{R}$, $\underline{u} \in \mathbb{R}^{d-1}$ refers to all components except the first one of $u \in \mathbb{R}^d$.

The coefficients are defined as follows

$$
\begin{aligned}
a &: \mathbb{R}^n \times \mathbb{M} \to \mathbb{R}^n \\
b &: \mathbb{R}^n \times \mathbb{M} \to \mathbb{R}^{n\times m} \\
g_1 &: \mathbb{R}^n \times \mathbb{M} \times \mathbb{R}^d \to \mathbb{R}^n \\
g_2 &: \mathbb{R}^n \times \mathbb{M} \times \mathbb{R}^d \to \mathbb{R}^n \\
\phi &: \mathbb{R}^n \times \mathbb{M} \times \mathbb{M} \times \mathbb{R}^{d-1} \to \mathbb{R}^n \\
\lambda &: \mathbb{R}^n \times \mathbb{M} \times \mathbb{M} \to \mathbb{R}_+ \\
c &: \mathbb{R}^n \times \mathbb{M} \times \mathbb{R}^d \to \mathbb{R}^N.
\end{aligned}
$$

Moreover, for all $k = 1, 2, \ldots, N$ we define measurable mappings $\Sigma_k : \mathbb{R}^n \times \mathbb{M} \to \mathbb{R}_+$ in a following manner

$$
\Sigma_k(x, e_i) = \begin{cases} \sum_{j=1}^{k} \lambda(x, e_i, e_j) & k > 0, \\ 0 & k = 0, \end{cases} \tag{2.31}
$$

function $c(\cdot,\cdot,\cdot)$ by

$$
c(x, e_i, u) = \begin{cases} e_j - e_i & \text{if } u_1 \in (\Sigma_{j-1}(x, e_i), \Sigma_j(x, e_i)], \\ 0 & \text{otherwise }, \end{cases} \tag{2.32}
$$

and function $g_2(\cdot,\cdot,\cdot)$ by

$$
g_2(x, e_i, u) = \begin{cases} \phi(x, e_i, e_j, \underline{u}) & \text{if } u_1 \in (\Sigma_{j-1}(x, e_i), \Sigma_j(x, e_i)], \\ 0 & \text{otherwise }. \end{cases} \tag{2.33}
$$

Let U_θ denote the projection of the support of function $\phi(\cdot,\cdot,\cdot,\cdot)$ on $\underline{U} = \mathbb{R}^{d-1}$. The jump size of X_t and the new value of θ_t at the jump times generated by Poisson random measure p_2 are determined by the functions (2.32) and (2.33) correspondingly. There are three different situations possible:

(i) Simultaneous jump of X_t and θ_t

$$
\begin{cases} c(\cdot,\cdot,u) \neq 0 & \text{if } u_1 \in (\Sigma_{j-1}(x, e_i), \Sigma_j(x, e_i)],\ i, j = 1, \ldots, N \text{ and } j \neq i, \\ g_2(\cdot,\cdot,u) \neq 0 & \text{if } u_1 \in (\Sigma_{j-1}(x, e_i), \Sigma_j(x, e_i)],\ i, j = 1, \ldots, N \text{ and } \underline{u} \in U_\theta. \end{cases}
$$

(ii) Switch of θ_t only

$$
\begin{cases} c(\cdot,\cdot,u) \neq 0 & \text{if } u_1 \in (\Sigma_{j-1}(x, e_i), \Sigma_j(x, e_i)],\ i, j = 1, \ldots, N \text{ and } j \neq i, \\ g_2(\cdot,\cdot,u) = 0 & \text{if } u_1 \in (\Sigma_{j-1}(x, e_i), \Sigma_j(x, e_i)],\ i, j = 1, \ldots, N \text{ and } \underline{u} \notin U_\theta. \end{cases}
$$

(iii) Jump of X_t only

$$
\begin{cases} c(\cdot,\cdot,u) = 0 & \text{if } u_1 \in (\Sigma_{j-1}(x, e_j), \Sigma_j(x, e_j)],\ j = 1, \ldots, N, \\ g_2(\cdot,\cdot,u) \neq 0 & \text{if } u_1 \in (\Sigma_{j-1}(x, e_j), \Sigma_j(x, e_j)],\ j = 1, \ldots, N, \text{ and } \underline{u} \in U_\theta. \end{cases}
$$

We make the following assumptions on the coefficients of SDE (2.29)–(2.30).

(A1) There exists a constant l such that for all $i = 1,2,\ldots,N$

$$|a(x,e_i)|^2 + |b(x,e_i)|^2 + \int_{\mathbb{R}^d} |g_1(x,e_i,u)|^2 m_1(du) \leq l(1+|x|^2).$$

(A2) For any $r > 0$ one can specify constant l_r such that for all $i = 1,2,\ldots,N$

$$|a(x,e_i) - a(y,e_i)|^2 + |b(x,e_i) - b(y,e_i)|^2 + \int_{\mathbb{R}^d} |g_1(x,e_i,u) - g_1(y,e_i,u)|^2 m_1(du) \leq l_r|x-y|^2$$

for $|x| \leq r$, $|y| \leq r$.

(A3) Function c satisfies (2.31), (2.32), and for $i,j = 1,2,\ldots,N$, $\lambda(e_i,e_j,\cdot)$ are bounded and measurable, $\lambda(e_i,e_j,\cdot) \geq 0$.

(A4) Function g_2 satisfies (2.31), (2.33), and for all $t > 0$, $i,j = 1,\ldots,N$

$$\int_0^t \int_{\mathbb{R}^d} |\phi(x,e_i,e_j,\underline{u})| p_2(ds,du) < \infty, \quad P\text{-a.s.}$$

THEOREM 2.6 *Assume (A1)–(A4). Let p_1, p_2, W, X_0 and θ_0 be independent. Then SDE (2.29)–(2.30) has a unique strong solution which is a semimartingale.*

PROOF See Theorem 4.1 in [15]. ■

In order to explicitly show the hybrid jump behavior as a strong solution to an SDE, Blom [2] has developed an approach to prove that solution of (2.29)–(2.30) is indistinguishable from the solution of the following set of Equations:

$$d\theta_t = \sum_{i=1}^{N} (e_i - \theta_{t-}) p_2\left(dt, (\Sigma_{i-1}(X_{t-},\theta_{t-}), \Sigma_i(X_{t-},\theta_{t-})] \times \mathbb{R}^{d-1}\right), \tag{2.34}$$

$$dX_t = a(X_t,\theta_t)dt + b(X_t,\theta_t)dW_t + \int_{\mathbb{R}^d} g_1(X_{t-},\theta_{t-},u) q_1(dt,du) + \int_{\mathbb{R}^d} \phi(X_{t-},\theta_{t-},\theta_t,\underline{u}) p_2(dt, (0,\Sigma_N(X_{t-},\theta_{t-})] \times d\underline{u}). \tag{2.35}$$

THEOREM 2.7 *Assume (A1)–(A4). Let p_1, p_2, W, X_0 and θ_0 be independent. Then SDE (2.34)–(2.35) has a unique strong solution which is a semimartingale.*

PROOF The proof consists of showing that the solution of (2.34)–(2.35) is indistinguishable from the solution of (2.29)–(2.30). Subsequently Theorem 2.7 is the consequence of Theorem 2.6.

Indeed, rewriting of (2.34) yields (2.30):

$$
\begin{aligned}
d\theta_t &= \sum_{i=1}^{N}(e_i - \theta_{t-})p_2\left(dt, (\Sigma_{i-1}(X_{t-},\theta_{t-}), \Sigma_i(X_{t-},\theta_{t-})] \times \mathbb{R}^{d-1}\right) \\
&= \int_{\mathbb{R}^d} \sum_{i=1}^{N}(e_i - \theta_{t-}) I_{(\Sigma_{i-1}(X_{t-},\theta_{t-}),\Sigma_i(X_{t-},\theta_{t-})]}(u_1)p_2(dt, du_1 \times d\underline{u}) \\
&= \int_{\mathbb{R}^d} c(X_{t-},\theta_{t-},u)p_2(dt,du).
\end{aligned}
$$

Next, since the first three right hand terms of (2.35) and (2.29) are equal, it remains to show that the fourth right hand term in (2.35) yields the fourth right hand term in (2.29) up to indistinguishability:

$$
\begin{aligned}
&\int_{\mathbb{R}^d} \phi(X_{t-},\theta_{t-},\theta_t,\underline{u})p_2\big(dt, (0,\Sigma_N(X_{t-},\theta_{t-})] \times d\underline{u}\big) \\
&= \int_{(0,\infty)}\int_{\mathbb{R}^{d-1}} \phi(X_{t-},\theta_{t-},\theta_t,\underline{u}) I_{(0,\Sigma_N(X_{t-},\theta_{t-})]}(u_1)p_2(dt,du_1 \times d\underline{u}) \\
&= \int_{(0,\infty)}\int_{\mathbb{R}^{d-1}} \phi(X_{t-},\theta_{t-},\theta_t,\underline{u}) \times \\
&\qquad \times \sum_{i=1}^{N} I_{(\Sigma_{i-1}(X_{t-},\theta_{t-}),\Sigma_i(X_{t-},\theta_{t-})]}(u_1)p_2(dt,du_1 \times d\underline{u}) \\
&= \int_{(0,\infty)}\int_{\mathbb{R}^{d-1}} \sum_{i=1}^{N} \big[\phi(X_{t-},\theta_{t-},\theta_t,\underline{u}) \times \\
&\qquad \times I_{(\Sigma_{i-1}(X_{t-},\theta_{t-}),\Sigma_i(X_{t-},\theta_{t-})]}(u_1)\big] p_2(dt,du_1 \times d\underline{u}) \\
&= \int_{(0,\infty)}\int_{\mathbb{R}^{d-1}} \sum_{i=1}^{N} \big[\phi(X_{t-},\theta_{t-},\theta_{t-} + \Delta\theta_t,\underline{u}) \times \\
&\qquad \times I_{(\Sigma_{i-1}(X_{t-},\theta_{t-}),\Sigma_i(X_{t-},\theta_{t-})]}(u_1)\big] p_2(dt,du_1 \times d\underline{u}) \\
&= \int_{(0,\infty)}\int_{\mathbb{R}^{d-1}} \sum_{i=1}^{N} \big[\phi(X_{t-},\theta_{t-},\theta_{t-} + (e_i - \theta_{t-}),\underline{u}) \times \\
&\qquad \times I_{(\Sigma_{i-1}(X_{t-},\theta_{t-}),\Sigma_i(X_{t-},\theta_{t-})]}(u_1)\big] p_2(dt,du_1 \times d\underline{u}) \\
&= \int_{(0,\infty)}\int_{\mathbb{R}^{d-1}} \sum_{i=1}^{N} \big[\phi(X_{t-},\theta_{t-},e_i,\underline{u}) \times \\
&\qquad \times I_{(\Sigma_{i-1}(X_{t-},\theta_{t-}),\Sigma_i(X_{t-},\theta_{t-})]}(u_1)\big] p_2(dt,du_1 \times d\underline{u}) \\
&= \int_{\mathbb{R}^d} g_2(X_{t-},\theta_{t-},u)p_2(dt,du).
\end{aligned}
$$

This completes the proof. ■

REMARK 2.6 We notice the interesting aspect that the presence of θ_t in ϕ (Equation (2.35)) explicitly shows that jump of $\{X_t\}$ depends on the switch from θ_{t-} to θ_t, i.e., it is a hybrid jump. ■

2.5 Instantaneous Hybrid Jumps at a Boundary

Up to now we have considered $\mathbb{R}^n \times \mathbb{M}$-valued processes the jumps and switches of which are driven by Poisson random measure. In this section we will consider $\mathbb{R}^n \times \mathbb{M}$-valued processes which also have instantaneous jumps and switches when hitting boundaries of some given sets. In order to simplify the analysis we assume that the purely discontinuous martingale term is equal to zero (i.e., we take $g_1 \equiv 0$).

First we define a particular sequence of processes. Suppose for each $e_i \in \mathbb{M}$, $i = 1, \dots, N$ there is an open connected set $E^i \subset \mathbb{R}^n$, with boundary ∂E^i. Let

$$E = \{x \,|\, x \in E^i, \text{ for some } i = 1, \dots, N\} = \bigcup_{i=1}^{N} E^i,$$

$$\partial E = \{x \,|\, x \in \partial E^i, \text{ for some } i = 1, \dots, N\} = \bigcup_{i=1}^{N} \partial E^i.$$

The interior of the set E is the jump "destination" set. Suppose that the function g_2, defined by (2.33), in addition to requirement (A4) has the following property:

(B1) $(x + \phi(x, e_i, \underline{u})) \in E^i$ for each $x \in E^i$, $\underline{u} \in \mathbb{R}^{d-1}$, $i = 1, \dots, N$.

Similarly as in [3, pp. 38–39], we consider an increasing sequence of stopping times τ_n^E and a sequence of jump-diffusions $\{X_t^n \,;\, t \geq \tau_{n-1}^E\}$, $n = 1, 2, \dots$, governed by the following SDE (in integral form):

$$X_t^n = X_{\tau_{n-1}^E}^n + \int_{\tau_{n-1}^E}^t a(X_s^n, \theta_s^n) ds + \int_{\tau_{n-1}^E}^t b(X_s^n, \theta_s^n) dW_s \tag{2.36}$$

$$+ \int_{\tau_{n-1}^E}^t \int_{\mathbb{R}^d} g_2(X_{s-}^n, \theta_{s-}^n, u) p_2(ds, du),$$

$$\theta_t^n = \theta_{\tau_{n-1}^E}^n + \int_{\tau_{n-1}^E}^t \int_{\mathbb{R}^d} c(X_{s-}^n, \theta_{s-}^n, u) p_2(ds, du), \tag{2.37}$$

$$X_{\tau_n^E}^{n+1} = g^x(X_{\tau_n^E}^n, \theta_{\tau_n^E}^n, \beta_{\tau_n^E}), \tag{2.38}$$

$$\theta_{\tau_n^E}^{n+1} = g^\theta(X_{\tau_n^E}^n, \theta_{\tau_n^E}^n, \beta_{\tau_n^E}). \tag{2.39}$$

More specifically, the stopping times are defined as follows.

$$\tau_k^E \triangleq \inf\{t > \tau_{k-1}^E : X_t^k \in \partial E\}, \tag{2.40}$$

$$\tau_0^E \triangleq 0 \tag{2.41}$$

$k = 1,2,\ldots,N$, i.e., $\tau_0^E < \tau_1^E < \cdots < \tau_k^E < \ldots$ a.s.,

$$g^x : \partial E \times \mathbb{M} \times V \to \mathbb{R}^n, \tag{2.42}$$

$$g^\theta : \partial E \times \mathbb{M} \times V \to \mathbb{M}, \tag{2.43}$$

and $\{\beta_t, t \in [0,\infty)\}$ is the sequence of V-valued (one may take $V = \mathbb{R}^d$) i.i.d. random variables distributed according to some given distribution. The initial values X_0^1 and θ_0^1 are some prescribed random variables.

REMARK 2.7 Assumption (B1) ensures that the sequence of stopping times (2.40) is well defined and the boundary ∂E can be hit only by the continuous part

$$X_t^{c,n} = X_{\tau_{n-1}^E}^n + \int_{\tau_{n-1}^E}^t a(X_s^n, \theta_s^n)ds + \int_{\tau_{n-1}^E}^t b(X_s^n, \theta_s^n)dW_s \tag{2.44}$$

of the processes $\{X_t^n\}$, $n = 1,2,\ldots$, between the jumps and/or switching times generated by Poisson random measure p_2. ■

In order to prove existence and uniqueness, we define the process $\{X_t, \theta_t\}$ as follows.

$$\begin{cases} X_t(\omega) = \sum_{n=1}^\infty X_t^n(\omega) I_{\left[\tau_{n-1}^E(\omega), \tau_n^E(\omega)\right)}(t) \\ \theta_t(\omega) = \sum_{n=1}^\infty \theta_t^n(\omega) I_{\left[\tau_{n-1}^E(\omega), \tau_n^E(\omega)\right)}(t) \end{cases} \tag{2.45}$$

provided there exist solutions $\{X_t^n, \theta_t^n\}$ of SDE (2.36)–(2.39). On the open set E, process $\{X_t, \theta_t\}$ (provided it exists) evolves according to SDE (2.29)–(2.30) or (2.34)–(2.35). At times τ_k^E there is a jump and/or switching determined by the mappings g^x and g^θ correspondingly, i.e., $X_{\tau_k^E} \neq X_{\tau_k^E-}$ and/or $\theta_{\tau_k^E} \neq \theta_{\tau_k^E-}$.

To ensure the existence of a strong unique solution of (2.45) we need assumption (B1) and the following:

(B2) $d(\partial E, g^x(\partial E, \mathbb{M}, V)) > 0$, i.e., $\{X_t\}$ may jump only inside of open set E.

(B3) Process (2.45) hits the boundary ∂E a.s. finitely many times on any finite time interval.

THEOREM 2.8 *Assume (A1)–(A4) and (B1)–(B3). Let W, p_2, $\{\beta_t, t \in [0,\infty)\}$, X_0 and θ_0 be independent. Then process (2.45) exists for every $t \in \mathbb{R}_+$, it is strongly unique and it is a semimartingale.*

PROOF See Theorem 5.2 in [15]. ■

2.6 Related SDE Models on Hybrid State Spaces

In this section we compare stochastic hybrid models developed by Blom [2], Blom *et al.* [3], and Ghosh and Bagchi [9] with the models presented in Sections 2.4 and 2.5. We will use the same notations and definitions of coefficients as in Sections 2.4 and 2.5. Table 2.2 lists the models we are dealing within this section.

Table 2.2: List of models and their main features.

	θ	$X1$	$X2$	$\theta \& X2$	B
HB1 [2]	✓	-	✓	✓	-
HB2 [3]	✓	-	✓	✓	✓
GB1 [9]	✓	-	✓	✓	-
GB2 [9]	✓	-	-	-	✓
KB1 [15]	✓	✓	✓	✓	-
KB2 [15]	✓	-	✓	✓	✓

The conventions used in Table 2.2 have the following meaning:

HB1 refers to switching hybrid-jump diffusion of Blom [2];

HB2 refers to switching hybrid-jump diffusion with hybrid jumps at the boundary of Blom *et al.* [3];

GB1 refers to switching jump diffusion of Ghosh and Bagchi [9];

GB2 refers to switching diffusion with hybrid jumps at the boundary of Ghosh and Bagchi [9];

KB1 refers to switching hybrid-jump diffusion developed in Section 2.4;

KB2 refers to switching hybrid-jump diffusion with hybrid jumps at the boundary developed in Section 2.5.

θ stands for independent random switching of θ_t;

$X1$ stands for independent random jump of X_t generated by compensated Poisson random measure;

$X2$ stands for independent random jump of X_t generated by Poisson random measure;

$\theta \& X2$ stands for simultaneous jump of X_t and θ_t generated by Poisson random measure;

B stands for simultaneous jump of X_t and θ_t at the boundary.

Stochastic hybrid model HB1 [2] forms a subset of KB1. The difference is that HB1 assumes a zero martingale measure q_1 in (2.29) or (2.34). Thanks to [16], Blom [2] also develops a verifiable version of condition (A4):

(A4′) For any $k \in \mathbb{N}$, there exists a constant N_k such that for each $i, j \in \{1, 2, \ldots N\}$

$$\sup_{|x| \leq k} \int_{\mathbb{R}^{d-1}} |\phi(x, e_i, e_j, \underline{u})| \bar{\mu}(d\underline{u}) \leq N_k.$$

Stochastic hybrid model HB2 [3] equals KB2; [3] also develops the verifiable version (A4′) of (A4). In order to explain the relation with GB1 and GB2 we first specify these stochastic hybrid models developed in [9].

2.6.1 Stochastic Hybrid Model GB1 of Ghosh and Bagchi

Now, let us consider the model GB1 of Ghosh and Bagchi [9].

The evolution of $\mathbb{R}^n \times \mathbb{M}$-valued Markov process $\{X_t, \theta_t\}$ is governed by the following equations:

$$dX_t = a(X_t, \theta_t)dt + b(X_t, \theta_t)dW_t + \int_{\mathbb{R}} g(X_{t-}, \theta_{t-}, u)p(dt, du), \tag{2.46}$$

$$d\theta_t = \int_{\mathbb{R}} h(X_{t-}, \theta_{t-}, u)p(dt, du). \tag{2.47}$$

Here:

(i) for $t = 0$, X_0 is a prescribed $\mathbb{R}^n$-valued random variable.

(ii) for $t = 0$, θ_0 is a prescribed $\mathbb{M}$-valued random variable.

(iii) W is an n-dimensional standard Wiener process.

(iv) $p(dt, du)$ is a Poisson random measure with intensity $dt \times m(du)$, where m is the Lebesgue measure on $\mathbb{R}$. p is assumed to be independent of W.

The coefficients are defined as:

$$\begin{aligned} a &: \mathbb{R}^n \times \mathbb{M} \to \mathbb{R}^n \\ b &: \mathbb{R}^n \times \mathbb{M} \to \mathbb{R}^{n \times n} \\ g &: \mathbb{R}^n \times \mathbb{M} \times \mathbb{R} \to \mathbb{R}^n \\ h &: \mathbb{R}^n \times \mathbb{M} \times \mathbb{R} \to \mathbb{R}^N. \end{aligned}$$

Function h is defined as:

$$h(x, e_i, u) = \begin{cases} e_j - e_i & \text{if } u \in \Delta_{ij}(x) \\ 0 & \text{otherwise,} \end{cases} \tag{2.48}$$

where for $i,j \in \{1,\dots,N\}$, $i \neq j$, $x \in \mathbb{R}^n$, $\Delta_{ij}(x)$ are the intervals of the real line defined as:

$$\begin{aligned}
\Delta_{12}(x) &= [0, \lambda_{12}(x)) \\
\Delta_{13}(x) &= [\lambda_{12}(x), \lambda_{12}(x) + \lambda_{13}(x)) \\
&\vdots \\
\Delta_{1N}(x) &= \Big[\textstyle\sum_{j=2}^{N-1} \lambda_{1j}(x), \sum_{j=2}^{N} \lambda_{1j}(x)\Big) \\
\Delta_{21}(x) &= \Big[\textstyle\sum_{j=2}^{N} \lambda_{1j}(x), \sum_{j=2}^{N} \lambda_{1j}(x) + \lambda_{21}(x)\Big)
\end{aligned}$$

and so on. In general,

$$\Delta_{ij}(x) = \Big[\sum_{i'=1}^{i-1} \sum_{\substack{j'=1 \\ j' \neq i'}}^{N} \lambda_{i'j'}(x) + \sum_{\substack{j'=1 \\ j' \neq i}}^{j-1} \lambda_{ij'}(x), \sum_{i'=1}^{i-1} \sum_{\substack{j'=1 \\ j' \neq i'}}^{N} \lambda_{i'j'}(x) + \sum_{\substack{j'=1 \\ j' \neq i}}^{j} \lambda_{ij'}(x)\Big).$$

For fixed x these are disjoint intervals, and the length of $\Delta_{ij}(x)$ is $\lambda_{ij}(x)$, $\lambda_{ij} : \mathbb{R}^n \to \mathbb{R}$, $i,j = 1,\dots,N$, $i \neq j$.

Let K_1 be the support of $g(\cdot,\cdot,\cdot)$ and let U_1 be the projection of K_1 on $\mathbb{R}$. It is assumed that U_1 is bounded. Let K_2 denote the support of $h(\cdot,\cdot,\cdot)$ and U_2 the projection of K_2 on $\mathbb{R}$. By definition of c, U_2 is a bounded set. One can define function $g(\cdot,\cdot,\cdot)$ so that the sets U_1 and U_2 form three nonempty sets: $U_1 \setminus U_2$, $U_1 \cap U_2$ and $U_2 \setminus U_1$ (see Figure 2.1). Then, we have the following:

(i) For $u \in U_1 \cap U_2$

$$\begin{cases} g(\cdot,\cdot,u) \neq 0 \\ h(\cdot,\cdot,u) \neq 0 \end{cases}$$

i.e., simultaneous jumps of X_t and switches of θ_t are possible.

(ii) For $u \in U_2 \setminus U_1$

$$\begin{cases} g(\cdot,\cdot,u) = 0 \\ h(\cdot,\cdot,u) \neq 0 \end{cases}$$

i.e., only random switches of θ_t are possible.

(iii) For $u \in U_1 \setminus U_2$

$$\begin{cases} g(\cdot,\cdot,u) \neq 0 \\ h(\cdot,\cdot,u) = 0 \end{cases}$$

i.e., only random jumps of X_t are possible.

Ghosh and Bagchi [9] proved that under the following conditions there exists an a.s. unique strong solution of SDE (2.46)–(2.47).

(D1) For each $e_i \in \mathbb{M}$, $i = 1,\dots,N$, $a(\cdot,e_i)$ and $b(\cdot,e_i)$ are bounded and Lipschitz continuous.

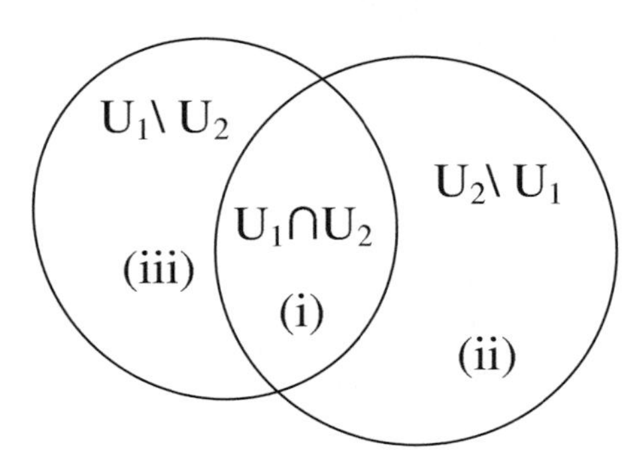

FIGURE 2.1: $U_1 \cup U_2$ is the projection of set $K_1 \cup K_2$ on $\mathbb{R}$.

(D2) For all $i, j \in \{1, \ldots, N\}$, $i \neq j$, functions $\lambda_{ij}(\cdot)$ are bounded and measurable, $\lambda_{ij}(\cdot) \geq 0$ for $i \neq j$ and $\sum_{j=1}^{N} \lambda_{ij}(\cdot) = 0$ for any $i \in \{1, \ldots, N\}$.

(D3) U_1, the projection of support of $g(\cdot,\cdot,\cdot)$ on $\mathbb{R}$, is bounded.

2.6.2 Stochastic Hybrid Model GB2 of Ghosh and Bagchi

Next, we present the GB2 model of Ghosh and Bagchi [9]. The state of the system at time t, denoted by (X_t, θ_t), takes values in $\bigcup_{n=1}^{\infty}(S_n \times \mathbb{M}_n)$, where $\mathbb{M}_n = \{e_1, e_2, \ldots, e_{N_n}\}$ and $S_n \subset \mathbb{R}^{d_n}$. Between the jumps of X_t the state equations are of the form

$$dX_t = a^n(X_t, \theta_t)dt + b^n(X_t, \theta_t)dW_t^n, \tag{2.49}$$

$$d\theta_t = \int_{\mathbb{R}} h^n(X_{t-}, \theta_{t-}, u)p(dt, du), \tag{2.50}$$

where for each $n \in \mathbb{N}$

$$a^n : S_n \times \mathbb{M}_n \to \mathbb{R}^{d_n}$$
$$b^n : S_n \times \mathbb{M}_n \to \mathbb{R}^{d_n \times d_n}$$
$$h^n : S_n \times \mathbb{M}_n \times \mathbb{R} \to \mathbb{R}^{N_n}.$$

Function h^n is defined in a similar way as (2.48) with rates $\lambda_{ij}^n : S_n \to \mathbb{R}$, $\lambda_{ij}^n \geq 0$ for $i \neq j$, and $\sum_{j=1}^{N_n} \lambda_{ij}^n(\cdot) = 0$ for any $i \in \{1, \ldots, N\}$. W^n is a standard d_n-dimensional Wiener process, and p is a Poisson random measure on $\mathbb{R}_+ \times \mathbb{R}$ with the intensity $dt \times m(du)$ as in the previous section.

For each $n \in \mathbb{N}$, let $A_n \subset S_n$, $D_n \subset S_n$. The set A_n is the set of instantaneous jump, whereas D_n is the destination set. It is assumed that for each $n \in \mathbb{N}$, A_n and D_n are closed sets, $A_n \cap D_n = \varnothing$ and $\inf_n d(A_n, D_n) > 0$, where $d(\cdot,\cdot)$ denotes the distance between two sets. If at some random time X_t hits A_n, then it executes an instantaneous jump. The destination of (X_t, θ_t) at this juncture is determined by a map

$$g_n : A_n \times \mathbb{M}_n \to \cup_{m\in\mathbb{N}}(D_m \times \mathbb{M}_m).$$

After reaching the destination, the process $\{X_t, \theta_t\}$ follows the same evolutionary mechanism over and over again.

Let $\{\eta_t\}$ be an $\mathbb{N}$ valued process defined by

$$\eta_t = n \text{ if } (X_t, \theta_t) \in S_n \times \mathbb{M}_n. \tag{2.51}$$

The $\{\eta_t\}$ is a piecewise constant process that changes from n to m when (X_t, θ_t) jumps from the regime $S_n \times \mathbb{M}_n$ to the regime $S_m \times \mathbb{M}_m$. Thus η_t is an indicator of a regime and a change in η_t means a switching in the regimes in which $\{X_t, \theta_t\}$ evolves.

Let

$$\begin{aligned}
\tilde{S} &= \{(x, e_i, n) | x \in S_n, e_i \in \mathbb{M}_n\},\\
\tilde{A} &= \{(x, e_i, n) | x \in A_n, e_i \in \mathbb{M}_n\},\\
\tilde{D} &= \{(x, e_i, n) | x \in D_n, e_i \in \mathbb{M}_n\}.
\end{aligned}$$

Then $\{X_t, \theta_t, \eta_t\}$ is an $\tilde{S}$-valued process, the set $\tilde{A}$ is the set where jumps occur and $\tilde{D}$ is the destination set for this process. The sets $\cup_n(S_n \times \mathbb{M}_n)$, $\cup_n(A_n \times \mathbb{M}_n)$, and $\cup_n(D_n \times \mathbb{M}_n)$ can be embedded in $\tilde{S}$, $\tilde{A}$, and $\tilde{D}$ respectively.

Let d^0 denote the injection map of $\cup_n(D_n \times \mathbb{M}_n)$ into $\tilde{D}$. Define the maps $\tilde{g}_1$, $\tilde{g}_2$, and $\tilde{h}$ as follows:

$$\begin{aligned}
&\tilde{g}_i : \tilde{A} \to \tilde{D},\, i = 1, 2,\\
&\tilde{h} : \tilde{A} \to \mathbb{N},
\end{aligned}$$

such that $\tilde{g}_1(x, e_i, n)$, $\tilde{g}_2(x, e_i, n)$ and $\tilde{h}(x, e_i, n)$ are the first, second and third component in $d^0(g_n(x, e_i))$ respectively. Let τ_{m+1} be the stopping time defined by

$$\tau_{m+1} = \inf\{t > \tau_m | X_{t-}, \theta_{t-}, \eta_{t-} \in \tilde{A}\}.$$

Now the equations for $\{X_t, \theta_t, \eta_t\}$ can be written as follows:

$$\begin{aligned}
dX_t &= \Big(a(X_t, \theta_t, \eta_t) + \sum_{m=0}^{\infty} [\tilde{g}_1(X_{\tau_m-}, \theta_{\tau_m-}, \eta_{\tau_m-}) - X_{\tau_m-}]\delta(t - \tau_m)\Big)dt \qquad (2.52)\\
&\quad + b(X_t, \theta_t, \eta_t)dW_t^{\eta_t},\\
d\theta_t &= \int_{\mathbb{R}} h(X_{t-}, \theta_{t-}, \eta_{t-}, u)p(dt, du) \qquad (2.53)\\
&\quad + \sum_{m=0}^{\infty} [\tilde{g}_2(X_{\tau_m-}, \theta_{\tau_m-}, \eta_{\tau_m-}) - \theta_{\tau_m-}]\delta(t - \tau_m)dt,\\
d\eta_t &= \sum_{m=0}^{\infty} [\tilde{h}(X_{\tau_m-}, \theta_{\tau_m-}, \eta_{\tau_m-}) - \eta_{\tau_m-}]I_{\{\tau_m \le t\}}, \qquad (2.54)
\end{aligned}$$

where δ is the Dirac measure and $a(x,e_i,n) = a^n(x,e_i)$, $b(x,e_i,n) = b^n(x,e_i)$, and $h(x,e_i,n,u) = h^n(x,e_i,u)$.

To ensure the existence of an a.s. unique strong solution of SDE (2.52)–(2.54), Ghosh and Bagchi [9] adopted the following assumptions:

(E1) For each $n \in \mathbb{N}$ and $e_i \in \mathbb{M}_i$, $a^n(\cdot,e_i)$ and $b^n(\cdot,e_i)$ are bounded and Lipschitz continuous.

(E2) For each $n \in \mathbb{N}$, $i,j = 1,\ldots,M_n$, $i \neq j$, functions $\lambda^n_{ij}(\cdot)$ are bounded and measurable, $\lambda^n_{ij}(\cdot) \geq 0$ for $i \neq j$ and $\sum_{j=1}^N \lambda^n_{ij}(\cdot) = 0$ for any $i \in \{1,\ldots,N\}$.

(E3) The maps g_n, $n \in \mathbb{N}$, are bounded and uniformly continuous.

(E4) $\inf_n d(A_n,D_n) > 0$.

2.6.3 Hierarchy Between Stochastic Hybrid Models

In this subsection we discuss the differences between the models and determine the hierarchy of these models. This hierarchy is organized on the basis of the behaviors of the processes, e.g., different types of jumps, and not on the assumptions applied to the models. We summarize this hierarchy of models in Figure 2.2.

First, let us compare GB1 and HB1 (=KB1 with $g_1 = 0$). Both models allow either independent or simultaneous jumps and switches of X_t and θ_t. However, there are some differences in assumptions imposed on the coefficients and in construction of the jump and switching coefficients. The first two terms (i.e., the drift and the diffusion term) in (2.29) and in (2.46) are identical. However, to assure the existence of a strong unique solution of SDE (2.46)–(2.47), Ghosh and Bagchi [9] assume that the drift and the diffusion coefficients are bounded, i.e., condition (D1). To prove the similar result for SDE (2.29)–(2.30) more general growth condition (A1) is adopted. The construction of the "switching" terms (2.30) and (2.47) is almost identical with some minor differences in defining the "rate" intervals. The conditions on the "rate" functions $\lambda(e_i,e_j,\cdot)$ and $\lambda_{ij}(\cdot)$ are the same, i.e., these functions are assumed to be bounded and measurable for all $i,j = 1,\ldots,N$, i.e., conditions (A3) and (D2).

There is a substantial difference in the construction of the g_2 jump part of X_t in the HB1/KB1 and GB1 models. In GB1 the jumps of X_t are described by a stochastic integral of function g with respect to a Poisson random measure $p(dt,du)$ with intensity $dt \times m(du)$, where m is the Lebesgue measure on $U = \mathbb{R}$. In order to satisfy the existence and uniqueness of solution, U_1, the projection of support of function g on $U = \mathbb{R}$, must be bounded, i.e., condition (D3). In HB1/KB1 the g_2 jumps of X_t are also defined by a stochastic integral driven by Poisson random measure $p_2(dt,du)$ but with intensity $dt \times m(du_1) \times \bar{\mu}(\underline{u})$, where m is the Lebesgue measure on $U_1 = \mathbb{R}$ and $\bar{\mu}$ is a probability measure on $\underline{U} = \mathbb{R}^{d-1}$. The integrand function g_2, which determines the jump size of X_t, compared to function g, has an extra argument

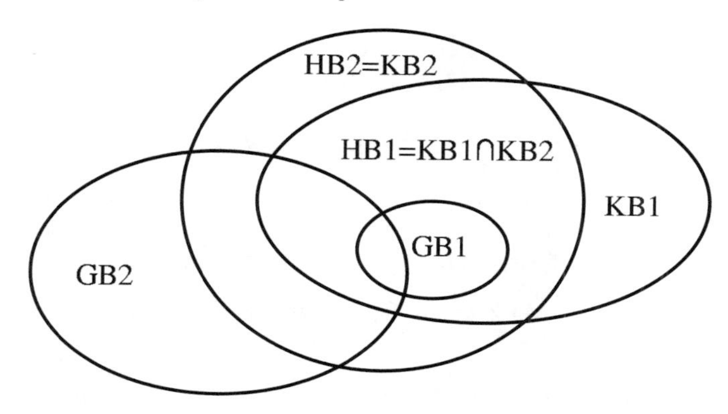

FIGURE 2.2: The hierarchy between stochastic hybrid models; the sets HB2=KB2 and GB2 fall within the set of Generalized Stochastic Hybrid Processes [4]. KB1 provides complementary modelling power in allowing processes that have infinite variation in jumps on a finite time interval.

$\underline{u} \in \underline{U} = \mathbb{R}^{d-1}$, and, since the intensity of p_2 with respect to $\underline{u}$ is a probability measure $\bar{\mu}$ (which is always finite), the projection of support of g_2 on $\underline{U} = \mathbb{R}^{d-1}$ can be unbounded. This gives some extra freedom in modelling the jumps of X_t component. It is only required that function g_2 must satisfy condition (A4) or the verifiable (A4′). From this follows that model HB1/KB1 includes model GB1 as a special case (GB1 $\subset$ HB1 $\subset$ KB1).

Models KB2 and GB2 have some similarities. Let us see what are the main differences between SDE (2.36)–(2.39) and SDE (2.52)–(2.54). Solutions of SDE (2.52)–(2.54) are the $\bigcup_{n=1}^{\infty}(S_n \times \mathbb{M}_n)$-valued switching diffusions with hybrid jumps at the boundary. Before hitting the boundary $\{X_t, \theta_t\}$ evolves as an $(S_n \times \mathbb{M}_n)$-valued switching diffusion in some regime $\eta_t = n \in \mathbb{N}$. The drift and the diffusion coefficients and the mapping determining a new starting point of the process after the hitting the boundary can be different for every different regime $n \in \mathbb{N}$.

Solutions of SDE (2.36)–(2.39) are the $(\mathbb{R}^n \times \mathbb{M})$-valued switching-jump diffusions with hybrid jumps at the boundary. The dimension of the state space and the coefficients of SDE are fixed. Hence, on this specific point, model GB2 is more general. However the jump term in KB2, see Equation (2.36), is more general than the jump term in GB2, see Equation (2.52).

Now let us have a look at conditions (E1)–(E4). Condition (E1) implies that our local conditions (A1) and (A2) for SDE (2.29)–(2.30) are definitely satisfied. Conditions (E2) and (E3) imply that conditions (A3) and (A4) for SDE (2.29)–(2.30) are satisfied. Condition (E4) implies that (B1) and (B2) adopted to SDE (2.36)–(2.39) are satisfied. It ensures that after the jump the process starts inside of some open set, but not on a boundary. Condition (B3) of SDE (2.36)–(2.39) is missing for GB2 [9].

In general GB2 is not a subclass of KB2 (or HB2) since in GB2 the state of the system (X_t, θ_t) takes values in $\bigcup_{k=1}^{\infty}(S_k \times \mathbb{M}_k)$, where $\mathbb{M}_k = \{e_1, e_2, \ldots, e_{N_k}\}$ and

$S_k \subset \mathbb{R}^{d_k}$ may be different for different k's. If $(S_k \times \mathbb{M}_k) = (\mathbb{R}^n \times \mathbb{M})$ for all $k \in \mathbb{N}$ then obviously GB2 $\subset$ KB2 (=HB2).

2.7 Markov and Strong Markov Properties

In this section we prove Markov and strong Markov properties for model HB2=KB2 (Section 2.5).

Assume we are given the following objects:

- A measurable space $(S, \mathscr{S})$.
- Another measurable space $(\Omega, \mathscr{G})$ and a family of σ-algebras $\{\mathscr{G}_t^s, 0 \leq s \leq t \leq \infty\}$, such that $\mathscr{G}_t^s \subset \mathscr{G}_v^u \subset \mathscr{G}$ provided $0 \leq u \leq s \leq t \leq v$; $\mathscr{G}_t^s$ denotes a σ-algebra of events on time interval $[s,t]$; we write $\mathscr{G}_t$ in place of $\mathscr{G}_t^0$ and $\mathscr{G}^s$ in place of $\mathscr{G}_\infty^s$.
- A probability measure $P_{s,x}$ for each pair $(s,x) \in [0,\infty) \times S$ on $\mathscr{G}^s$.
- A function (stochastic process) $\xi_t(\omega) = \xi(t,\omega)$ defined on $[0,\infty) \times \Omega$ with values in S.

The system consisting of these four objects will be denoted by $\{\xi_t, \mathscr{G}_t^s, P_{s,x}\}$ [10].

DEFINITION 2.7 *A system of objects $\{\xi_t, \mathscr{G}_t^s, P_{s,x}\}$ is called a Markov process provided:*

(i) for each $t \in [0,\infty)$ $\xi_t(\omega)$ is measurable mapping of $(\Omega, \mathscr{G})$ into $(S, \mathscr{S})$;

(ii) for arbitrary fixed s,t and B $(0 \leq s \leq t, B \in \mathscr{S})$ the function $P(s,x,t,B) = P_{s,x}(\xi_t \in B)$ is $\mathscr{S}$-measurable with respect to x;

(iii) $P_{s,x}(\xi_s = x) = 1$ for all $s \geq 0$ and $x \in S$;

(iv) $P_{s,x}(\xi_u \in B \mid \mathscr{G}_t^s) = P_{t,\xi_t}(\xi_u \in B)$ for all $s,t,u, 0 \leq s \leq t \leq u < \infty$, $x \in S$ and $B \in \mathscr{S}$.

The measure $P_{s,x}$ should be considered as a probability law which determines the probabilistic properties of the process $\xi_t(\omega)$ given that it starts at point x at the time s. Condition (iv) in Definition 2.7 expresses the Markov property of the processes. Let $\mathbb{E}_{s,x}$ denote the expectation with respect to measure $P_{s,x}$. For $\mathscr{G}^s$-measurable random variable $\xi(\omega)$

$$\mathbb{E}_{s,x}[\xi(\omega)] = \int \xi(\omega) P_{s,x}(d\omega).$$

It is not difficult to show that the Markov property (iv) in Definition 2.7 can be rewritten in terms of expectations as follows:

$$\mathbb{E}_{s,x}[f(\xi_u) \mid \mathscr{G}_t^s] = \mathbb{E}_{t,\xi_t}[f(\xi_u)], \; 0 \le s \le t \le u < \infty,$$

where f is an arbitrary $\mathscr{S}$-measurable bounded function.

Next, let us show that process

$$\begin{cases} X_t(\omega) = \sum_{n=1}^{\infty} X_t^n(\omega) I_{\left[\tau_{n-1}^E(\omega), \tau_n^E(\omega)\right)}(t) \\ \theta_t(\omega) = \sum_{n=1}^{\infty} \theta_t^n(\omega) I_{\left[\tau_{n-1}^E(\omega), \tau_n^E(\omega)\right)}(t) \end{cases} \tag{2.55}$$

defined as a concatenation of solutions $\{X_t^n, \theta_t^n\}$ of the system of SDE (2.36)–(2.39) (see Sections 2.4 and 2.5) is Markov. We follow the approach used in [11]. Let $\xi_t^{s,\eta} = (X_t^{s,x}, \theta_t^{s,\theta})$ denote the process (2.55) on $[s,\infty)$ satisfying initial condition $\xi_s^{s,\eta} = \eta = (X_s^{s,x}, \theta_s^{s,\theta})$. Note that now $S = \mathbb{R}^n \times \mathbb{M}$ and $\mathscr{S} = \mathscr{B}_{\mathbb{R}^n \times \mathbb{M}}$ is the σ-algebra of Borel sets on $\mathbb{R}^n \times \mathbb{M}$. Assume that conditions of Theorem 2.8 are satisfied. Let $\mathscr{F}_t^s$, $s < t$ be the σ-algebras generated by $\{W_u - W_s, p_2([s,u], dz), \beta_u, u \in [s,t]\}$, $\mathscr{F}_t^0 = \mathscr{F}_t$, $\mathscr{F}_\infty^s = \mathscr{F}^s$. For $s \le t$ the σ-algebras $\mathscr{F}_s$ and $\mathscr{F}^s$ are independent. Process $\xi_t^{s,\eta}$ is $\mathscr{F}^s$-measurable, hence, it is independent of σ-algebra $\mathscr{F}_s$. Let η_s be an arbitrary $\mathbb{R}^n \times \mathbb{M}$-valued $\mathscr{F}_s$ measurable random variable. Then ξ_t^{s,η_s}, $t \ge s$, is unique $\mathscr{F}_t$-measurable process on $[s,\infty)$ satisfying the initial condition $\xi_s^{s,\eta_s} = \eta_s$. Since for $u < s$ process $\xi_t^{u,y}$ is $\mathscr{F}_t$-measurable on $[s,\infty)$ with initial condition $\xi_s^{u,y}$ then the following equality holds

$$\xi_t^{u,y} = \xi_t^{s,\xi_s^{u,y}}, \; u < s < t. \tag{2.56}$$

Let φ be a bounded measurable function on $\mathbb{R}^n \times \mathbb{M}$, let ζ_s be an arbitrary bounded $\mathscr{F}_s$-measurable quantity. The independence of $\mathscr{F}_s$ and $\mathscr{F}^s$ and the Fubini theorem imply that measure P on $\mathscr{F}_\infty$ is a product of measures P_s and P^s, where P_s is a restriction of P on $\mathscr{F}_s$, where P^s is a restriction of P on $\mathscr{F}^s$, and

$$\mathbb{E}[\varphi(\xi_t^{u,y})\zeta_s] = \mathbb{E}[\varphi(\xi_t^{s,\xi_s^{u,y}})\zeta_s] = \mathbb{E}\left[\zeta_s(\mathbb{E}[\varphi(\xi_t^{s,x})])_{x=\xi_s^{u,y}}\right].$$

Since $\xi_s^{u,y}$ is $\mathscr{F}_s$-measurable then $\mathbb{E}[\varphi(\xi_t^{u,y}) \mid \mathscr{F}_s] = \left[\mathbb{E}[\varphi(\xi_t^{s,x})]\right]_{x=\xi_s^{u,y}}$. Let

$$P(s,x,t,B) = P(\xi_t^{s,x} \in B), \; B \in \mathscr{B}_{\mathbb{R}^n \times \mathbb{M}}, \tag{2.57}$$

here $\mathscr{B}_{\mathbb{R}^n \times \mathbb{M}}$ is the σ-algebra of Borel sets on $\mathbb{R}^n \times \mathbb{M}$. Then, by taking $\varphi = I_B$, we obtain

$$P(\xi_t^{u,y} \in B \mid \mathscr{F}_s) = P(s, \xi_s^{u,y}, t, B). \tag{2.58}$$

If ξ_t is an arbitrary process defined by (2.55), by the same reasoning with help of which equalities (2.56) and (2.58) have been obtained, one can show that $\xi_t = \xi_t^{s,\xi_s}$ for $s < t$ and that

$$P(\xi_t \in B \mid \mathscr{F}_s) = P(s, \xi_s, t, B).$$

Hence, the process defined by (2.55) is a Markov process with transition probability $P(s,x,t,B)$ defined by (2.58). To be precise, we have shown that the system of objects $\{(X_t,\theta_t),\mathscr{F}_t^s,P_{s,(x,\theta)}\}$, where $P_{s,(x,\theta)}\big((X_t,\theta_t)\in B\big) = P(s,(x,\theta),t,B) = P\big((X_t^{s,x},\theta_t^{s,\theta})\in B\big)$, $B\in\mathscr{B}_{\mathbb{R}^n\times\mathbb{M}}$, is a Markov process.

Next, we prove the Markov property

$$P_{s,x}(\xi_u\in B\mid\mathscr{G}_t^s) = P_{t,\xi_t}(\xi_u\in B),\ \ s\le t\le u$$

remains valid also when a fixed time moment t is replaced by a stopping time.

Let $\{\xi_t(\omega),\mathscr{G}_t^s,P_{s,x}\}$ be a Markov process in the space $(S,\mathscr{S})$. Let $\mathscr{T}$ denote the σ-algebra of Borel sets on $[0,\infty)$.

DEFINITION 2.8 *A Markov process is called strong Markov if:*

(i) the transition probability $P(s,x,t,B)$ for a fixed B is a $\mathscr{T}\times\mathscr{S}\times\mathscr{T}$-measurable function of (s,x,t) on the set $0\le s\le t<\infty$, $x\in S$;

(ii) it is progressively measurable;

(iii) for any $s\ge 0$, $t\ge 0$, $\mathscr{S}$-measurable function $f(x)$ and arbitrary stopping time τ,

$$\mathbb{E}_{s,x}[f(\xi_{t+\tau})\mid\mathscr{G}_\tau^s] = \mathbb{E}_{\tau,\xi_\tau}[f(\xi_{t+\tau})]. \tag{2.59}$$

REMARK 2.8 For Equation (2.59) to be satisfied, it is necessary that the random variable $g(\xi_\tau,\tau,t+\tau) = \mathbb{E}_{\tau,\xi_\tau}[f(\xi_{t+\tau})]$ be $\mathscr{G}_\tau^s$-measurable. For this reason assumptions (i) and (ii) make part of the definition of the strong Markov property [10]. ■

Now we return to the process $\xi_t = (X_t,\theta_t)$ defined in Section 2.5. We have shown that it is a Markov process. The following theorem proves that it is a strong Markov process also.

THEOREM 2.9 *Assume (A1)–(A4) and (B1)–(B3). Let W, p_2, μ^E, X_0 and θ_0 be independent. Let $\mathscr{F}_t^s$, $s<t$ be the σ-algebras generated by $\{W_u - W_s, p_2(dz,[s,u]),\beta_u, u\in[s,t]\}$. For any bounded Borel function $f:\mathbb{R}^n\times\mathbb{M}\to\mathbb{R}$ and any $\mathscr{F}_t^s$-stopping time τ*

$$\mathbb{E}_{s,x}[f(\xi_{t+\tau})\mid\mathscr{F}_\tau^s] = \mathbb{E}_{\tau,\xi_\tau}[f(\xi_{t+\tau})].$$

PROOF Let $\{\sigma_k, k=0,1,\dots\}$ denote the ordered set of the stopping times $\{\tau_k^E, k=1,2,\dots\}$ and $\{\tau_k, k=0,1,\dots\}$. The latter set is the set of the stopping times generated by Poisson random measure p_2. Then on each time interval $[\sigma_{k-1},\sigma_k)$, $k=1,2,\dots$ process ξ_t evolves as a diffusion staring at point

$\xi_{\sigma_{k-1}}$ at the time σ_{k-1}. This means that on each time interval $[\sigma_{k-1}, \sigma_k)$ the strong Markov property holds. Let $\mathscr{F}^s_\tau$ be the σ-algebra generated by the $\mathscr{F}^s_t$-stopping time τ. The sets $\{\omega : \tau(\omega) \in [\sigma_{k-1}(\omega), \sigma_k(\omega))\}$, $k = 1, 2, \ldots$ are $\mathscr{F}^s_\tau$-measurable. Hence

$$\begin{aligned}
\mathbb{E}_{s,x}[f(\xi_{t+\tau}) \mid \mathscr{F}^s_\tau] &= \sum_{k=0}^{\infty} I_{[\sigma_{k-1},\sigma_k)}(\tau)\mathbb{E}_{s,x}\left[f(\xi_{t+\tau}) \mid \mathscr{F}^s_\tau\right] \\
&= \sum_{k=0}^{\infty} \mathbb{E}_{s,x}\left[I_{[\sigma_{k-1},\sigma_k)}(\tau) f(\xi_{t+\tau}) \mid \mathscr{F}^s_\tau\right] \\
&= \sum_{k=0}^{\infty} \mathbb{E}_{\tau,\xi_\tau}\left[I_{[\sigma_{k-1},\sigma_k)}(\tau) f(\xi_{t+\tau})\right] \\
&= \mathbb{E}_{\tau,\xi_\tau}\left[\sum_{k=0}^{\infty} I_{[\sigma_{k-1},\sigma_k)}(\tau) f(\xi_{t+\tau})\right] \\
&= \mathbb{E}_{\tau,\xi_\tau}\left[f(\xi_{t+\tau})\right].
\end{aligned}$$

This completes the proof. ■

REMARK 2.9 The approach taken in the proof of Theorem 2.9 was initially developed for switching diffusion processes by Vera Minina, Twente University. ■

2.8 Concluding Remarks

We have given an overview of stochastic hybrid processes as strongly unique solutions to stochastic differential equations on hybrid state space. These SDEs are driven by Brownian motion and Poisson random measure. Our overview has shown several new classes of stochastic hybrid processes each of which goes significantly beyond the well known class of jump-diffusions with Markov switching coefficients, whereas semimartingale and strong Markov properties have been shown to hold true. The main phenomena covered by these extensions are:

- Hybrid jumps, i.e., continuous valued jumps that happen simultaneously with a mode switch, and the size of which depends of the mode value prior and after the switch;
- Instantaneous jump reflection at the boundary, i.e., upon hitting a given measurable boundary of the Euclidean valued set, the continuous valued process component jumps instantaneously away from the boundary;
- The continuous valued process component may jump so frequently that it is no longer a process of finite variation;

- Feasible combinations of these phenomena within one SDE such that its solution still is a semimartingale strong Markov process.

For each of the extensions, our overview provides the specific conditions on the SDE under which there exist strongly unique semimartingale solutions. We also presented a novel approach to prove strong Markov property for general stochastic hybrid processes.

References

[1] Bensoussan, A. and J.L. Menaldi (2000). Stochastic hybrid control. *J. Math. Analysis and Applications* **249**, 261–268.

[2] Blom, H.A.P. (2003). Stochastic hybrid processes with hybrid jumps. *Proc. IFAC Conf. Analysis and Design of Hybrid Systems* (ADHS 2003). Eds: S. Engell, H. Guéguen, J. Zaytoon. Saint-Malo, Brittany, France, June 16–18, 2003.

[3] Blom, H.A.P., G.J. Bakker, M.H.C. Everdij and M.N.J. van der Park (2003). *Stochastic analysis background of accident risk assessment for Air Traffic Management*. Hybridge Report D2.2. National Aerospace Laboratory NLR. http://www.nlr.nl/public/hosted-sites/hybridge.

[4] Bujorianu, M. and J. Lygeros (2004). General stochastic hybrid systems: Modelling and optimal control. *Proc. IEEE Conference on Decision and Control.* Bahamas.

[5] Davis, M.H.A. (1984). Piecewise-deterministic markov processes: A general class of non-diffusion stochastic models. *J.R. Statist. Soc. B* **46**(3), 353–388.

[6] Davis, M.H.A. (1993). *Markov Models and Optimization*. Chapman & Hall. London.

[7] Ghosh, M.K., A. Arapostathis and S.I. Marcus (1993). Optimal control of switching diffusions with application to flexible manufacturing systems. *SIAM J. Control Optimization* **31**, 1183–1204.

[8] Ghosh, M.K., A. Arapostathis and S.I. Marcus (1997). Ergodic control of switching diffusions. *SIAM J. Control Optimization* **35**, 1952–1988.

[9] Ghosh, M.K. and A. Bagchi (2004). Modeling stochastic hybrid systems. *System Modeling and Optimization. Proceedings of the* 21st *IFIP TC7 Conference* (J. Cagnol and J.P. Zolèsio, Eds.). Sophia Antipolis, France. pp. 269–279.

[10] Gihman, I.I. and A.V. Skorohod (1975). *The theory of stochastic processes II.* Springer-Verlag. Berlin.

[11] Gihman, I.I. and A.V. Skorohod (1982). *Stochastic differential equations and their applications*. Naukova Dumka. Kiev. in Russian.

[12] Hu, J., J. Lygeros and S. Sastry (2000). Towards a theory of stochastic hybrid systems. *Lecture notes in Computer Science*. Vol. 1790. pp. 160–173.

[13] Jacod, J. and A.N. Shiryaev (1987). *Limit Theorems for Stochastic Processes*. Springer. Berlin.

[14] Jacod, J. and A.V. Skorokhod (1996). Jumping Markov processes. *Ann. Inst. Henri Poincare* **32**(1), 11–67.

[15] Krystul, J. and H.A.P. Blom (2005). Generalized stochastic hybrid processes as strong solutions of stochastic differential equations. *Hybridge report D2.3*, http://www.nlr.nl/public/hosted-sites/hybridge/.

[16] Lepeltier, J.P. and B. Marchal (1976). Probleme des martingales et equations differentielles stochastiques associees a un operateur integro-differentiel. *Ann. Inst. Henri Poincare* **12**(Section B), 43–103.

[17] Marcus, S.I. (1978). Modeling and analysis of stochastic differential equations driven by point processes. *IEEE Trans. Information Theory* **24**, 164–172.

[18] Mariton, M. (1990). *Jump Linear Systems in Automatic Control*. Marcel Dekker. New York.

[19] Pola, G., M. Bujorianu, J. Lygeros and M. D. Di Benedetto (2003). Stochastic hybrid models: An overview. *Proc. IFAC Conference on Analysis and Design of Hybrid Systems ADHS03*. Saint Malo, France. p. 44-50.

[20] Schouten, F.A. Van Der Duyn and A. Hordijk (1983). Average optimal policies in markov decision drift processes with applications to a queueing and a replacement model. *Adv. Appl. Prob.* **15**,p. 274–303.

[21] Snyder, D.L. (1975). *Random Point Processes*. Wiley. New York.

[22] Wonham, W.M. (1970). *Random differential equations in control theory. Probability Analysis in Applied Mathematics*. Vol. 2. Academic Press.

Chapter 3

Compositional Modelling of Stochastic Hybrid Systems

Stefan Strubbe
University of Twente

Arjan van der Schaft
University of Groningen

3.1 Introduction 47
3.2 Semantical Models 48
3.3 Communicating PDPs 55
3.4 Conclusions 75
References 76

3.1 Introduction

Stochastic hybrid systems often have a complex structure, meaning that they consist of many interacting components. Think for example of an Air Traffic Management system where multiple aircraft, multiple humans, etc., are involved. These systems are too complex to be modelled in a monolithic way. Therefore, for these systems there is a need for compositional modelling techniques, where the system can be modelled in a stepwise manner by first modelling all individual components and secondly by connecting these components to each other.

In this chapter we present the framework of CPDPs (Communicating Piecewise Deterministic Markov Processes), which is a compositional modelling framework for stochastic hybrid systems of the PDP type. (For the PDP model we refer to [4] or [3].) In this framework each component of a complex stochastic hybrid system can be modelled as a single CPDP and all these component CPDPs can be connected through a composition operator. As we will see, connecting two or more CPDP components results in another CPDP. In other words, the class of CPDP is closed under the composition operation. CPDP is an automaton framework (like the models from [1] and [11]). Another framework for compositional modelling of PDP-type systems is [5, 6], which is a Petri-net framework.

The framework of CPDPs can be seen as an extension of the established framework of IMCs (Interactive Markov Chains, [7]). The main extension of CPDPs with respect to IMCs is the same as the extension of a Piecewise Deterministic Markov Process with respect to a continuous-time Markov chain: it allows for a general continuous dynamics in the continuous state variables, while the jump rates of the Poisson processes may depend on these continuous state variables, and the continuous state variables are stochastically reset at event times. As a result, CPDPs cover quite a large class of stochastic hybrid systems as encountered in applications, although *diffusions* can*not* be included.

A CPDP is a syntactical object. To make clear how a CPDP behaves, we need to give a formal semantics of CPDP. In Section 3.2 we introduce some semantical models and we explain the behavior of these models. Then, in Section 3.3 we introduce the CPDP model and we give its semantics in terms of the semantical models of Section 3.2. By giving these semantics, we make clear how the CPDP behaves. At the end of Section 3.3, the CPDP model is extended to the so-called value-passing CPDP model. In this extended model there are richer interaction possibilities: CPDP components can now send information to each other concerning the continuous variables.

The contents of this chapter is based on the papers [13] and [16] and on the thesis [12]. For more material on this subject, for further explanation of the material of this chapter, and for the proofs of the theorems of this chapter we refer to the thesis [12]. A different introduction to the framework of CPDPs, on a less general level and with an emphasis on the description of CPDPs as an extension to the existing framework of Interactive Markov Chains, can be found in [17].

3.2 Semantical Models

Semantical models are used to capture/express the behavior of a syntactical model. Semantical models are also used to compare syntactical models with each other. For example, if two syntactical objects have the same semantics, they can be regarded equivalent.

In this chapter we consider two syntactical and four semantical models. The syntactical models are: Piecewise Deterministic Markov Processes (PDPs) and Communicating Piecewise Deterministic Markov Processes (CPDPs). The semantical models are: Transition Mechanism Structures (TMSs), Non-deterministic Transition Systems (NTSs), Continuous Flow Spontaneous Jump Systems (CFSJSs), and Forced Transition Systems (FTSs). We can distinguish different levels for the semantical and syntactical models that we use. If the behavior of a semantical model M_1 can be expressed within the semantical model M_2, then we say that M_1 is a higher level semantical model than M_2. From high to low we consider the following levels.

- Syntactical level: PDP, CPDP
- High semantical level: CFSJS, NTS

- Intermediate semantical level: FTS
- Low semantical level: TMS

In this section we define all semantical models and we show how these models are related to each other, i.e., how lower semantical objects express the behavior of higher semantical objects.

All semantical models in this section will be used to capture a certain part of the behavior of PDP or CPDP-type systems. The final definition of the semantics of CPDPs in Section 3.3.2 will be done in terms only of a CFSJS and an NTS.

3.2.1 Transition Mechanism Structure

A Transition Mechanism Structure (TMS) gives us the random variables and the flow maps that are necessary to determine execution paths of a stochastic (hybrid) system. Once we know the TMS of a system, we can directly determine the stochastic process and the stochastic execution paths of the system. The stochastic process or the execution paths can be used to analyze the systems (stochastic) behavior.

A TMS consists of two parts: 1. a transition mechanism which determines the time of a transition and the target state of the transition, 2. a flow map which determines the continuous flow between two transitions. The semantical model TMS is formally defined as follows.

DEFINITION 3.1 *A* Transition Mechanism Structure *(TMS) is a tuple* (E, ξ_0, ϕ, TM). *E is a Borel state space, ξ_0 is the initial state. ϕ is a flow map, i.e., the process evolves from state ξ_0 at time zero to state $\phi(t,\xi)$ at time t if no transitions occur in the interval $[0,t]$, etc. TM is a* transition mechanism *on E. A* transition mechanism *on a Borel space E is a pair (T,Q) with $T: E \to RV(\bar{\mathbb{R}}_+, \mathscr{B}(\bar{\mathbb{R}}_+))$, where $RV(\Omega, \mathscr{F})$ denotes the set of all random variables (defined on any probability space) taking values in the measurable space $(\Omega, \mathscr{F})$ and where $\bar{\mathbb{R}}_+ := \mathbb{R}_+ \cup \{\infty\}$, and with $Q: E \to Prob(E)$. Here $Prob(E)$ denotes the set of all probability measures on the measurable space $(E, \mathscr{B}(E))$.*

Given a state $\xi \in E$, the transition mechanism $TM = (T,Q)$ determines a transition-time t and a transition target state ξ' by drawing a sample t from the random variable $T(\xi)$ followed by drawing a sample ξ' from the probability measure $Q(\phi(t,\xi))$. We also say that with this procedure we have drawn the sample (t,ξ') from the transition mechanism $TM(\xi)$. If the sample ∞ is drawn from $T(\xi)$, then this is not followed by drawing a sample from Q. We then say that the sample $(\infty, \emptyset)$ is drawn from $TM(\xi)$.

An execution path of a TMS (E, ξ_0, ϕ, TM) is generated as follows. Draw a sample (t_1, ξ_1) from $TM(\xi_0)$. For $t \in [0, t_1[$ the execution path has value $\phi(t, \xi_0)$. Draw a sample (t_2, ξ_2) from $TM(\xi_1)$. For $t \in [t_1, t_1 + t_2[$, the execution path has value $\phi(t - t_1, \xi_1)$. Draw a sample (t_3, ξ_3) from $TM(\xi_2)$, etc.

TMS forms the lowest semantical level that we use. The TMS of a system can be derived from the CFSJS and the FTS semantics of the system. This derivation is done in Section 3.2.4.

3.2.2 Continuous Flow Spontaneous Jump System (CFSJS)

For both the syntactical models PDP and CPDP, we have that transitions, where the state instantaneously jumps to another state, can happen in two ways: 1. spontaneously (with some probability distribution), 2. forced (when the state reaches some "forbidden" area and is forced to jump to a another state). In between transitions, the state of a PDP/CPDP evolves continuously. The part of the PDP/CPDP system behavior concerning the continuous evolution and the spontaneous transitions is captured by a CFSJS. The part concerning the forced transitions is captured by an FTS.

DEFINITION 3.2 *A CFSJS is a tuple $(E, \xi_0, \phi, \lambda, Q)$. The state space E is a Borel space. ξ_0 is the initial state, $\phi : \mathbb{R}_+ \times E \rightarrow E$ is the flow map, $\lambda : E \rightarrow \mathbb{R}_+$ is the jump rate and $Q : E \rightarrow Prob(E)$ is the transition measure.*

The jump rate $\lambda(\xi)$ of a CFSJS at state ξ determines the probability of a spontaneous transition "near" state ξ as follows: if the system is at state ξ at time t, then the probability that a spontaneous transition occurs in the interval $[t, t + \Delta t]$ equals $\lambda(\xi)\Delta t + o(\Delta t)$, where $o(\Delta t)$ denotes a function such that $\lim_{\Delta t \rightarrow 0} \frac{o(\Delta t)}{\Delta t} = 0$. In other words, for Δt small enough, the probability that a spontaneous transition occurs in the interval $[t, t + \Delta t]$ equals approximately $\lambda(\xi)\Delta t$. (This means that if the process is at state ξ at time $\hat{t}$ and the next jump happens at time $\hat{t} + t$, then t is determined by a Poisson process with intensity $\lambda(\phi(t, \xi))$, see [9].)

If a systems behavior is completely captured by a CFSJS, i.e., if there are no forced transitions, then the CFSJS completely determines the stochastic executions of the system. By determining the TMS of a CFSJS, we indirectly determine the stochastic process/executions of the CFSJS.

DEFINITION 3.3 *The TMS (Transition Mechanism Structure) of a CFSJS $(E, \xi_0, \phi, \lambda, Q)$ is defined as $(E, \xi_0, \phi, (T, Q))$, where, for $\xi \in E$, the* survivor function *$\Psi_{T(\xi)}(t)$ of $T(\xi)$, is defined as*

$$\Psi_{T(\xi)}(t) = \mathrm{e}^{-\int_0^t \lambda(\phi(s,\xi))\mathrm{d}s}. \tag{3.1}$$

The survivor function *$\Psi_{T(\xi)}(t)$ is by definition equal to $\mathrm{P}(\mathrm{T}(\xi) > \mathrm{t})$ and thus expresses the probability that $T(\xi)$ "survives" be the time instant t, or, in other words, expresses the probability that a transition does not occur until time t.*

We show that (3.1) indeed expresses that if at time zero, i.e., the time of the previous transition, the process is at state $\hat{\xi}$ and at time t the process is at state ξ, then, given that no jump occurred in the interval $[0, t]$, the probability that a spontaneous transition occurs in the interval $[t, t+\Delta t]$ equals $\lambda(\xi)\Delta t + o(\Delta t)$.

$$\mathrm{P}(T(\hat{\xi}) \in [t, t+\Delta t] \mid T(\hat{\xi}) > t) = \frac{\psi_{T(\hat{\xi})}(t) - \psi_{T(\hat{\xi})}(t+\Delta t)}{\psi_{T(\hat{\xi})}(t)} =$$

$$1 - \mathrm{e}^{-\int_0^{t+\Delta t} \lambda(\phi(s,\hat{\xi}))\mathrm{d}s + \int_0^t \lambda(\phi(s,\hat{\xi}))\mathrm{d}s} = 1 - \mathrm{e}^{-\int_0^{\Delta t} \lambda(\phi(s+t,\hat{\xi}))\mathrm{d}s},$$

which, after Taylor expansion, equals $\lambda(\xi)\Delta t + o(\Delta t)$.

3.2.2.1 Memoryless Property of the Jump Times

Let X be a TMS with transition mechanism (T, Q) and flow map ϕ. We can execute X as described in Section 3.2.1. Suppose that during such an execution we lose at some time $\hat{t}$, while the process is at state $\xi_{\hat{t}}$, the information of the last drawn sample from T. Can we now continue the execution path from $\xi_{\hat{t}}$ in a correct way, or do we have to start a new execution path from ξ_0? Let t_l denote the time of the previous transition (before $\hat{t}$) and let ξ_{t_l} be the state of the execution path at time t_l, which is the target state of the transition at time t_l. If t_l and ξ_{t_l} are known, then it is correct to continue the stochastic execution from $\hat{t}$ as follows: draw a sample $\tilde{t}$ from $\tilde{T}$, where $\tilde{T}$ is a random variable such that

$$\mathrm{P}(\tilde{T} > t) = \mathrm{P}(T(\xi_{t_l}) > \hat{t} - t_l + t | T(\xi_{t_l}) > \hat{t} - t_l).$$

Now let the execution path flow from state $\xi_{\hat{t}}$ to state $\phi(\tilde{t}, \xi_{\hat{t}})$ and switch at state $\phi(\tilde{t}, \xi_{\hat{t}})$ according to the measure $Q(\phi(\tilde{t}, \xi_{\hat{t}}))$. From the new state we again draw a new sample from the transition mechanism, etc.

Now we show that we can determine $\mathrm{P}(\tilde{T} > t)$ without knowing t_l and ξ_{t_l}, and consequently we can conclude that we can correctly continue the execution path from state $\hat{\xi}$ without having any information except that the process is at state $\hat{\xi}$.

The transition mechanism (T, Q) of a CFSJS has a special structure expressed by the following property.

$$\mathrm{P}(T(\xi) > \hat{t} + t | T(\xi) > \hat{t}) = \mathrm{P}(T(\phi(\hat{t}, \xi)) > t). \tag{3.2}$$

This property expresses the fact that the jump times are memoryless. Because of this property we have

$$\mathrm{P}(\tilde{T} > t) = P(T(\hat{\xi}) > t).$$

Thus, if during the execution of a CFSJS, we lose the information of the last drawn sample before time $\hat{t}$ and state $\xi_{\hat{t}}$, then because of property (3.2), we can continue the execution by considering $\xi_{\hat{t}}$ as a state right after some switch. This means that we draw a sample $\tilde{t}$ from $T(\xi_{\hat{t}})$, followed by drawing a sample from $Q(\phi(\tilde{t}, \xi_{\hat{t}}))$, etc. As we will see in the next section, this observation makes it possible that the behavior of a system that consists of two CFSJSs executed at the same time can be expressed as a single CFSJS.

3.2.2.2 Representing Two Parallel CFSJSs as a Single CFSJS

Let $X = (E_X, \xi_{X,0}, \phi_X, \lambda_X, Q_X)$ and $Y = (E_Y, \xi_{Y,0}, \phi_Y, \lambda_Y, Q_Y)$ be two CFSJSs. Assume that at time t_0 both the processes X and Y are started. Let $\xi_X : \mathbb{R}_+ \to E_X$ be an execution path generated by the TMS of X and let $\xi_Y : \mathbb{R}_+ \to E_Y$ be an execution path generated by the TMS of Y. Then we call $\xi : \mathbb{R}_+ \to E_X \times E_Y$, where $\xi(t) = (\xi_X(t), \xi_Y(t))$, an execution path of the simultaneous execution of X and Y on the combined state space $E_X \times E_Y$. We show that these combined execution paths can be generated by the TMS of a single CFSJS denoted as $X|Y$. In Section 3.3 we need this result when two components (i.e., two CPDPs) that are executed in parallel, need to be represented as a single component (i.e., as a single CPDP). The state space of $X|Y$ is the product space $E_X \times E_Y$.

Suppose that after the start at t_0, X switches for the first time at t_1 at state $\xi_{X,1}$ and Y does not switch before t_1. Then at t_1, the state of X is reset by the measure $Q(\xi_{X,1})$. Let $\xi_{Y,1}$ be the state of Y at time t_1. The state of Y is not reset at time t_1, but from Section 3.2.2.1 we know that the stochastic behavior of Y will not change if we reset the state of Y at time t_1 with probability one to the same state. (Equivalently, the reset measure is the Dirac measure concentrated at the current state.) Then for the execution of Y, the state does not change at time t_1, but a new sample is drawn from the transition mechanism at state $\xi_{Y,1}$, which does not influence the stochastic execution according to Section 3.2.2.1.

We define a transition mechanism (T, Q) on the state space $E_X \times E_Y$ such that generating an execution path of (T, Q) is equal to generating a combined execution path for X and Y as described above.

Let for all $(\xi_X, \xi_Y) \in E_X \times E_Y$ the random variable $T(\xi_X, \xi_Y)$ be equal to

$$\min\{T_X(\xi_X), T_Y(\xi_Y)\}.$$

Then T determines the jump time of either X or Y. It can be seen that the survivor function of $T(\xi_X, \xi_Y)$ equals

$$\Psi_{T(\xi_X,\xi_Y)}(t) = \mathrm{e}^{-\int_0^t (\lambda_X(\phi(s,\xi_X)) + \lambda_Y(\phi(s,\xi_Y)))\mathrm{d}s}. \tag{3.3}$$

If a switch happens at combined state (ξ_X, ξ_Y), then it can be seen that the probability that this switch is a switch of X is equal to $\frac{\lambda_X(\xi_X)}{\lambda_X(\xi_X)+\lambda_Y(\xi_Y)}$ and the probability that this switch is a switch of Y is equal to $\frac{\lambda_Y(\xi_Y)}{\lambda_X(\xi_X)+\lambda_Y(\xi_Y)}$.

If X switches at state (ξ_X, ξ_Y), the reset measure $Q_X(\xi_X) \times Id(\xi_Y)$ is used and if Y switches at state (ξ_X, ξ_Y), the reset measure $Id(\xi_X) \times Q_Y(\xi_Y)$ is used. Then we get for Q

$$Q(\xi_X, \xi_Y) = \frac{\lambda_X(\xi_X)}{\lambda_X(\xi_X) + \lambda_Y(\xi_Y)} Q_X(\xi_X) \times Id(\xi_Y) + \tag{3.4}$$

$$\frac{\lambda_Y(\xi_Y)}{\lambda_X(\xi_X) + \lambda_Y(\xi_Y)} Id(\xi_X) \times Q_Y(\xi_Y).$$

Define CFSJS $X|Y$ as $(E_X \times E_Y, (\xi_{X,0}, \xi_{Y,0}), (\phi_X, \phi_Y), \lambda, Q)$, where

$$\lambda(\xi_X, \xi_Y) = \lambda_X(\xi_X) + \lambda_Y(\xi_Y).$$

The TMS of $X|Y$ equals (T, Q) and therefore CFSJS $X|Y$ generates the same execution paths (with the same probabilities) as the combination of execution paths of X and Y.

If $|$ denotes the operator that maps two CFSJSs to the combined CFSJS, then it can be seen that $|$ is associative and the combination of X,Y and Z can be expressed as either $(X|Y)|Z$ or $X|(Y|Z)$.

3.2.3 Forced Transition Structure (FTS)

If a system has forced transitions, then the behavior of the system concerning these forced transitions can be captured as an FTS.

DEFINITION 3.4 *An FTS is a tuple $(E, \mathscr{T})$, where the state space E is a Borel space, and $\mathscr{T} \subset E \times Prob(E)$ is the transition relation. For each $\xi \in E$, there exists at most one measure m such that $(\xi, m) \in \mathscr{T}$. If a state ξ is such that there exists an m such that $(\xi, m) \in \mathscr{T}$, then we call ξ an* enabled state *of the FTS.*

If a system X has corresponding FTS $(E, \mathscr{T})$, then if $(\xi, m) \in \mathscr{T}$ means that if X reaches state ξ at some time t, then X is forced to switch at this state and the target state of the switch is determined by measure m.

3.2.4 CFSJS Combined with FTS

The behavior of a PDP and, under certain conditions, the behavior of a CPDP can be captured as a combination of a CFSJS and an FTS. In fact, this combination means that the process runs as the CFSJS until an enabled state of the FTS is reached. Then the forced transition is executed, and the CFSJS execution continues from the state right after the forced transition. We now show how this combination of CFSJS and FTS behaves in terms of TMS.

Let (X_C, X_F), where $X_C = (E, \xi_0, \phi, \lambda, Q)$ is an CFSJS and $X_F = (E, \mathscr{T})$ is an FTS, be the combined semantics of a system X with state space E. For each $\xi \in E$ we define $t_*(\xi)$ as

$$t_*(\xi) := \begin{cases} \inf\{t \geq 0 | \phi(t, \xi) \text{ is an enabled state of } X_F\} \\ \infty \text{ if no such time exists.} \end{cases}$$

Thus, $t_*(\xi)$ is the maximum time before a jump surely occurs from the moment that the process is in state ξ. Either a jump occurs before time $t_*(\xi)$ because of the CFSJS part or a forced jump happens at time $t_*(\xi)$.

The transition mechanism structure of (X_C, X_F) is then equal to $(E, \xi_0, \phi, (T, \tilde{Q}))$, where $\tilde{Q}(\xi)$ equals $Q(\xi)$ if ξ is not an enabled state of X_F and $\tilde{Q}(\xi)$ equals m if ξ is an enabled state of X_F, where m is such that $(\xi, m) \in \mathscr{T}$. The survivor function of T (whose definition we take from [4]) equals

$$\Psi_{T(\xi)}(t) = I_{(t<t_*(\xi))} \mathrm{e}^{-\int_0^t \lambda(\phi(s,\xi))\mathrm{d}s}. \tag{3.5}$$

In [12] it is described how the semantics of a PDP or a CPDP (that has no non-determinism) can be given as a CFSJS together with an FTS. In this chapter the CPDPs do have non-determinism. Then the semantics can not be expressed by a CFSJS together with an FTS, but instead will be expressed by a CFSJS together with an NTS.

3.2.5 Non-deterministic Transition System (NTS)

Besides spontaneous and forced transitions, we can also distinguish *non-deterministic transitions* . We call a ξ-enabled transition *non-deterministic* if, when a process reaches state ξ, the process has the potential to execute the transition but it is not forced to execute the transition. In other words, it is not determined whether the process should execute the transition. Forced transitions are clearly not non-deterministic since the process has no choice but is forced to execute a ξ-enabled forced transition when its state reaches ξ. Execution of spontaneous transitions is determined by random variables, spontaneous transitions are not non-deterministic therefore. We now define the semantical model NTS (Non-deterministic Transition Structure), whose transitions are non-deterministic.

DEFINITION 3.5 *An NTS is a tuple $(E, \Sigma, \mathscr{T})$. The state space E is a Borel space. Σ is a set of labels. $\mathscr{T} \subset E \times \Sigma \times Prob(E)$ is the transition relation.*

(At this level of generality Σ is an arbitrary set, although in most situations it will actually be a finite set.) We write $\xi \xrightarrow{\sigma} m$ for $(\xi, \sigma, m) \in \mathscr{T}$. If a system X has corresponding NTS $(E, \Sigma, \mathscr{T})$, then $\xi \xrightarrow{\sigma} m$ has the meaning that if X reaches state ξ at some time t, then X has the possibility/potential to jump at this point on action σ and the target state of the jump is determined by measure m. Whether the transition is really executed at state ξ is not-determined. If there are multiple ξ-enabled transitions, then the process has, at state ξ, the potential to execute either one of them or to execute none of them. In the concurrent processes literature non-determinism is often used in a stricter sense than here. A ξ-enabled transition with label σ is then called non-deterministic if there is another ξ-enabled transition with the same label σ.

Actions σ are used for interaction between systems. In Section 3.3 we show how this interaction is established in the context of CPDPs.

If a system has non-deterministic transitions, then the system is open in the sense that its behavior can be influenced by other systems. Therefore, a system with non-deterministic transitions, if the non-determinism is not resolved by for example a scheduler, can not be stochastically executed. This means that we cannot determine the TMS of such a system.

We now give two simple examples that show how the behavior of a stochastic system can be captured in terms of CFSJS, FTS, and NTS. In the first example the behavior is captured as a CFSJS together with an FTS, in the second example the behavior is captured as a CFSJS together with an NTS.

EXAMPLE 3.1 The state x of system X takes value in $\mathbb{R}$ and evolves continuously as described by the ordinary differential equation $\dot{x} = 1$. The initial state is $x_0 = 0$. A spontaneous transition may happen and the time of such a transition is exponentially distributed with parameter λ. The target state of such a transition is chosen with uniform distribution in $[0,1]$. If x reaches the value 1, a transition is forced to happen.

The behavior of X can be captured as a CFSJS together with an FTS. The CFSJS equals $(\mathbb{R}, 0, \phi, \lambda, Q)$, where $\phi(t,x) = x + t$ for all states x and Q equals $U[0,1]$, i.e., the uniform distribution on $[0,1]$, for all states x. The FTS equals $(\mathbb{R}, \mathscr{T}_F)$, with $\mathscr{T}_F = \{(1, U[0,1])\}$.

The TMS that corresponds to the combination of this CFSJS and FTS equals $(\mathbb{R}, 0, \phi, (T, Q))$, where for all states $x < 1$

$$\mathrm{P}(T(x) > t) = I_{(t<1-x)} \mathrm{e}^{-\lambda t}.$$

■

EXAMPLE 3.2 The state x of system X takes value in $\mathbb{R}$ and evolves continuously as described by the ordinary differential equation $\dot{x} = 1$. The initial state $x_0 = 0$. A spontaneous transition may happen and the time of such a transition is exponentially distributed with parameter λ. The target state of such a transition is chosen with uniform distribution in $[0,1]$. If $x \geq 1$, a transition is allowed but not forced to happen. The action of the transition is τ.

The behavior of X can be captured as a CFSJS together with an NTS. The CFSJS equals the CFSJS of Example 3.1. The NTS equals $(\mathbb{R}, \{\tau\}, \mathscr{T}_N)$, with $\mathscr{T}_N = \{(x, \tau, U[0,1]) | x \geq 1\}$. Note that because of the presence of non-determinism, the behavior of the CFSJS together with the NTS cannot be captured as a TMS. ■

3.3 Communicating PDPs

With PDPs we can model a broad class of stochastic systems. However, because PDPs do not allow modelling in a compositional way, the modelling process becomes nearly impossible if systems have a (very) complex structure. In this section we introduce the automaton model CPDP (communicating PDP), which makes it possible to model PDP-type systems in a compositional way.

3.3.1 Definition of the CPDP Model

DEFINITION 3.6 *A CPDP is a tuple* $(L, V, \nu, W, \omega, F, G, \Sigma, \mathscr{A}, \mathscr{P}, \mathscr{S})$, *where*

- L *is a set of locations.*
- V *is a set of state variables. With* $d(v)$ *for* $v \in V$ *we denote the dimension of variable* v. $v \in V$ *takes its values in* $\mathbb{R}^{d(v)}$.
- W *is a set of output variables. With* $d(w)$ *for* $w \in W$ *we denote the dimension of the variable* w. $w \in W$ *takes its values in* $\mathbb{R}^{d(w)}$.
- $\nu : L \to 2^V$ *maps each location to a subset of* V, *which is the set of state variables of the corresponding location. We call the valuation space of the variables of* $\nu(l)$ *the state space of location* l.
- $\omega : L \to 2^W$ *maps each location to a subset of* W, *which is the set of output variables of the corresponding location. We call the valuation space of the variables of* $\omega(l)$ *the output space of location* l.
- F *assigns to each location* l *and each* $v \in \nu(l)$ *a mapping from* $\mathbb{R}^{d(v)}$ *to* $\mathbb{R}^{d(v)}$, *i.e.,* $F(l,v) : \mathbb{R}^{d(v)} \to \mathbb{R}^{d(v)}$. $F(l,v)$ *is the vector field that determines the evolution of* v *for location* l *(i.e.,* $\dot{v} = F(l,v)$ *for location* l*).*
- G *assigns to each location* l *and each* $w \in \omega(l)$ *a mapping from* $\mathbb{R}^{d(v_1)+\cdots+d(v_m)}$ *to* $\mathbb{R}^{d(w)}$, *where* v_1 *till* v_m *are the state variables of location* l. $G(l,w)$ *determines the output equation of* w *for location* l *(i.e.,* $w = G(l,w)$*).*
- Σ *is the set of communication labels.* $\bar{\Sigma}$ *denotes the "passive" mirror of* Σ *and is defined as* $\bar{\Sigma} = \{\bar{a} | a \in \Sigma\}$.
- $\mathscr{A}$ *is a set of active transitions and consists of five-tuples* (l, a, l', G, R), *denoting a transition from location* $l \in L$ *to location* $l' \in L$ *with communication label* $a \in \Sigma$, *guard* G *and reset map* R. G *is a subset of the* valuation space *of* l, *which notion is introduced in Section 3.3.1.1. The reset map* R *assigns to each point in* G *for each variable* $v \in \nu(l')$ *a probability measure on* $\mathbb{R}^{d(v)}$, *i.e.,* $R(g,v) \in Prob(\mathbb{R}^{d(v)})$ *for all* $g \in G$ *and all* $v \in \nu(l')$.
- $\mathscr{P}$ *is a set of passive transitions of the form* $(l, \bar{a}, l', R)$. R *is defined on the valuation space of* l *as the* R *of an active transition is defined on the guard space.*

- *$\mathscr{S}$ is a set of spontaneous transitions and consists of four-tuples (l, λ, l', R), denoting a transition from location $l \in L$ to location $l' \in L$ with jump-rate λ and reset map R. The jump rate λ (i.e., the Poisson rate of the Poisson process of the spontaneous transition) is a mapping from the state space of l to $\mathbb{R}_+$. R is defined on the state space of l as it is done for passive transitions.*

The graphical notation of a CPDP is as follows. The locations are pictured as circles (see for example Figure 3.1). The differential and output equations that belong to a location l are written inside the circle of l. A transition from location l to location l' is drawn as an arrow from l to l'. The communication label from Σ, the reset map and the guard of the transition are written above (or next to) the arrow. A passive transition is pictured as an active transition, except that the label is now from $\bar{\Sigma}$ and there is no guard. A spontaneous transition is pictured as an arrow with a little box in the middle. This notation is chosen in line with the notation used for IMCs (Interactive Markov Chains, cf. [7]. The jump rate and reset map of a spontaneous transition are written above or next to this little box.

EXAMPLE 3.3 In Figure 3.1 we see the CPDP X, which models a flying aircraft. The initial location of X is l_1. The initial state x_0 in l_1 represents the position and velocity of the aircraft at initial time t_0. Since the aircraft flies in three dimensional space, the state space of variable x is $\mathbb{R}^6$. Location l_1 represents a flying mode. This means that in l_1 the aircraft is somewhere up in the sky and not at the ground. The dynamics of the aircraft in flying mode l_1 is determined by the vector field f_1. In this model we do not discriminate between state and output, therefore in all locations the output is chosen to be a copy of the state, i.e., $y = x$. Location l_2 represents a non-nominal flying mode. In this mode the aircraft is flying while there is a defect. (This mode represents for example that the navigation system is not working properly.) In this non-nominal flying mode the dynamics is determined by vector field f_2. In location l_1, the time till the defect occurs, is exponentially distributed with parameter λ_1. This is expressed by the spontaneous transition from l_1 to l_2 with jump rate λ_1. When this transition is executed at some state $(l_1, \{x = x_1\})$, i.e., when the defect occurs at this state, the state of l_2 is reset by $R_1(\{x = x_1\})$, which is a probability measure on $\mathbb{R}^6$, the state space of x. Since the position and velocity of the aircraft do not change when a switch from l_1 to l_2 happens, $R_1(\{x = x_1\})$ equals the Dirac measure at x_1, which assigns probability one to the singleton set $\{x = x_1\}$. We call R_1 an identity reset map. In the non-nominal mode l_2, repair activities are undertaken. The time till the defect is repaired is exponentially distributed with parameter λ_2. This is expressed by the spontaneous transition from l_2 to l_1 which has reset map R_2, which also equals the identity reset map.

If the aircraft approaches the airport, the flying mode will change to a landing mode where the aircraft prepares for landing. These modes are rep-

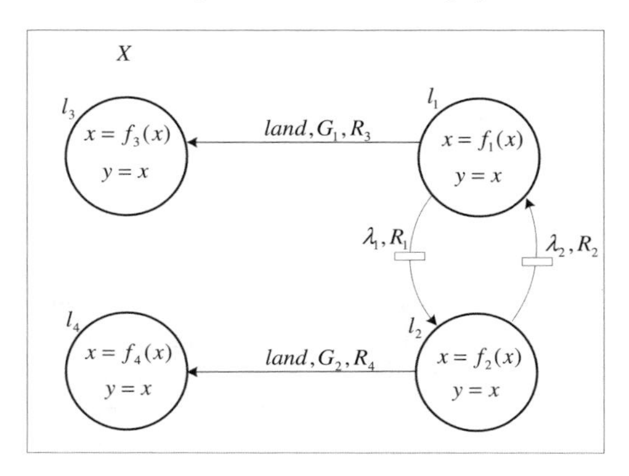

FIGURE 3.1: Landing aircraft modelled as CPDP.

resented by l_3 and l_4, where l_3 represents nominal landing and l_4 represents landing while there still is an unrepaired defect. l_1 switches to l_3 as soon as the altitude drops below a certain level h. Let x_1 till x_6 be the components of variable x and let x_3 denote the altitude of the aircraft. The guard G_1 of the transition from l_1 to l_3 is equal to $\{x = (x_1, x_2, \cdots, x_6) | x_3 \leq h\}$. This expresses that this transition is allowed to be executed as soon as $x \in G_1$, i.e., as soon as $x_3 \leq h$. In fact, we want the transition executed at the first time instant where $x \in G_1$. Later we will see that we can express this by assuming maximal progress, which means, roughly said, that active transitions should be executed as soon as the guard is satisfied. When this active transition is executed at some time t_2, then the identity reset map R_3 (also an identity reset map) resets the state and the discrete event *land* is executed at time t_2. The discrete event (and its transition) is executed instantaneously, i.e., does not consume time. Later we will see that these discrete events can be used for communication between CPDPs. In location l_3 the continuous dynamics (expressing the landing phase) is determined by vector field f_3. In the same way the transition from l_2 to l_4 with identity reset map R_4 represents the transition from flying mode to landing mode, but now while there is a defect present.

In this example the jump rates λ_1 and λ_2 are constants, expressing exponentially distributed times. However, the CPDP model also allows λ_1 and λ_2 to depend on the state, and therefore indirectly depend on the time. If for example λ_1 depends on x_3 such that $\lambda_1(x_1, x_2, \cdots, x_6) > \lambda_1(x_1', x_2', \cdots, x_6')$ when $x_3 > x_3'$, then this would express that the rate of switching is larger for great altitudes, i.e., at great altitudes it is more likely that a defect occurs than at small altitudes. ■

One feature of the CPDP model, the passive transition, is not explained in the above example. The meaning of passive transitions becomes apparent in the context of communication between multiple CPDPs and is explained and illustrated in Section 3.3.3.

3.3.1.1 The State and Output Space of a CPDP

The state of a CPDP is hybrid; it consists of a location on the one hand and of values for the continuous variables on the other hand.

DEFINITION 3.7 *Let X be a CPDP with location set L, set of state variables V, set of output variables W, and for each $l \in L$ the sets of active state and output variables $\nu(l) \subset V$ and $\omega(l) \subset W$. The (hybrid) state space of X is defined as*

$$\{(l, val) \mid l \in L, val \in vs(l)\},$$

where $vs(l)$ denotes the valuation space of location l, which in case $\nu(l) = \{v_1, v_2, \cdots, v_n\}$, is given as

$$\mathbb{R}^{d(v_1)} \times \cdots \times \mathbb{R}^{d(v_n)}$$

and in case $\nu(l) = \emptyset$ is defined as $\{0\}$. The output space of X is defined as

$$\{(l, val) \mid l \in L, val \in os(l)\},$$

where $os(l)$ denotes the output space of location l, which in case $\omega(l) = \{w_1, \cdots, w_m\}$ is defined as $\mathbb{R}^{d(\mathbf{w}_1)} \times \cdots \times \mathbb{R}^{d(\mathbf{w}_m)}$ and in case $\omega(l) = \emptyset$ is given as $\{0\}$. The output value 0 is used for CPDP states where no output is defined.

EXAMPLE 3.4 The state space of CPDP X of Figure 3.1 equals

$$\{(l, val) \mid l \in \{l_1, l_2, l_3, l_4\}, val \in \mathbb{R}^6\}.$$

The output space of X equals $\mathbb{R}^6$. ■

REMARK 3.1 We allow that for CPDP locations l we have $\nu(l) = \emptyset$, i.e., we allow that locations do not have continuous variables attached to it. We call these locations *empty locations*. According to Definition 3.7, the valuation space of an empty location l equals $\{0\}$ and therefore l contributes one state to the state space of the CPDP; the state $(l, 0)$. The guard of an active transition α with origin location l is then equal to $\{0\}$, which means that at an empty location, active transitions are always enabled. Spontaneous transitions at empty locations assign a constant λ to the single state of the valuation space. This means that the jump time of such a spontaneous transition is exponentially distributed with parameter λ. A reset map of a transition whose target location is an empty location assigns probability one to the single state 0 of the valuation space of that empty location.

We also allow for CPDP locations such that $\omega(l) = \emptyset$. This means that no output dynamics is defined for such locations. The output at states of these locations will later be defined as 0. ∎

Let $\alpha = (l, a, l', G, R)$ be an active transition. Then we we define the mappings *oloc* (origin location), *lab* (label), *tloc* (target location), *guard*, and *rmap* (reset map) as: $oloc(\alpha) = l$, $lab(\alpha) = a$, $tloc(\alpha) = l'$, $guard(\alpha) = G$, $rmap(\alpha) = R$. These mappings, except for *guard*, are also defined in the same way for passive transitions. *oloc*, *tloc*, and *rmap* are also defined in the same way for spontaneous transitions. Furthermore, let $\xi = (l, val)$ be some hybrid state. Then $loc(\xi) := l$ maps ξ to its discrete part, and $val(\xi) := val$ maps ξ to its continuous part.

We define the flow map $\phi : \mathbb{R}_+ \times E \to E$ of a CPDP $X = (L, V, \nu, W, \omega, F, G, \Sigma, \mathscr{A}, \mathscr{P}, \mathscr{S})$ with state space E. $\phi(t, \xi)$ is determined by the differential equations

$$\dot{x}_1 = F(l, x_1),\ \dot{x}_2 = F(l, x_2), \cdots\ \dot{x}_n = F(l, x_n), \tag{3.6}$$

where $l = loc(\xi)$ and $\nu(l) = \{x_1, x_2, \cdots, x_n\}$. Thus, for $t \geq 0$ and $\xi = (l, \{x_1 = r_1, \cdots, x_n = r_n\}) \in E$, $\phi(t, \xi)$ equals $\xi' = (l, \{x_1 = r'_1, \cdots, x_n = r'_n\})$, where $r_1, \cdots, r_n$ are the solutions of (3.6) for $x_1 \cdots x_n$ at time t where $x_1 \cdots x_n$ at time zero have values $r_1 \cdots r_n$. For empty locations l we define the flow map as $\phi(t, (l, 0)) := (l, 0)$ for all $t \geq 0$.

3.3.2 Semantics of CPDPs

Let $X = (L, V, \nu, W, \omega, F, G, \Sigma, \mathscr{A}, \mathscr{P}, \mathscr{S})$ be a CPDP with state space E, flow map ϕ and initial state ξ_0. We define the semantics of X as the combination of a CFSJS and an NTS. Let X_C denote the CFSJS of X and let X_N denote the NTS of X. Then, $X_C = (E, \xi_0, \phi, \lambda, Q)$, where

$$\lambda(l, val) := \sum_{\alpha \in \mathscr{S}_{l\to}} \lambda_\alpha(val),$$

where $\mathscr{S}_{l\to}$ denotes the set of all spontaneous transitions with origin location l, and for all $A \in \mathscr{B}(E)$,

$$Q(l, val)(A) = \sum_{\alpha \in \mathscr{S}_{l\to}} \frac{\lambda_\alpha(val)}{\lambda(l, val)} R_\alpha(A)$$

and $X_N = (E, \Sigma \cup \bar{\Sigma}, \mathscr{T})$, where

- $(\xi, a, m) \in \mathscr{T}$ if and only if there exists an $\alpha \in \mathscr{A}$ such that $lab(\alpha) = a$, $oloc(\alpha) = loc(\xi)$, $val(\xi) \in guard(\alpha)$, and $rmap(\alpha)(\xi) = m$.

- $(\xi, \bar{a}, m) \in \mathscr{T}$ if and only if there exists an $\alpha \in \mathscr{P}$ such that $lab(\alpha) = \bar{a}$, $oloc(\alpha) = loc(\xi)$, and $rmap(\alpha)(\xi) = m$.

Note that the CFSJS defined above expresses correctly that in each location there is a "race" between the spontaneous transitions enabled at that location just as the "race" between the spontaneous transitions of two CFSJSs that are running in parallel as described in Section 3.2.2.2. That the λ and Q of the CFSJS correctly express this "race" can, mutatis mutandis, also be found in Section 3.2.2.2.

EXAMPLE 3.5 The semantics of CPDP X of Figure 3.1 with state space E, flow map ϕ, and initial state ξ_0 is as follows. $X_C = (E, \xi_0, \phi, \lambda, Q)$, where for $\xi = (l, val) \in E$,

$$\lambda(\xi) = \begin{cases} \lambda_1 \text{ if } l = l_1, \\ \lambda_2 \text{ if } l = l_2, \\ 0 \quad \text{if } l \in \{l_3, l_4\} \end{cases}$$

and for all $B \in \mathscr{B}(E)$

$$Q(\xi)(B) = \begin{cases} R_1(\xi)(B) \text{ if } l = l_1, \\ R_2(\xi)(B) \text{ if } l = l_2, \\ \text{undefined if } l \in \{l_3, l_4\}. \end{cases}$$

$X_N = (E, \Sigma \cup \bar{\Sigma}, \mathscr{T})$, where

$$\mathscr{T} = \{(\xi, land, m) | \xi \in G_1, m = R_3(\xi)\} \cup \{(\xi, land, m) | \xi \in G_2, m = R_4(\xi)\}.$$

■

We cannot give the execution of a CPDP X with active/passive transitions in terms of a transition mechanism system, because it is not determined when transitions from $\mathscr{T}$ are executed. However, we can describe the execution of a general CPDP $X = (L, V, \nu, W, \omega, F, G, \Sigma, \mathscr{A}, \mathscr{P}, \mathscr{S})$ as follows. Let $X_C = (E, \xi_0, \phi, \lambda, Q)$ be the CFSJS of X and let $X_N = (E, \Sigma, \mathscr{T})$ be the NTS of X. Then, the execution of X can be seen as the execution of X_C while at every state ξ the process has the potential to switch with measure m if $(\xi, \sigma, m) \in \mathscr{T}$ for some $\sigma \in \Sigma \cup \bar{\Sigma}$.

3.3.2.1 Output Semantics

The CFSJS and NTS do not capture the complete behavior of a CPDP. At every state $\xi \in E$ of a CPDP, the CPDP also has an output value which lies in its output space E_O and which is determined by the output mapping $G : E \to E_O$. Therefore we could say that the complete behavior of a CPDP is captured by its CFSJS, its NTS and its output mapping.

3.3.3 Composition of CPDPs

Now we will define how CPDPs can be composed. The composition operator that we use, which can be seen as a generalization of the composition operator for Interactive Markov Chains from [7], will be denoted by $|_A^P|$. We do not have the space here to explain the full interaction-potential of this operator. We now give informally the main features of $|_A^P|$. For a full explanation we refer to [12] or [14].

First we discuss the distinction between active and passive transitions. Active transitions can be executed independently from passive transitions. Passive transitions can only be executed when they are triggered by active transitions in another component. If CPDPs X and Y are composed, and CPDP component X executes an a-transition, then this transition will trigger (if available) a passive $\bar{a}$-transition in component Y. We could also say that the $\bar{a}$-transition of Y observes the a-transition of X.

In $|_A^P|$, A, which is a subset of Σ, is the set of active events that should synchronize. This means that if CPDPs X and Y are composed through operator $|_A^P|$, and if $a \in A$, then an a-transition of X can be executed only if at the same time an a-transition of Y is executed (and vice versa). If CPDPs X and Y are composed through $|_A^P|$ and $a \in A$, then an a-transition of X cannot trigger a $\bar{a}$-transition of Y. In other words, the events from A are used for active-active synchronization and the events from $\Sigma \backslash A$ are used for active-passive synchronization.

P, which is a subset of $\bar{\Sigma}$, is the set of all passive events that should synchronize. Briefly said, an event $\bar{a} \in P$ is such that multiple $\bar{a}$-transitions can be triggered by a single a-transition. This means that an a-transition of X can trigger a passive $\bar{a}$-transition in all of the other components Y, Z, etc. If $\bar{a} \notin P$, then an a-transition of X can trigger a $\bar{a}$-transition in only one of the other components Y, Z, etc.

In the definition of composition of CPDPs, communication is expressed through synchronization of transitions and not through the sharing of continuous variables. Therefore, each component should have its own continuous variables, i.e., the intersection of the sets of continuous variables of the two components should be empty. If this is not the case, then the two components are not compatible for composition.

We now give the definition of composition of CPDPs. Afterwards we briefly explain the composition rules and therewith explain how interaction is expressed in this definition of composition.

DEFINITION 3.8 *Let* $X = (L_X, V_X, \nu_X, W_X, \omega_X, F_X, G_X, \Sigma, \mathscr{A}_X, \mathscr{P}_X, \mathscr{S}_X)$ *and* $Y = (L_Y, V_Y, \nu_Y, W_Y, \omega_Y, F_Y, G_Y, \Sigma, \mathscr{A}_Y, \mathscr{P}_Y, \mathscr{S}_Y)$ *be two CPDPs such that* $V_X \cap V_Y = W_X \cap W_Y = \emptyset$. *Then* $X|_A^P|Y$ *is defined as the CPDP* $(L, V, \nu, W, \omega, F, G, \Sigma, \mathscr{A}, \mathscr{P}, \mathscr{S})$, *where*

- $L = \{l_1|_A^P|l_2 \mid l_1 \in L_X, l_2 \in L_Y\}$.
- $V = V_X \cup V_Y$, $W = W_X \cup W_Y$.
- $\nu(l_1|_A^P|l_2) = \nu(l_1) \cup \nu(l_2)$, $\omega(l_1|_A^P|l_2) = \omega(l_1) \cup \omega(l_2)$.
- $F(l_1|_A^P|l_2, v)$ *equals* $F_X(l_1, v)$ *if* $v \in \nu_X(l_1)$ *and equals* $F_Y(l_2, v)$ *if* $v \in \nu_Y(l_2)$.
- $G(l_1|_A^P|l_2, w)$ *equals* $G_X(l_1, w)$ *if* $w \in \omega_X(l_1)$ *and equals* $G_Y(l_2, w)$ *if* $w \in \omega_Y(l_2)$.
- $\mathscr{A}$, $\mathscr{P}$ *and* $\mathscr{S}$ *are the least relations satisfying the rules r1, r2, r2′, r3, r3′, r4, r4′, r5, r6, r6′, r7 and r7′, defined below*

$$r1.\frac{l_1 \overset{a,G_1,R_1}{\longrightarrow} l_1', l_2 \overset{a,G_2,R_2}{\longrightarrow} l_2'}{l_1|_A^P|l_2 \overset{a,G_1\times G_2,R_1\times R_2}{\longrightarrow} l_1'|_A^P|l_2'}(a \in A).$$

$$r2.\frac{l_1 \overset{a,G_1,R_1}{\longrightarrow} l_1', l_2 \overset{\bar{a},R_2}{\longrightarrow} l_2'}{l_1|_A^P|l_2 \overset{a,G_1\times vs(l_2),R_1\times R_2}{\longrightarrow} l_1'|_A^P|l_2'}(a \notin A).$$

$$r2'.\frac{l_1 \overset{\bar{a},R_1}{\longrightarrow} l_1', l_2 \overset{a,G_2,R_2}{\longrightarrow} l_2'}{l_1|_A^P|l_2 \overset{a,vs(l_1)\times G_2,R_1\times R_2}{\longrightarrow} l_1'|_A^P|l_2'}(a \notin A).$$

$$r3.\frac{l_1 \overset{a,G_1,R_1}{\longrightarrow} l_1', l_2 \overset{\bar{a}}{\not\longrightarrow}}{l_1|_A^P|l_2 \overset{a,G_1\times vs(l_2),R_1\times Id}{\longrightarrow} l_1'|_A^P|l_2}(a \notin A).$$

$$r3'.\frac{l_1 \overset{\bar{a}}{\not\longrightarrow}, l_2 \overset{a,G_2,R_2}{\longrightarrow} l_2'}{l_1|_A^P|l_2 \overset{a,vs(l_1)\times G_2,Id\times R_2}{\longrightarrow} l_1|_A^P|l_2'}(a \notin A).$$

$$r4.\frac{l_1 \overset{\bar{a},R_1}{\longrightarrow} l_1'}{l_1|_A^P|l_2 \overset{\bar{a},R_1\times Id}{\longrightarrow} l_1'|_A^P|l_2}(\bar{a} \notin P), \quad r4'.\frac{l_2 \overset{\bar{a},R_2}{\longrightarrow} l_2'}{l_1|_A^P|l_2 \overset{\bar{a},Id\times R_2}{\longrightarrow} l_1|_A^P|l_2'}(\bar{a} \notin P)$$

$$r5.\frac{l_1 \overset{\bar{a},R_1}{\longrightarrow} l_1', l_2 \overset{\bar{a},R_2}{\longrightarrow} l_2'}{l_1|_A^P|l_2 \overset{\bar{a},R_1\times R_2}{\longrightarrow} l_1'|_A^P|l_2'}(\bar{a} \in P).$$

$$r6.\frac{l_1 \overset{\bar{a},R_1}{\longrightarrow} l_1', l_2 \overset{\bar{a}}{\not\longrightarrow}}{l_1|_A^P|l_2 \overset{\bar{a},R_1\times Id}{\longrightarrow} l_1'|_A^P|l_2}(\bar{a} \in P), \quad r6'.\frac{l_1 \overset{\bar{a}}{\not\longrightarrow}, l_2 \overset{\bar{a},R_2}{\longrightarrow} l_2'}{l_1|_A^P|l_2 \overset{\bar{a},Id\times R_2}{\longrightarrow} l_1|_A^P|l_2'}(\bar{a} \in P)$$

$$r7.\frac{l_1 \overset{\lambda_1,R_1}{\longrightarrow} l_1'}{l_1|_A^P|l_2 \overset{\hat{\lambda}_1,R_1\times Id}{\longrightarrow} l_1'|_A^P|l_2}, \quad r7'.\frac{l_2 \overset{\lambda_2,R_2}{\longrightarrow} l_2'}{l_1|_A^P|l_2 \overset{\hat{\lambda}_2,Id\times R_2}{\longrightarrow} l_1|_A^P|l_2'},$$

where $\hat{\lambda}_1$ *and* $\hat{\lambda}_2$ *are defined as* $\hat{\lambda}_1(\xi_1,\xi_2) := \lambda_1(\xi_1)$ *and* $\hat{\lambda}_2(\xi_1,\xi_2) := \lambda_2(\xi_2)$.

We briefly explain how the composition ruler r1 till r7 should be interpreted. r1 says that if $a \in A$ and both $l_1 \overset{a,G_1,R_1}{\longrightarrow} l_1'$ and $l_2 \overset{a,G_2,R_2}{\longrightarrow} l_2'$ are true, i.e., if X has an a-transition from location l_1 to location l_1' with guard G_1 and reset map R_1 and if Y has an a-transition from location l_2 to location l_2' with guard G_2 and reset map R_2, then CPDP $X|_A^P|Y$ has an a-transition from location $l_1|_A^P|l_2$ to location $l_1'|_A^P|l_2'$ with guard $G_1 \times G_2$ and with reset map $R_1 \times R_2$ (where in the latter $\times$ denotes the product probability measure). Rule r1 expresses that a-transitions with $a \in A$ should synchronize. In the same way rule $r2$ expresses that for $a \notin A$, an a-transition of X synchronizes with (i.e., triggers) a $\bar{a}$-transition of Y (and vice versa with rule r2′). Note that $vs(l_2)$ denotes the whole state space of location l_2. Rule r3 expresses that if no $\bar{a}$-transition is present in Y, then the a-transition of X will be executed on its own. Note that Id denotes the identity reset map (i.e., the Dirac probability measure). Rules r4 and r4$'$ express that for $\bar{a} \notin P$, passive $\bar{a}$-transitions do not synchronize and are therefore executed on their own. Rule $r5$ expresses that for $\bar{a} \in P$, $\bar{a}$-transitions synchronize. Rules r6 and r6$'$ express for $\bar{a} \in P$, that if one of the components does not have a $\bar{a}$-transition, then the other component can still execute its passive $\bar{a}$-transition. This expresses (in the context of three components) that an a-transition of X can trigger a $\bar{a}$-transition of Y also when Z does not have a $\bar{a}$-transition enabled. Rules $r7$ and $r7'$ express that all spontaneous transitions remain (unchanged) in the composition.

In [12], the composition operator $|_A^P|$ is also defined on the semantical level of NTS. Then the following result holds, which shows that a composed CPDP correctly expresses the behavior (as an NTS and CFSJS) of the interaction of the component CPDPs.

THEOREM 3.1 *Let X and Y be two CPDPs with semantics (X_C, X_N) and (Y_C, Y_N) respectively, where X_C and Y_C are CFSJSs and X_N and Y_N are NTSs. Let $(X|_A^P|Y_C, X|_A^P|Y_N)$ be the semantics of CPDP $X|_A^P|Y$. Then,*

$$(X|_A^P|Y_C, X|_A^P|Y_N) = (X_C|Y_C, X_N|_A^P|Y_N).$$

Also we have the following result.

THEOREM 3.2 *$|_A^P|$ for CPDPs is commutative for all A and P. $|_A^P|$ for CPDPs is associative if and only if for all events a we have $a \notin A \Rightarrow \bar{a} \in P$.*

EXAMPLE 3.6 The CPDP X of Figure 3.2 models a flying aircraft and has initial state $\xi_{X,0} = (l_1, x_0)$. Note that, for reasons of simplicity, the non-nominal locations from Figure 3.1 are not modelled here. CPDP Y_1 of Figure 3.2 models a control tower at an airport that can communicate with the aircraft modelled by X. Location l_3 is the location where Y_1 waits for a signal from X. The dynamics of l_3 is a clock dynamics expressing the time that Y has to wait before X sends a signal. Therefore, the initial valuation of initial location l_3 equals $\{x_1 = 0\}$. Location l_4 is the location where Y_1 "knows" that X has send a signal. The dynamics of this location is again a clock dynamics. If Y_1 enters location l_4, then the timer is reset to zero, which means that reset map R_2 assigns for each value of x_1 in l_3 probability one to the Borel set $\{\{x_1 = 0\}\}$.

We connect X and Y_1 via composition operator $|_A^P|$, where $A = \{land\}$ (P is not relevant here). This means that the signal/label *land* is used as a shared synchronization action between X and Y_1. Then, Y_1 can execute the *land* transition only when at the same time X executes its *land* transition. We want to model that X can execute its *land* transition independently from

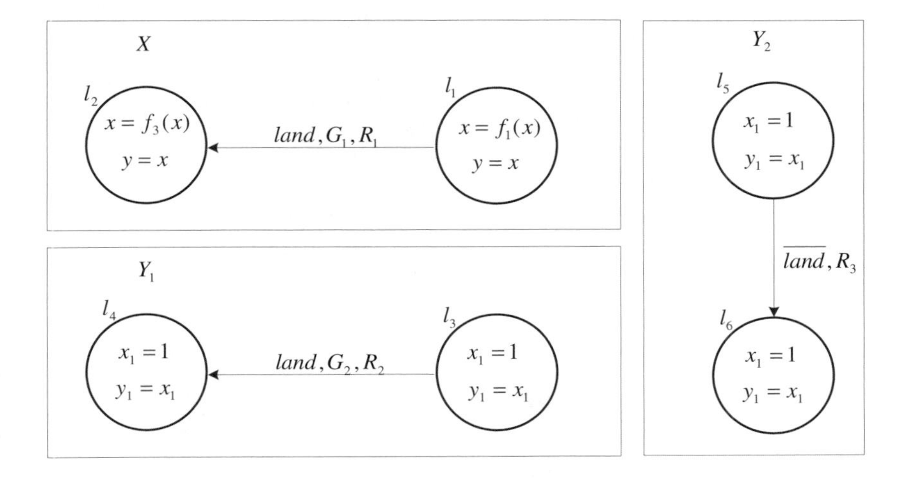

FIGURE 3.2: Landing aircraft and control tower modelled as interacting CPDPs.

Y. Once this happens, this transition should be communicated to Y. We can express this via the guards G_1 and G_2. G_1 equals G_1 from Example 3.3, expressing that this switch may happen as soon as the altitude of the aircraft drops under a certain level h. G_2 equals the whole valuation space of location l_3. This expresses that this transition can always be taken and consequently it expresses that this transition cannot block the *land* transition of X. We assume maximal progress. Then, the synchronized *land* transition is executed as soon as guard G_1 is satisfied. After the synchronized *land* transition, Y_1 is in location l_4. We could say that the information "X switched to landing mode," which is received by Y_1, is stored in the discrete component of the hybrid state of Y_1. In other words, discrete state l_4 of Y_2 has the meaning "X is in landing mode."

The CPDP $X|_A^P|Y_1$, which expresses the composite system of X interconnected with Y_1, is pictured in Figure 3.3. According to composition rule r1, the guard G_3 equals $G_1 \times G_2$ and the reset map R_4 equals $R_1 \times R_2$. If we look at the behavior of CPDP $X|_A^P|Y_1$ under maximal progress, then we will see that this CPDP indeed expresses the communication from X to Y_1 that we wanted to model: the initial hybrid state of $X|_A^P|Y_1$ equals $(l_1|l_3, \{x = x_0, x_1 = 0\})$. The continuous state variables x and x_1 evolve along vector fields f_1 and 1 respectively until guard G_3 is satisfied. G_3 is satisfied when the vertical position of x reaches the level h. Then, the *land* transition is executed and the state variables x and x_1 are reset by R_4, which means that x is reset by R_1 and x_1 is reset by R_2. Thus, we see that at the moment that X switches to landing mode, Y_1 switches to location l_4, which indeed establishes the communication that we intended.

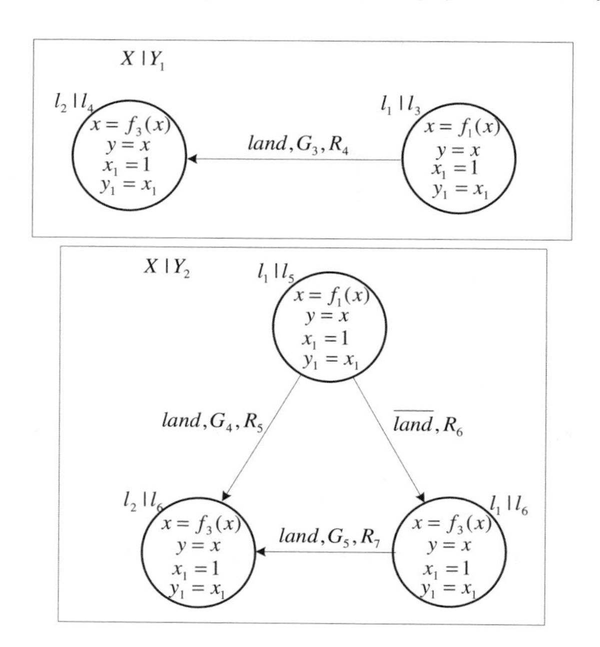

FIGURE 3.3: Composite CPDP of landing aircraft and control tower.

Now we show how the aircraft/control tower system can be modelled by using a passive transition. For this example, we think that modelling the communication with a passive transition is more natural, since there is a clear distinction between an active system (the aircraft which sends the information of the switch) and a passive system (the control tower which receives the information). Now, the control tower is modelled as the CPDP Y_2 of Figure 3.2. Y_2 is exactly the same as Y_1, except that the active transition is replaced by a passive transition with label $\overline{land}$. This passive transition expresses that as soon as a *land* signal is received (from X), the passive transition is executed and reset map R_3 (whose action equals the action of R_2) resets the timer x_1 to zero. Since *land* is not a synchronization action here, we connect X and Y_2 via $|_A^P|$, where $A = \emptyset$ (P is not relevant). The resulting CPDP $X|_A^P|Y_2$ is pictured in Figure 3.3. It can be seen from rule r2, that guard G_4 is equal to G_3 and reset map R_5 is equal to R_4. This means that as far as locations $l_1|l_3$ and $l_2|l_4$ / $l_1|l_5$ and $l_2|l_6$ are concerned, $X|_A^P|Y_1$ and $X|_A^P|Y_2$ have the same behavior. The difference between $X|_A^P|Y_1$ and $X|_A^P|Y_2$ lies in the fact that $X|_A^P|Y_2$ can switch to location $l_1|l_6$ via a passive transition, while $X|_A^P|Y_1$ cannot do this. The meaning of this switch to $l_1|l_6$ becomes apparent in a composition context with more than two components. A third component could by means of executing an active *land*-transition then trigger this passive transition. ■

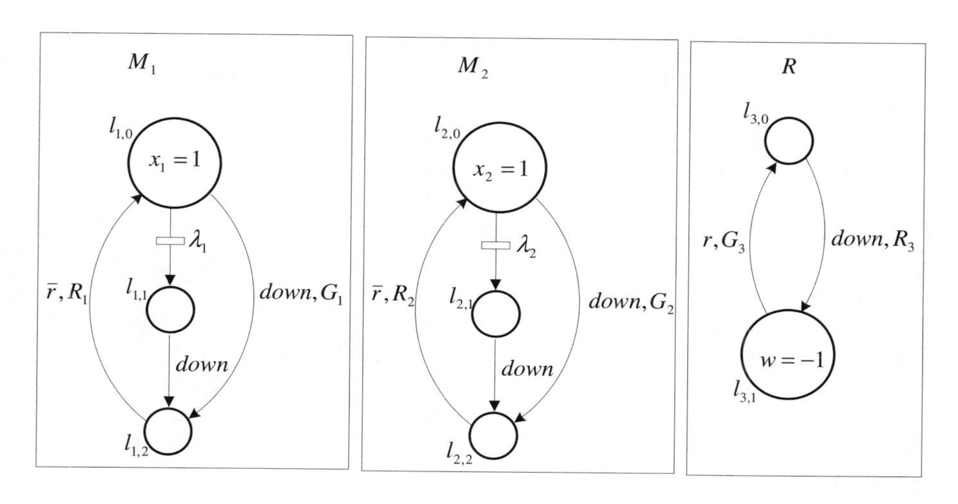

FIGURE 3.4: CPDP model of the repair shop system.

EXAMPLE 3.7 In Figure 3.4, a repair shop system is modelled as the composition of CPDPs M_1,M_2 and R. CPDPs M_1 and M_2 model two machines and CPDP R models a repair shop. M_1 initially starts in location $l_{1,0}$ with a clock dynamics for its state variable x_1. M_1 can break down with state dependent jump rate λ_1. This is modelled by the spontaneous transition to $l_{1,1}$. $l_{1,1}$ is an empty location, therefore the spontaneous transition to $l_{1,1}$ has a trivial reset map that assigns probability 1 to state $(l_{1,1}, 0)$. This reset map is not pictured in Figure 3.4. From $l_{1,1}$ an active transition is executed to $l_{1,3}$, with label *down*. We want to model that this *down*-transition is executed immediately after location $l_{1,1}$ is reached. Then, the *down* signal is executed exactly when M_1 breaks down. In the next chapter we will see that with maximal progress M_1 indeed models that no time is consumed in location $l_{1,1}$. If the machine does not break down via the spontaneous transition before s_1 time units, i.e., the maximal age of the machine, then the machine should be taken out of order to the repair shop. This is modelled by the *down*-transition from $l_{1,0}$ to $l_{1,2}$ with guard G_1 equal to $x_1 \geq s_1$. In location $l_{1,2}$, machine M_1 waits for an r signal. This is expressed by the passive $\bar{r}$-transition. This r signal will be sent by the repair shop, indicating that the machine has been repaired. Reset map R_1 resets state variable x_1 to zero, which expresses that the machine starts brand new. Machine M_2 is modelled likewise.

The repair shop CPDP R starts in empty location $l_{3,0}$. Here it waits until one of the machines needs to be repaired. The switch to repair mode $l_{3,1}$ is modelled by the active *down*-transition. We define *down* to be a synchronization action and therefore this *down*-transition synchronizes with either a *down*-transition of M_1 or a *down*-transition of M_2. Due to this synchronization, R switches to repair mode $l_{3,1}$ exactly when one of the machines need to be repaired. Reset map R_3 resets state variable w with a uniform distribution on the interval $[t_1, t_2]$, determining the time needed to repair the machine. In $l_{3,1}$, w counts down to zero, expressed by the dynamics $\dot{w} = -1$. If w has been counted down to zero, R switches back to $l_{3,0}$ where it waits for a new machine to be repaired. This switch is modelled by the active r transition. The guard G_3 of this transition equals $w = 0$. The passive $\bar{r}$ transitions of M_1 and M_2 can synchronize with this active r-transition, therefore these passive $\bar{r}$-transitions are executed exactly when the machine is repaired.

From the description above, we get that *down* is an interleaving action between M_1 and M_2, *down* is a synchronization action between R and M_1 or M_2, and r is an interleaving action between R and M_1 or M_2. The passive action $\bar{r}$ may be chosen interleaving or synchronizing. This choice does not influence the behavior since M_1 and M_2 will never visit their locations $l_{1,2}$ and $l_{2,2}$ at the same time (i.e., joint location $(l_{1,2}, l_{2,2})$ is never reached). This gives that the total repair shop system is modelled as the CPDP

$$(M_1|_{\emptyset}^{\emptyset}|M_2)|_{down}^{\emptyset}|R.$$

■

3.3.4 Value Passing CPDPs

In this section we extend the CPDP model to *value passing CPDPs*. For CPDPs, interaction is established through synchronization of transitions. This means that the information that one CPDP can obtain concerning other CPDPs in the composition is, first, which active actions are executed and, second, at which times these actions are executed. For example, via a passive $\bar{a}$-transition, a CPDP "knows" when another CPDP executes an a-transition. With value passing CPDPs we extend the CPDP model such that it is possible for one CPDP to obtain information about the values of the output variables of other CPDPs. The moments where this information is communicated from one CPDP to another CPDP are the moments where the transitions synchronize. In other words, this communication of output information is expressed through synchronization of transitions. This idea of passing values through synchronizing transitions is called *value passing* in the literature and has been developed for example for the specification languages LOTOS [2, 10] and CSP [8].

This section is organized as follows. First we define the value passing CPDP model and we give the CFSJS/NTS semantics of a value passing CPDP. Then we define the composition operator $|_A^P|$ for value passing CPDPs. As in the case of CPDPs, we will see that the behavior of two interacting value passing CPDPs X and Y is equal to the CFSJS and NTS of the value passing CPDP $X|_A^P|Y$. Finally, we give some examples illustrating the expressiveness of value passing in the context of value passing CPDPs.

3.3.4.1 Definition and Semantics of Value Passing CPDPs

DEFINITION 3.9 *A value-passing CPDP is a tuple $(L, V, W, \nu, \omega, F, G, \Sigma, \mathscr{A}, \mathscr{P}, \mathscr{S})$, where all elements except $\mathscr{A}$ are defined as in Definition 3.6 and where $\mathscr{A}$ is a finite set of active transitions that consists of six-tuples (l, a, l', G, R, vp), denoting a transition from location $l \in L$ to location $l' \in L$ with communication label $a \in \Sigma$, guard G, reset map R and value-passing element vp. G is a subset of the valuation space of l. vp can be equal to either $!Y$, $?U$ or $\emptyset$. For the case $!Y$, Y is an ordered tuple $(w_1, w_2, \cdots, w_m)$ where $w_i \in w(l)$ for $i = 1 \cdots m$. If for a transition we have $vp = !Y$ for some Y, then this means that in a synchronization with other transitions, this transition passes the values of the variables in Y to the other transition. For the case $?U$, we have $U \subset \mathbb{R}^n$ for some $n \in \mathbb{N}$. If for a transition we have $vp = ?U$, then this means that in a synchronization with another transition that has $vp = !Y$, this transition receives the values from the variables of Y as long as these values are contained in the set U. If the other transition wants to pass values that do not lie in U, then the synchronization will not take place, i.e., it is blocked by U. If a transition is not used for value passing (either output $!Y$ or input $?U$), then this transition has $vp = \emptyset$. The reset map R assigns to each point in $G \times U$ (for the case $vp = ?U$) or to each point in G (for the cases $vp = !Y$ and $vp = \emptyset$) for each state variable $v \in \nu(l')$ a probability measure on $\mathbb{R}^{d(v)}$. Active transitions α with $\omega(oloc(\alpha)) = \emptyset$, i.e., whose origin locations have no continuous variables, have value passing element $vp = \emptyset$.*

Let $X = (L, V, \nu, W, \omega, F, G, \Sigma, \mathscr{A}, \mathscr{P}, \mathscr{S})$ be a value passing CPDP with state space E, flow map ϕ and initial state ξ_0. We define the CFSJS and NTS semantics of X. Let X_C be the CFSJS of X and let X_N be the NTS of X. X_C is defined as in Section 3.3.2. $X_N = (E, \Sigma^{vp} \cup \Sigma \cup \bar{\Sigma}, \mathscr{T})$, where

- $\Sigma^{vp} := \{(a, r) | a \in \Sigma, r \in \mathbb{R}^n \text{ for some } n \in \mathbb{N}\}$.
- $(\xi, a, m) \in \mathscr{T}$ if and only if there exists an $\alpha \in \mathscr{A}$ such that $lab(\alpha) = a$, $oloc(\alpha) = loc(\xi)$, $val(\xi) \in guard(\alpha)$, $rmap(\alpha)(\xi) = m$ and $vp(\alpha) = \emptyset$.
- $(\xi, (a, r), m) \in \mathscr{T}$, with $r \in \mathbb{R}^n$, if and only if there exists an $\alpha \in \mathscr{A}$ such that $lab(\alpha) = a$, $oloc(\alpha) = loc(\xi)$, and $val(\xi) \in guard(\alpha)$ and

  (i) $vp(\alpha) = !(w_1, \cdots, w_k)$, where

  $$(G(loc(\xi), w_1)(val(\xi)), \cdots, G(loc(\xi), w_k)(val(\xi))) = r$$

  (i.e., the output for w_1 at ξ equals r), and $rmap(\alpha)(\xi) = m$, or

  (ii) $vp(\alpha) = ?U$ and $r \in U$ and $rmap(\alpha)(\xi, r) = m$
- $(\xi, \bar{a}, m) \in \mathscr{T}$ if and only if there exists an $\alpha \in \mathscr{P}$ such that $lab(\alpha) = \bar{a}$, $oloc(\alpha) = loc(\xi)$ and $rmap(\alpha) = m$.

EXAMPLE 3.8 Let X be a CPDP with one location l. At l there is continuous dynamics $\dot{x} = 1$ and the output map equals $y = x$. There is one active transition $\alpha = (l, a, l, G, R, vp)$ with guard G satisfied if $x \geq 1$, with reset map $R(\xi)(\{x = 0\}) = 1$, i.e., R resets x to 0 at all states $\xi = (l, \{x = r\})$ with $r \geq 1$, and with value passing element $vp = !y$.

The NTS of X, whose state space we denote by E, equals $(E, \Sigma^{vp} \cup \Sigma \cup \bar{\Sigma}, \mathscr{T})$ with $\Sigma = \{a\}$, $\Sigma^{vp} = \{(a, r) | r \in \mathbb{R}^n \text{ for some } n \in \mathbb{N}\}$ and

$$\mathscr{T} = \{((l, \{x = r\}), (a, r), m) | r \geq 1, m = \text{Dirac measure at} x = 0)\}.$$

If we have $vp = ?U$ for some $U \subset \mathbb{R}$ instead of $vp = !y$ and we have $R(\xi, r) = Id(\{x = r\}$, then we get

$$\mathscr{T} = \{((l, \{x = r\}), (a, r'), m) \mid r \geq 1, r' \in U, m = \text{Dirac measure at} x = r')\}.$$

In the latter case we have that the NTS has for states $\xi \in G$ for all $r' \in U$ a transition with label (a, r'). If another CPDP Y outputs value $r' \in U$ through an a-transition, then the NTS of Y has a transition with label (a, r'). In the NTS composition of the NTSs of X with Y, these (a, r') transitions synchronize, which expresses that X accepts the output r' of Y. X then resets its state to r' as expressed by the reset measure $Id(l, \{x = r'\})$. This idea of composition of value passing CPDPs is formally defined as follows. ■

3.3.4.2 Composition of Value Passing CPDPs

DEFINITION 3.10 *Let* $X = (L_X, V_X, \nu_X, W_X, \omega_X, F_X, G_X, \Sigma, \mathscr{A}_X, \mathscr{P}_X, \mathscr{S}_X)$ *and* $Y = (L_Y, V_Y, \nu_Y, W_Y, \omega_Y, F_Y, G_Y, \Sigma, \mathscr{A}_Y, \mathscr{P}_Y, \mathscr{S}_Y)$ *be two value passing CPDPs such that* $V_X \cap V_Y = W_X \cap W_Y = \emptyset$. *Then* $X|_A^P|Y$ *is defined as the CPDP* $(L, V, \nu, W, \omega, F, G, \Sigma, \mathscr{A}, \mathscr{P}, \mathscr{S})$, *where* L, V, ν, W, ω, F, G, Σ, $\mathscr{P}$, *and* $\mathscr{S}$ *are defined as in Definition 3.8 and* $\mathscr{A}$ *is the least relation satisfying the rules (note that the rules r1,r2,r2',r,r3' are the same as in the ordinary composition of CPDPs, cf. Definition 3.8.)*

$$r1.\frac{l_1 \overset{a,G_1,R_1}{\longrightarrow} l_1', l_2 \overset{a,G_2,R_2}{\longrightarrow} l_2'}{l_1|_A^P|l_2 \overset{a,G_1\times G_2,R_1\times R_2}{\longrightarrow} l_1'|_A^P|l_2'}(a \in A).$$

$$r2.\frac{l_1 \overset{a,G_1,R_1}{\longrightarrow} l_1', l_2 \overset{\bar{a},R_2}{\longrightarrow} l_2'}{l_1|_A^P|l_2 \overset{a,G_1\times vs(l_2),R_1\times R_2}{\longrightarrow} l_1'|_A^P|l_2'}(a \notin A).$$

$$r2'.\frac{l_1 \overset{\bar{a},R_1}{\longrightarrow} l_1', l_2 \overset{a,G_2,R_2}{\longrightarrow} l_2'}{l_1|_A^P|l_2 \overset{a,vs(l_1)\times G_2,R_1\times R_2}{\longrightarrow} l_1'|_A^P|l_2'}(a \notin A).$$

$$r3.\frac{l_1 \overset{a,G_1,R_1}{\longrightarrow} l_1', l_2 \overset{\bar{a}}{\not\longrightarrow}}{l_1|_A^P|l_2 \overset{a,G_1\times vs(l_2),R_1\times Id}{\longrightarrow} l_1'|_A^P|l_2}(a \notin A).$$

$$r3'.\frac{l_1 \not\xrightarrow{\bar{a}}, l_2 \xrightarrow{a,G_2,R_2} l_2'}{l_1|_A^P|l_2 \xrightarrow{a,vs(l_1)\times G_2,Id\times R_2} l_1|_A^P|l_2'}(a \notin A).$$

$$r1data.\frac{l_1 \xrightarrow{a,G_1,R_1,v_1} l_1', l_2 \xrightarrow{a,G_2,R_2,v_2} l_2'}{l_1|_A^P|l_2 \xrightarrow{a,G_1|G_2,R_1\times R_2,v_1|v_2} l_1'|_A^P|l_2'}(a \in A, v_1|v_2 \neq \perp).$$

$$r2data.\frac{l_1 \xrightarrow{a,G_1,R_1,v_1} l_1'}{l_1|_A^P|l_2 \xrightarrow{a,G_1\times val(l_2),R_1\times Id,v_1} l_1'|_A^P|l_2}(a \notin A).$$

$$r2data'.\frac{l_2 \xrightarrow{a,G_2,R_2,v_2} l_2'}{l_1|_A^P|l_2 \xrightarrow{a,val(l_1)\times G_2,Id\times R_2,v_2} l_1|_A^P|l_2'}(a \notin A),$$

where $l_1 \xrightarrow{a,G_1,R_1,v_1} l_1'$ *means* $(l_1,a,l_1',G_1,R_1,v_1) \in \mathscr{A}_X$ *with* $v_1 \neq \emptyset$, $l_1 \xrightarrow{a,G_1,R_1} l_1'$ *means* $(l_1,a,l_1',G_1,R_1,\emptyset)$, *and* $v_1|v_2$ *is defined as:*

- $v_1|v_2 :=!Y$ *if* $v_1 =!Y$ *and* $v_2 :=?U$ *and* $dim(U)=dim(Y)$ *or if* $v_2 =!Y$ *and* $v_1 :=?U$ *and* $dim(U)=dim(Y)$,
- $v_1|v_2 :=?(U_1 \cap U_2)$ *if* $v_1 =?U_1$ *and* $v_2 =?U_2$ *and* $dim(U_1)=dim(U_2)$,
- $v_1|v_2 := \perp$ *otherwise, where* $\perp$ *means that* v_1 *and* v_2 *are not compatible.*

Furthermore, $G_1|G_2$ *is, only when* $v_1|v_2 \neq \perp$, *defined as:*

- $G_1|G_2 := (G_1 \cap U) \times G_2$ *if* $v_1 =!Y$ *and* $v_2 =?U$,
- $G_1|G_2 := G_1 \times (G_2 \cap U)$ *if* $v_1 =?U$ *and* $v_2 =!Y$,
- $G_1|G_2 := G_1 \times G_2$ *if* $v_1 =?U_1$ *and* $v_2 =?U_2$.

Here we define $G \cap U$ *as the set of all states in* G *whose output values lie in* U.

THEOREM 3.3 *Let* X *and* Y *be two value passing CPDPs with semantics* (X_C, X_N) *and* (Y_C, Y_N) *respectively. Let* $(X|_A^P|Y_C, X|_A^P|Y_N)$ *be the semantics of value passing CPDP* $X|_A^P|Y$. *Assume that there do not exist value-passing transitions* $(l_1,a,l_1',G_1,R_1,!(w_1,\cdots,w_k)) \in \mathscr{A}_X$ *and* $(l_2,a,l_2',G_2,R_2,!(\tilde{w}_1,\cdots,\tilde{w}_l)) \in \mathscr{A}_Y$ *such that* $a \in A$ *and there exist* $\xi_1 \in G_1$ *and* $\xi_2 \in G_2$ *such that* $(G_X(\xi_1,w_1),\cdots, G_X(\xi_1,w_k)) = (G_Y(\xi_2,\tilde{w}_1),\cdots,G_Y(\xi_2,\tilde{w}_l))$. *Then,*

$$(X|_A^P|Y_C, X|_A^P|Y_N) = (X_C|Y_C, X_N|_A^P|Y_N).$$

REMARK 3.2 The assumption in Theorem 3.3 says that there may not be two value passing output transitions with the same label (in A) and with

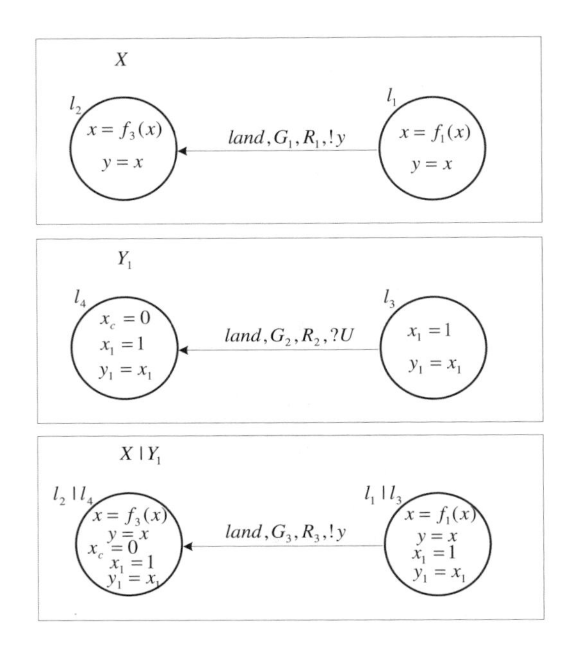

FIGURE 3.5: Value passing CPDPs.

the same output value for some states. Rule $r1data$ expresses that two value passing output transitions can not synchronize, which is in line with the philosophy that at any moment only one component can determine the output, while multiple components may receive this value via value passing input transitions. If the assumption is not satisfied, then on the level of composition of NTSs, there will be synchronized transition that comes from these two output transitions, while the NTS of the composition does not have this synchronized transition because of rule $r1data$. ■

THEOREM 3.4 *$|_A^P|$ for value passing CPDPs is commutative for all A and P. $|_A^P|$ for value passing CPDPs is associative if and only if for all events a we have $a \notin A \Rightarrow \bar{a} \in P$.*

EXAMPLE 3.9 In Figure 3.5 we see the value passing CPDPs X and Y_1. X and Y are the same as the X and Y_1 of Figure 3.2, except that here the active transitions are value passing active transitions. More specific, at the moment of switching to landing mode, the aircraft X sends the value of its state (position and velocity) to the control tower Y_1.

Sending the state information is modelled as the value $!y$ for the value

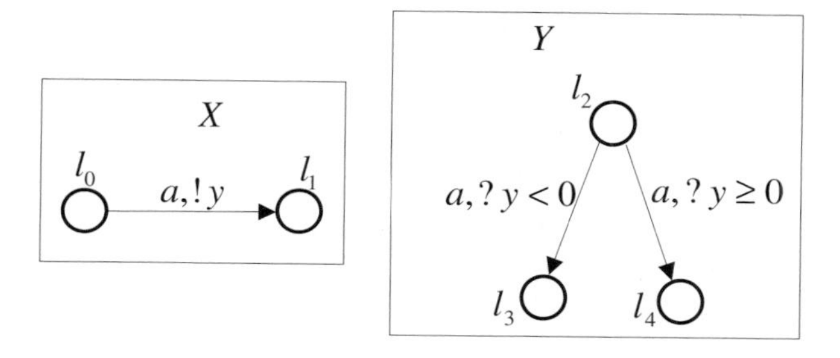

FIGURE 3.6: Value passing used to express scheduling.

passing part of the transition. y is the only output variable of X and is a copy of x and contains therefore the exact information of the state. The value passing element of the transition in Y_1 equals $?U$, where $U = \mathbb{R}^6$. This means that this transition can receive all six dimensional real values. Note that if we would have $r \notin U$ for some $r \in \mathbb{R}^6$, then the transition of X would be blocked at state $\{x = r\}$.

Location l_4 of Y_1 has the new state variable x_c. This variable is used to store the information received from X. At the moment that X switches to l_2, Y_1 will switch to l_4 and the value of y, communicated by X, will be stored in x_c. Storing received data is done via the reset maps and in the case of Figure 3.5 it is expressed as $R(\{x_1 = r_1, x = r_2\})(\{\{x_1 = r_1, x_c = r_2\}\}) = 1$. Note that this indeed expresses that x_1 will not change by the switch and x_c holds the value of y after the switch.

■

3.3.4.3 Expressiveness of Value Passing

In Example 3.9 we have seen that value passing can express sending/receiving of the value of output variables. There are more types of communication that can be expressed by using value passing. We give two more examples which show two more types of communication: scheduling via value passing, constraint conjunction via value passing.

EXAMPLE 3.10 In this example we show how one CPDP can schedule transitions of another CPDP. In Figure 3.6, we see two CPDPs, X and Y, which are pictured without the details concerning state/output dynamics, guards and reset maps. CPDP X can switch from location l_0 to location l_1. With this switch, the value of output variable y is communicated over channel a. This value of y can be received by Y at initial location l_2. Y uses this information to schedule its two transitions at location l_2. If the value of y is smaller than zero, then the transition to location l_3 is taken, otherwise

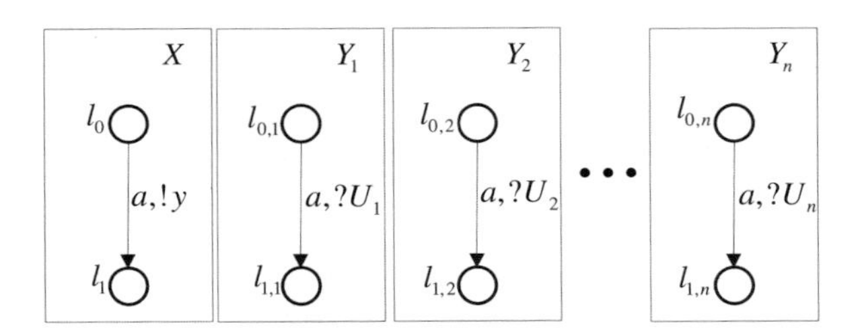

FIGURE 3.7: Value passing used for constraint conjunction.

the transition to location l_4 is taken. In Figure 3.6, $?y<0$ actually stands for $?\{r\in\mathbb{R}|r<0\}$, and $?y\geq 0$ stands for $?\{r\in\mathbb{R}|r\geq 0\}$. In fact, these value passing transitions of Y can receive any one dimensional value that is communicated over channel a. This means that, if we compose Y with a component that sends at some time the value of some one dimensional variable y_2 over channel a, then Y can receive this value. We specifically write $y<0$, to clarify that we intend that this transition is used to receive values of variable y of CPDP X.

In the composition $X|_A^P|Y$, with $A=\{a\}$ and P not relevant since no passive transitions are involved, X schedules the transitions of Y through the values of y. This method can, for example, be applied to systems where one component can perform different strategies, while the specific strategy that is chosen depends on the output variables of some other component. ■

EXAMPLE 3.11 In Figure 3.7, CPDP X can switch from initial location l_0 to location l_1. The guard of this transition (not pictured) equals the whole valuation space of l_0. If X would be executed as a stand alone CPDP, it would, because of maximal progress, switch immediately to l_1. In this example we show how other CPDPs, Y_1 till Y_n, can independently put constraints on the execution time of the active transition of X. For $i=1\cdots n$, U_i is the constraint put by CPDP Y_i on the execution time of the transition of X. Let y be one dimensional. Then, if $n=2$, U_1 equals $y\geq -1$ and U_2 equals $y\leq 1$, then in $X|_A^P|Y_1|_A^P|Y_2$, with $A=\{a\}$ and P not relevant, the guard on the a-transition from location $l_0|l_{0,1}|l_{0,2}$ to location $l_1|l_{1,1}|l_{1,2}$ is equal to the part of the valuation space where $y\in[-1,1]$. ■

3.4 Conclusions

In conclusion we summarize some aspects of the CPDP model, and describe which types of systems and which types of communication between those systems can be captured with the theory of this chapter.

A CPDP models a system with multiple locations. In each location, the continuous state of the CPDP has dynamics determined by some ordinary differential equation. The CPDP can jump from one location to another by means of a spontaneous transition or by means of a non-deterministic (or active) transition. A spontaneous transition is determined by some probability distribution. A non-deterministic (or active) transition can happen only if the continuous state lies inside the guard of that transition. However, if the process enters the guard of an active transition, then the process is not forced to execute the transition, but it is allowed to execute the transition.

Two CPDPs can communicate via the synchronization of transitions. If a is a synchronization action, then active a-transitions of the CPDPs should synchronize. This means that if one CPDP has an a-transition enabled (i.e., is inside the guard of some a-transition), and the other CPDP has no a-transition enabled, then this other CPDP blocks the enabled a-transitions of the first CPDP. We call this kind of communication *blocking interaction*. The other kind of communication that can be expressed is called *broadcasting interaction*. This happens if an active a-transition of one CPDP triggers a passive $\bar{a}$-transition of the other CPDP. Then the other CPDP "observes" that the first CPDP executes an a-transition. Thus, communication/interaction for CPDPs means that CPDPs can get knowledge about the execution of transitions in other CPDPs. Although two (or more) CPDPs cannot have shared continuous variables (as is the case in some other compositional hybrid systems frameworks), it is still possible that information concerning the continuous variables is communicated from one CPDP to the other. For this we need value-passing CPDPs, where active transitions of one CPDP can pass values (which come from the continuous variables) to active transitions in other CPDPs. These passed values can then influence the reset maps of the transitions that received these values and in that way one CPDP can get knowledge about the continuous variables of other CPDPs.

CPDPs have non-determinism. It is not determined when active transitions have to be executed and it is not determined which transition is executed at states where multiple transitions are enabled. In [12], the maximal progress assumption is used to resolve the first type of non-determinism: an active transition is executed as soon as the guard area of some transition is entered. In [12], the second type of non-determinism is resolved by defining a scheduler which probabilistically chooses which transition will be executed. Then, it is shown in [12], that a scheduled CPDP behaves under maximal progress as a PDP and an algorithm is given to transform such a scheduled CPDP into a PDP. With this equivalence result, scheduled CPDPs can be analyzed through PDP analysis techniques.

Also in [12] (or [15]), a notion of bisimulation is defined for CPDP, and an algorithm is given for finding bisimilation relations. Through bisimulation the state space of a CPDP can be reduced without changing the stochastic behavior of the CPDP.

References

[1] R. Alur, C. Coucoubetis, N. Halbwachs, T. Henzinger, P. Ho, X. Nicolin, A. Olivero, J. Sifakis, and S. Yovine. The algorithmic analysis of hybrid systems. *Theoretical Computer Science*, 138:3–34, 1995.

[2] T. Bolognesi and E. Brinksma. Introduction to the ISO specification language LOTOS. *Comp. Networks and ISDN Systems*, 14:25–59, 1987.

[3] M. H. A. Davis. Piecewise Deterministic Markov Processes: a general class of non-diffusion stochastic models. *Journal Royal Statistical Soc. (B)*, 46:353–388, 1984.

[4] M. H. A. Davis. *Markov Models and Optimization.* Chapman & Hall, London, 1993.

[5] M. H. C. Everdij and H. A. P. Blom. Petri-nets and hybrid-state Markov processes in a power-hierarchy of dependability models. In *Preprints Conference on Analysis and Design of Hybrid Systems ADHS 03*, pages 355–360, 2003.

[6] M. H. C. Everdij and H. A. P. Blom. Piecewise deterministic Markov processes represented by dynamically coloured Petri nets. *Stochastics*, 77(1):1–29, 2005.

[7] H. Hermanns. *Interactive Markov Chains*, volume 2428 of *Lecture Notes in Computer Science*. Springer, Berlin, 2002.

[8] C.A.R. Hoare. *Communicating Sequential Processes.* Prentice-Hall, Upper Saddle River, NJ, USA, 1985.

[9] J.F.C. Kingman. *Poisson Processes.* Oxford Clarendon Press, Oxford, 1993.

[10] M. Haj-Hussein, L. Logrippo, and M. Faci. An introduction to LOTOS: Learning by examples. *Comp. Networks and ISDN Systems*, 23(5):325–342, 1992.

[11] N. A. Lynch, R. Segala, and F. W. Vaandrager. Hybrid I/O automata. *Information and Computation*, 185(1):105–157, 2003.

[12] S. N. Strubbe. Compositional Modelling of Stochastic Hybrid Systems. *Phd. Thesis*, Twente University, 2005.

[13] S. N. Strubbe, A. A. Julius, and A. J. van der Schaft. Communicating Piecewise Deterministic Markov Processes. In *Preprints Conference on Analysis and Design of Hybrid Systems ADHS 03*, pages 349–354, 2003.

[14] S. N. Strubbe and R. Langerak. A composition operator with active and passive actions. Proc. 25th IFIP WG 6.1 International Conference on Formal Techniques for Networked and Distributed Systems, Taipei, 2005.

[15] S. N. Strubbe and A. J. van der Schaft. Bisimulation for Communicating Piecewise Deterministic Markov Processes (cpdps). In *Hybrid Systems: Computation and Control*, volume 3414 of *Lecture Notes in Computer Science*, pages 623–639. Springer, Berlin, 2005.

[16] S. N. Strubbe and A. J. van der Schaft. Stochastic semantics and value-passing for communicating piecewise deterministic Markov processes. Proc. Conf. Decision and Control, Seville, 2005.

[17] S. N. Strubbe and A. J. van der Schaft. Communicating Piecewise Deterministic Markov Processes. In H.A.P. Blom and J. Lygeros, editors, *Stochastic hybrid systems: theory and safety applications*, volume 337 of *Lecture Notes in Control and Informations Sciences*, pages 65–104. Springer, Berlin, 2006.

Chapter 4

Stochastic Model Checking

Joost-Pieter Katoen
RWTH Aachen

4.1 Introduction 79
4.2 The Discrete-time Setting 81
4.3 The Continuous-time Setting 87
4.4 Bisimulation and Simulation Relations 94
4.5 Epilogue 100
References 102

4.1 Introduction

When a program is run on a computer one of the most important considerations is whether it will work correctly. The fundamental question "when and why does software not work as expected?" has been the subject of intense research in computer science for decades. [1] The origins of a sound mathematical approach toward program correctness can be traced back to Turing in 1949 [59]. Early attempts to assess the correctness of computer programs were based on mathematical proof rules [37, 51, 3]. Proofs of realistic programs, though, are rather lengthy for programs of realistic size and require a large amount of human ingenuity.

In the early 1980s an alternative to using proof rules was proposed, independently by researchers in Europe [53] and the US [19], that given a (finite) model of a program systematically checks whether it satisfies a given property. This breakthrough was the first step towards the *automated verification* of concurrent programs. Typical questions treated by "*model checking*" are:

- safety: e.g., does a given mutual exclusion algorithm guarantee exclusive access to the shared resource?
- liveness: e.g., will a transmitted packet eventually arrive at the destination?

[1] According to Hoare, one of the pioneers in program verification and currently at Microsoft's laboratory, up to three-quarters of the 400 billion US dollar spent annually employing computer programmers in the US goes on debugging.

- fairness: e.g., will a repetitive attempt to carry out a transaction be eventually granted?

Over the last two decades, model checking has received a lot of attention and is subject of study of a rapidly growing research community [21, 20].

How does model checking work? Given a model of the system (the "possible behavior") and a specification of the property to be considered (the "desirable behavior"), model checking is a technique that systematically checks the validity of the property in the model. Models are typically nondeterministic finite-state automata, consisting of a finite set of states and a set of transitions that describe how the system evolves from one state into another. These automata are usually composed of concurrent entities and are often generated from a high-level description language such as Petri nets, process algebra [50], PROMELA [38] or Statecharts [30]. Properties are typically specified in temporal logic such as CTL (Computation Tree Logic) [19], an extension of propositional logic that allows one to express properties that refer to the relative order of events. Statements can either be made about states or about paths, i.e., sequences of states that model an evolution of the system.

The basis of model checking is a systematic, usually exhaustive, state-space exploration to check whether the property is satisfied in each state of the model, thereby using effective methods (such as symbolic data structures, partial-order reduction or clever hashing techniques) to combat the state-space explosion problem. Due to unremitting improvements of underlying algorithms and data structures together with hardware technology improvements, model-checking techniques that a decade ago only worked for simple examples, are nowadays applicable to more realistic designs. State-of-the-art model checkers can handle state spaces of about 10^9 states using off-the-shelf technology. Using clever algorithms and tailored data structures, larger state spaces (up to 10^{476} states [55]) can be handled for specific problems and restricted types of correctness properties, namely so-called reachability properties.

4.1.1 Stochastic Model Checking

Whereas model checking focuses on the absolute correctness, in practice such rigid notions are hard, or even impossible, to guarantee. Instead, systems are subject to various phenomena of stochastic nature, such as message loss or garbling, unpredictable environments, faults, and delays. Correctness thus is of a less absolute nature. Accordingly, instead of checking whether system failures are impossible a more realistic aim is to establish, for instance, whether "the chance of shutdown occurring is at most 0.01%." Similarly, the question whether a distributed computation terminates becomes "does it eventually terminate with probability 1?" These queries can be checked using *stochastic* model checking, an automated technique for stochastic models in which state transitions encode the probability of making a transition between states rather than just the existence of such transition.

Stochastic model checking is based on conventional model checking, since it relies on reachability analysis of the underlying transition system, but must also entail the calculation of the actual likelihoods through appropriate numerical methods. In addition to the qualitative statements made by conventional model checking, this provides the possibility to make quantitative statements about the system. Stochastic model checking uses extensions of temporal logics with probabilistic operators, affording the expression of these quantitative statements. Prominent examples of such extensions for CTL are PCTL [29] and CSL [4, 8].

Stochastic model checking is typically based on discrete-time and continuous-time Markov chains (DTMCs and CTMCs, respectively), or Markov decision processes (MDPs). Whereas Markov chains are fully stochastic, MDPs allow for nondeterminism. The former models are intensively used in performance and dependability analysis, whereas MDPs are of major importance in stochastic operations research [57, 52] and automated planning in AI [15]. Extensions of model checking to stochastic models originate from the mid 1980s [31, 60], first focusing on 0-1 probabilities, but later also considering quantitative properties [23]. During the last decade, these methods have been extended, refined and improved, and – most importantly – been supported by software tools [48, 35]. Currently, model checkers for DTMCs, CTMCs and MDPs do exist, and have been applied to several case studies ranging from randomized distributed algorithms [47] to dependability analysis of workstation clusters [33, 18]. With the currently available technology, models of $10^7 - 10^8$ states can be successfully checked [18, 44]. This number can be increased significantly by applying aggressive abstraction techniques such as symmetry reduction, (bi)simulation relations, or three-valued logics.

4.1.2 Topic of this Survey

This chapter surveys the state-of-the-art in model checking of fully probabilistic models, i.e., stochastic models without nondeterminism. Discrete- and continuous-time models are covered as well as their extensions with costs. Syntax and semantics of (core fragments of) temporal logics are provided and the algorithms for checking (conditional) reachability properties are considered. This encompasses simple probabilistic reachability, i.e., what is the probability to reach a set of goal states, as well as time-bounded probabilistic reachability where in addition the goal state should be reached within a given deadline (which can be either discrete or continuous). For cost-extended models, we consider cost- and time-bounded reachability. For these models the following question is central: what is the probability to reach a given set of goal states within a given time-bound and a given bound on the cost? Bisimulation and simulation relations are defined and their relationship to the temporal logics covered in this chapter is established.

4.2 The Discrete-time Setting

We first consider discrete-time Markov chains and equip these with costs.

4.2.1 Discrete-time Markov Chains

DTMCs. Let AP be a fixed, finite set of *atomic propositions*. The atomic propositions will be used to label states in a Markov chain, and are used to express the most elementary properties of a state. Such properties could be, e.g., "the discrete variable x lies between 0 and 201," or "the number of active processes in this state equals 7." It is assumed that the validity of an atomic proposition in a state can easily be determined (e.g., by inspection).

A (labelled) DTMC $\mathcal{D}$ is a tuple $(S, \mathbf{P}, L)$ where S is a finite set of *states*, $\mathbf{P} : S \times S \rightarrow [0,1]$ is a *probability matrix* such that $\sum_{s' \in S} \mathbf{P}(s,s') = 1$ for all $s \in S$, and $L : S \rightarrow 2^{AP}$ is a *labelling* function which assigns to each state $s \in S$ the set $L(s)$ of atomic propositions that are valid in s. A path through a DTMC is a sequence[2] of states $\sigma = s_0\, s_1\, s_2 \ldots$ with $\mathbf{P}(s_i, s_{i+1}) > 0$ for all i. Let $Path^{\mathcal{D}}$ denote the set of all paths in DTMC $\mathcal{D}$. $\sigma[i]$ denotes the $(i{+}1)$th state of σ, i.e., $\sigma[i] = s_i$. Let $\Pr_s$ denote the unique probability measure on sets of paths that start in state s. This probability measure is defined in the standard way, see e.g., [46].

PCTL. Properties over DTMCs are expressed using an extension of temporal logic. Let $a \in AP$, probability $p \in [0,1]$, k be a natural number (or ∞) and $\bowtie$ a binary comparison operator in $\{\leq, \geq\}$. The syntax of Probabilistic CTL (PCTL) [29] is defined by the grammar:

$$\Phi ::= \mathrm{tt} \;\Big|\; a \;\Big|\; \Phi \wedge \Phi \;\Big|\; \neg\Phi \;\Big|\; \mathcal{P}_{\bowtie p}(\Phi\, \mathcal{U}^{\leq k}\, \Psi).$$

Thus, a formula in PCTL is built up from the basic formulas tt (true) and atomic proposition a, can be obtained by combining two PCTL formulas by $\wedge$ (conjunction), prefixing a PCTL-formula with $\neg$ (negation), or by a so-called until-formula (denoted $\mathcal{U}$) that is contained in a $\mathcal{P}$-context which has as parameters a probability and a binary comparison operator. The other usual boolean connectives such as disjunction, implication and equivalence are derived in the usual way, e.g., $\Phi \vee \Psi = \neg(\neg\Phi \wedge \neg\Psi)$. The formula $\Phi\, \mathcal{U}^{\leq k}\, \Psi$ states a property over paths; a path $s_0\, s_1\, s_2 \ldots$ satisfies this formula if within k steps a Ψ-state is reached, and when all preceding states satisfy Φ. That is, when $\sigma[j]$ satisfies Ψ with $j \leq k$, and $\sigma[i]$ satisfies Ψ, for all indices i such that $i < j$. The unbounded until formula that is standard in temporal logics is obtained by taking k equal to ∞, i.e., $\Phi \mathcal{U} \Psi = \Phi\, \mathcal{U}^{\leq\infty}\, \Psi$. For the sake of simplicity, we do not consider the next operator.

The semantics of PCTL is defined by [29] a binary relation, denoted $\models$, between states of the DTMC and PCTL formulas. The fact that $(s, \Phi) \in \models$ is denoted as $s \models \Phi$, and denotes that the PCTL-formula Φ holds in state s. The relation $\models$ is defined by structural induction on the formula Ψ in the following way (where iff stands for if and only if):

$$\begin{array}{lll} s \models \mathrm{tt} & \text{for all } s \in S & \\ s \models a & \text{iff } a \in L(s) & \\ s \models \neg\Phi & \text{iff } s \not\models \Phi & \end{array} \qquad \begin{array}{ll} s \models \Phi \wedge \Psi & \text{iff } s \models \Phi \wedge s \models \Psi \\ s \models \mathcal{P}_{\bowtie p}(\Phi\, \mathcal{U}^{\leq k}\, \Psi) & \text{iff } Prob^{\mathcal{D}}(s, \Phi\, \mathcal{U}^{\leq k}\, \Psi) \bowtie p \end{array}$$

[2] In this chapter, we do not dwell upon distinguishing finite and infinite paths.

$\mathcal{P}_{\bowtie p}(\Phi\, \mathcal{U}^{\leq k}\, \Psi)$ asserts that the probability measure of the paths that start in s and that satisfy $\Phi\, \mathcal{U}^{\leq k}\, \Psi$ meets the bound $\bowtie p$. Here,

$$Prob^{\mathcal{D}}(s, \Phi\, \mathcal{U}^{\leq k}\, \Psi) = \Pr_s \left\{ \sigma \in Path^{\mathcal{D}} \mid \sigma \models \Phi\, \mathcal{U}^{\leq k}\, \Psi \right\}.$$

Formula $\Phi\, \mathcal{U}^{\leq k}\, \Psi$ asserts that Ψ will be satisfied within k steps and that all preceding states satisfy Φ, i.e.,

$$\sigma \models \Phi\, \mathcal{U}^{\leq k}\, \Psi \ \text{ iff } \exists j \leq k.\, (\sigma[j] \models \Psi \wedge \forall i < j.\, \sigma[i] \models \Phi).$$

The hop-constraint $\leq k$ can easily be generalised towards arbitrary intervals. The same applies to cost- and real-time constraints that are considered later. For the sake of brevity, we refrain from going into the details of these generalizations.

Let us illustrate the expressiveness of PCTL by means of an abstract example. Suppose that the states in the DTMC under consideration are forbidden (= *illegal*) states, *goal* states and others. Assume that *illegal* and *goal* are atomic propositions. The logic PCTL allows to express, e.g.,

- with probability ≥ 0.92, a goal state is reached via legal states only:

$$\mathcal{P}_{\geq 0.92}\,((\neg\, illegal)\ \mathcal{U}\ goal)$$

- ... in maximally 137 steps: $\mathcal{P}_{\geq 0.92}\left((\neg\, illegal)\ \mathcal{U}^{\leq 137}\ goal\right)$
- ... once there, remain almost always for at least the next 31 steps:

$$\mathcal{P}_{\geq 0.92}\left((\neg\, illegal)\ \mathcal{U}^{\leq 137}\ \mathcal{P}_{\geq 0.9999}\left(\Box^{\leq 31}\ goal\right)\right)$$

where $\mathcal{P}_{\geq p}(\Box^{\leq k}\Phi) = \mathcal{P}_{\leq 1-p}(\Diamond^{\leq k}\neg\,\Phi)$ and $\Diamond^{\leq k}\Phi$ stands for tt $\mathcal{U}^{\leq k}\, \Phi$. The formula $\Diamond^{\leq k}\Phi$ thus denotes that eventually a Φ-state will be reached within k steps and $\Box^{\leq k}\Phi$ asserts that the next k states all satisfy Φ.

Verifying hop-constrained probabilistic reachability. PCTL model checking [29] is carried out in the same way as verifying CTL [21] by recursively computing the set $Sat(\Phi) = \{s \in S \mid s \models \Phi\}$. This is done by means of a bottom-up recursive algorithm over the parse tree of Φ. Checking bounded until-formulas amounts to computing the least solution[3] of the following set of equations: $Prob^{\mathcal{D}}(s, \Phi\, \mathcal{U}^{\leq k}\, \Psi)$ equals 1 if $s \in Sat(\Psi)$,

$$Prob^{\mathcal{D}}(s, \Phi\, \mathcal{U}^{\leq k}\, \Psi) = \sum_{s' \in S} \mathbf{P}(s,s') \cdot Prob^{\mathcal{D}}(s', \Phi\, \mathcal{U}^{\leq k-1}\, \Psi) \tag{4.1}$$

if $s \in Sat(\Phi \wedge \neg\Psi)$ and $k > 0$, and equals 0 otherwise. This probability can be computed as the solution of a regular system of linear equations by standard means such

[3]Strictly speaking, the function $s \mapsto Prob^{\mathcal{D}}(s, \Phi\, \mathcal{U}^{\leq k}\Psi)$ is the least fixpoint of a higher-order function on $(S \to [0,1]) \to (S \to [0,1])$ where the underlying partial order on $S \to [0,1]$ is defined for $F_1, F_2 : S \to [0,1]$ by $F_1 \leq F_2$ if and only if $F_1(s) \leq F_2(s)$ for all $s \in S$.

as Gaussian elimination or can be approximated by an iterative approach (fixed point computation). The following alternative recipe is of interest to the stochastic models treated later in this chapter and has the same time complexity.

For DTMC $\mathcal{D} = (S, \mathbf{P}, L)$ and PCTL formula Φ, let DTMC $\mathcal{D}[\Phi] = (S, \mathbf{P}', L)$ where if $s \not\models \Phi$, then $\mathbf{P}'(s,s') = \mathbf{P}(s,s')$ for all $s' \in S$, and if $s \models \Phi$, then $\mathbf{P}'(s,s) = 1$ and $\mathbf{P}'(s,s') = 0$ for all $s' \neq s$. We have $\mathcal{D}[\Phi][\Psi] = \mathcal{D}[\Phi \vee \Psi]$. Let $\pi^{\mathcal{D}}(s,k)(s')$ denote the probability of being in state s' after exactly k steps in DTMC $\mathcal{D}$ when starting in s, i.e., $\pi^{\mathcal{D}}(s,k)(s') = \Pr_s\left\{\sigma \in Path^{\mathcal{D}} \mid \sigma[k] = s'\right\}$. This is known as the transient probability of state s' after k steps and can be obtained by $(\alpha_s \cdot \mathbf{P}^k)(s')$ where α_s is a probability vector that is one in state s, and 0 otherwise. It now follows that for any DTMC $\mathcal{D}$:

$$Prob^{\mathcal{D}}(s, \Phi\, \mathcal{U}^{\leq k}\, \Psi) = \sum_{s' \models \Psi} \pi^{\mathcal{D}[\neg\Phi \vee \Psi]}(s,k)(s'). \tag{4.2}$$

Note that $\mathcal{D}[\neg\Phi \vee \Psi] = \mathcal{D}[\neg(\Phi \wedge \Psi)][\Psi]$, i.e., all $\neg(\Phi \vee \Psi)$-states and all Ψ-states in $\mathcal{D}$ are made absorbing. That is, the only transitions available in these states are self-loops with probability one. The former is correct since $\Phi\, \mathcal{U}^{\leq k}\, \Psi$ is violated as soon as some state is visited that neither satisfies Φ nor Ψ. The latter is correct since, once a Ψ-state in $\mathcal{D}$ has been reached (along a Φ-path) in at most k steps, then $\Phi\, \mathcal{U}^{\leq k}\, \Psi$ holds, regardless of which states will be visited later on. This modification of the system dynamics is closely related to the one in Chapter 5 of this volume for the numerical computation of reachability probabilities in a more general class of stochastic hybrid systems.

Determining the set of states that satisfy $\Phi\, \mathcal{U}^{\leq k}\, \Psi$ thus amounts to computing $(\mathbf{P}^{\mathcal{D}[\neg\Phi \vee \Psi]})^k \cdot \underline{\iota}_\Psi$, where $\underline{\iota}_\Psi$ characterises $Sat(\Psi)$, i.e., $\iota_\Psi(s) = 1$ if $s \models \Psi$, and 0 otherwise. As iterative squaring is not attractive for stochastic matrices due to fill in [56], the product is typically computed in an iterative fashion: $\mathbf{P} \cdot (\ldots (\mathbf{P} \cdot \underline{\iota}_\Psi))$.

4.2.2 Rewards

DMRM. A discrete-time Markov reward model (DMRM) $\mathcal{D}_r$ is a tuple $(\mathcal{D}, r)$ where $\mathcal{D}$ is a DTMC and $r : S \rightarrow \mathbb{R}_{\geq 0}$ is a *reward* assignment function. The quantity $r(s)$ indicates the reward that is earned on leaving state s. Note that rewards could also be attached to edges in a DTMC, but this does not increase expressivity. A path through a DMRM is a path through its DTMC, i.e., sequence of states $\sigma = s_0\, s_1\, s_2 \ldots$ with $\mathbf{P}(s_i, s_{i+1}) > 0$ for all i. The probability measure on sets of paths is defined as for DTMCs.

PRCTL. Let $r \in \mathbb{R}_{\geq 0}$ be a nonnegative reward bound, k a natural number, $p \in [0,1]$ and $a \in AP$. The syntax of Probabilistic Reward CTL (PRCTL) [2] is defined by the following grammar:

$$\Phi ::= \mathrm{tt} \;\Big|\; a \;\Big|\; \Phi \wedge \Phi \;\Big|\; \neg\Phi \;\Big|\; \mathcal{P}_{\bowtie p}(\Phi\, \mathcal{U}^{\leq k}_{\leq r}\, \Phi) \;\Big|\; \mathcal{E}^{=k}_{\leq r}(\Phi).$$

Note that the binary until-operator is now equipped with two bounds: one on the maximum number (k) of allowed hops to reach the goal states, and one on the maximum allowed cumulated reward (r) before reaching the goal states. Formula $\mathcal{E}^{=k}_{\leq r}(\Phi)$

asserts that the expected cumulated reward in Φ-states until the k-th transition is at most r. Thus, in order to check the validity of this formula for a given path, all visits to Φ-state are considered in the first k steps and the total reward that is obtained in these states; the reward earned in other states is not relevant, this also applies to Φ-states that are visited after having visited k states in the path. Whenever the expected value of this quantity over all paths that start in state s is at most r, state s is said to satisfy $\mathcal{E}_{\leq r}^{=k}(\Phi)$. The formal definition follows below. Other operators involving rewards that could be considered can be found in [2].

The semantics of the state-formulas of PRCTL that are common with PCTL is identical to the semantics for PCTL as presented above. Formula $\Phi\, \mathcal{U}_{\leq r}^{\leq k} \Psi$ asserts that Ψ will be satisfied within k steps, that all preceding states satisfy Φ, and that the cumulated reward until reaching the Ψ-state is at most r. Thus, for path σ we have:

$$\sigma \models \Phi\, \mathcal{U}_{\leq r}^{\leq k} \Psi \text{ iff } \exists j \leq k.\, \Big(\sigma[j] \models \Psi \wedge (\forall i < j.\, \sigma[i] \models \Phi) \,\wedge\, \textstyle\sum_{i=0}^{j-1} r(\sigma[i]) \leq r\Big)\,.$$

Similar as for PCTL:

$$s \models \mathcal{P}_{\bowtie p}(\Phi\, \mathcal{U}_{\leq r}^{\leq k} \Psi) \text{ if and only if } Prob^{\mathcal{D}_r}(s, \Phi\, \mathcal{U}_{\leq r}^{\leq k} \Psi) \bowtie p$$

where $Prob^{\mathcal{D}_r} = Prob^{\mathcal{D}}$. The semantics of the expected cumulated reward operator is defined by:

$$s \models \mathcal{E}_{\leq r}^{=k}(\Phi) \text{ if and only if } \sum_{i=0}^{k-1} \sum_{s' \models \Phi} \pi(s,i)(s') \cdot r(s') \leq r.$$

Note that Φ plays the role of a state selector: only in states that satisfy Φ, the reward is considered. Rewards in the other states are ignored.

Multiple rewards. The logic PRCTL can easily be enhanced such that properties over models equipped with multiple reward structures can be treated. Suppose $\mathcal{C} = (\mathcal{D}, r_1, \ldots, r_k)$ is a DMRM with $k > 0$ reward assignment functions, and let $0 < j \leq k$. The reward operators of PRCTL can be generalized in a straightforward manner such that constraints on all k reward structures can be expressed in a single formula. For instance, the formula $\mathcal{E}_{\leq r_1 \ldots \leq r_k}^{=k}(\Phi)$ expresses that the expected cumulative reward in Φ-states until the k-th transition meets the upper bounds r_i of the i-th reward (for $0 < i \leq k$). The bounded-until operator can be generalised in a similar manner. Note that the hop-constraint (k) can also be considered as a reward in this setting.

Verifying hop- and reward-bounded probabilistic reachability.

Checking of the bounded until-operator in PRCTL amounts to computing the least solution of the following set of linear equations: $Prob^{\mathcal{D}_r}(s, \Phi\, \mathcal{U}_{\leq r}^{\leq k}\, \Psi)$ equals 1 if $s \in Sat(\Psi)$,

$$Prob^{\mathcal{D}_r}(s, \Phi\, \mathcal{U}_{\leq r}^{\leq k} \Psi) = \textstyle\sum_{s' \in S} \mathbf{P}(s,s') \cdot Prob^{\mathcal{D}_r}(s', \Phi\, \mathcal{U}_{\leq r - r(s)}^{\leq k-1}\, \Psi) \tag{4.3}$$

if $s \in Sat(\Phi \wedge \neg \Psi)$, $k > 0$, and $r(s) \geq r$, and equals 0 otherwise.

Let $\pi_r^{\mathcal{D}_r}(s,k)(s') = \Pr_s\left\{\sigma \in Path^{\mathcal{D}} \mid \sigma[k] = s' \wedge \sum_{i=0}^{k-1} r(\sigma[i]) \leq r\right\}$. Then for any DMRM $\mathcal{D}_r$:

$$Prob^{\mathcal{D}_r}(s, \Phi\, \mathcal{U}^{\leq k}_{\leq r} \Psi) = \sum_{s' \models \Psi} \pi_r^{\mathcal{D}_r[\neg\Phi \vee \Psi]}(s,k)(s') \qquad (4.4)$$

where for formula Φ, the DMRM $\mathcal{D}_r[\Phi]$ is defined as $(\mathcal{D}[\Phi], r')$ with DTMC $\mathcal{D}[\Phi]$ as before and $r'(s) = r(s)$ if $s \not\models \Phi$ and 0 otherwise. That is, all states that are made absorbing obtain a zero reward.

The mathematical characterization (4.4) has a strong resemblance with the result for DTMCs, cf. Equation (4.2). Nevertheless, computing the transient reward probabilities $\pi_r^{\mathcal{D}_r}(s,k)(s')$ is more involved than computing transient probabilities $\pi^{\mathcal{D}}$. We sketch two algorithms. The first algorithm is based on the following recursion scheme [58]. Assume the rewards are either natural or rational numbers. Let $p_r(s,k)$ be the probability to be in state s at the k-th step while having incurred an accumulated cost exactly r. Then:

$$\pi_r^{\mathcal{D}_r}(s,k)(s') = \sum_{i=1}^{r} p_i(s',k).$$

Let $p_i(s',1) = 1$ if $s' = s$ and $i = r(s)$ and 0 otherwise. Then for $k \geq 0$:

$$p_i(s,k{+}1) = \sum_{s' \in S} p_{i-r(s')}(s',k) \cdot \mathbf{P}(s',s).$$

Alternatively, we exploit the following adaptation of the path graph generation algorithm [54]. The basic idea is to unfold the DMRM while keeping track of the cumulative reward so far. In the i-th step, only Φ-successors of state s are "unfolded" if $i < k{-}1$, and if $i = k{-}1$, only Ψ-successors of state s are considered. Vertices that have the same cumulated reward are grouped. The groups have the form $(R, \{(s_1,p_1),\ldots,(s_m,p_m)\}) \in V_h$ where h is the unfolding depth, the root vertex is $(0,\{s,1\})$ (only element in V_0), $\sum_i p_i$ is the probability to gain R reward in h transitions, and p_i is the probability to reach s_i when starting from s. The unfolding is stopped on reaching depth k. On termination, the total probability of reward r can easily be obtained from the groups of vertices.

Checking the operator on expected cumulated reward amounts to solving a system of linear equations. The quantity $\sum_{i=0}^{k-1} \sum_{s' \models \Phi} \pi(s,i)(s') \cdot r(s')$ can be characterized as the smallest solution of the following system of linear equations:

$$H(s,k) = \begin{cases} 0 & \text{if } n = 0 \\ r(s) + \sum_{s' \in S} \mathbf{P}(s,s') \cdot H(s',k{-}1) & \text{if } s \in Sat(\Phi) \wedge k > 0 \\ \sum_{s' \in S} \mathbf{P}(s,s') \cdot H(s',k{-}1) & \text{if } s \notin Sat(\Phi) \wedge k > 0 \end{cases} .$$

4.3 The Continuous-time Setting

4.3.1 Continuous-time Markov Chains

CTMCs. A (labelled) CTMC $\mathcal{C}$ is a tuple $(S, \mathbf{R}, L)$ where S and L are as for DTMCs, and $\mathbf{R} : S \times S \to \mathbb{R}_{\geq 0}$ is the *rate matrix*. The exit rate $E(s) = \sum_{s' \in S} \mathbf{R}(s, s')$ denotes that the probability of taking a transition from s within t time units equals $1 - e^{-E(s) \cdot t}$. If $\mathbf{R}(s, s') > 0$ for more than one state s', a race between the outgoing transitions from s exists. That is, the probability $\mathbf{P}(s, s')$ of moving from s to s' in a single step equals the probability that the delay of going from s to s' "finishes before" the delays of any other outgoing transition from s. The probability of moving from state s to a state s' is $\mathbf{P}(s, s') = \frac{\mathbf{R}(s,s')}{E(s)}$. The probability of making a transition from state s to s' within time t is given by:

$$\frac{\mathbf{R}(s, s')}{E(s)} \cdot (1 - e^{-E(s) \cdot t}).$$

The time-abstract behaviour of a CTMC is described by its embedded DTMC. For CTMC $\mathcal{C} = (S, \mathbf{R}, L)$, the *embedded* DTMC is given by $emb(\mathcal{C}) = (S, \mathbf{P}, L)$, where $\mathbf{P}(s, s') = \mathbf{R}(s, s')/E(s)$ if $E(s) > 0$, and $\mathbf{P}(s, s) = 1$ and $\mathbf{P}(s, s') = 0$ for $s \neq s'$ if $E(s) = 0$.

A path in a CTMC is an alternating sequence $\sigma = s_0\, t_0\, s_1\, t_1\, s_2 \ldots$ with $\mathbf{R}(s_i, s_{i+1}) > 0$ and $t_i \in \mathbb{R}_{>0}$ for all i. The time stamps t_i denote the amount of time spent in state s_i. Let $Path^{\mathcal{C}}$ denote the set of paths through $\mathcal{C}$. $\sigma @ t$ denotes the state of σ occupied at time t, i.e., $\sigma @ t = \sigma[i]$ with i the smallest index such that $t \leq \sum_{j=0}^{i} t_j$. Let $\Pr_s$ denote the unique probability measure on sets of paths that start in s [10].

CSL. Let a, p and $\bowtie$ be as before and $t \in \mathbb{R}_{\geq 0}$ (or ∞). The syntax of Continuous Stochastic Logic (CSL) [4, 10] is:

$$\Phi ::= \mathrm{tt} \;\Big|\; a \;\Big|\; \Phi \wedge \Phi \;\Big|\; \neg \Phi \;\Big|\; \mathcal{P}_{\bowtie p}(\Phi\, \mathcal{U}^{\leq t}\, \Phi).$$

The semantics of CSL for the boolean operators is identical to that for PCTL. For the time-bounded until-formula:

$$s \models \mathcal{P}_{\bowtie p}(\Phi\, \mathcal{U}^{\leq t}\, \Phi) \text{ if and only if } Prob^{\mathcal{C}}(s, \Phi\, \mathcal{U}^{\leq t}\, \Psi) \bowtie p.$$

$Prob^{\mathcal{C}}(\cdot)$ is defined in a similar way as for DTMCs:

$$Prob^{\mathcal{C}}(s, \Phi\, \mathcal{U}^{\leq t}\, \Psi) = \Pr_s \left\{ \sigma \in Path^{\mathcal{C}} \mid \sigma \models \Phi\, \mathcal{U}^{\leq t}\, \Psi \right\}.$$

It is not difficult to establish that the set indicated on the right-hand side is measurable. The operator $\mathcal{U}^{\leq t}$ is the real-time variant of the PCTL operator $\mathcal{U}^{\leq k}$ for natural k; $\Phi\, \mathcal{U}^{\leq t}\, \Psi$ asserts that Ψ will be satisfied at some time instant in the interval $[0, t]$ and that at all preceding time instants Φ holds:

$$\sigma \models \Phi\, \mathcal{U}^{\leq t}\, \Psi \text{ if and only if } \exists x \leq t.\, (\sigma @ x \models \Psi \wedge \forall y < x.\, \sigma @ y \models \Phi).$$

Note that the standard until operator is obtained by taking t equal to ∞.

Model checking time-bounded until properties. CSL model checking [10, 13] is performed in the same way as for CTL [21] and PCTL [29], by recursively computing the set $Sat(\Phi)$. For the boolean operators this is exactly as for CTL and for unbounded until (i.e., $t{=}\infty$) this is exactly as for PCTL. Checking time-bounded until formulas is based on determining the least solution of the following set of Volterra integral equations: $Prob^{\mathcal{C}}(s, \Phi\, \mathcal{U}^{\leq t}\, \Psi)$ equals 1 if $s \in Sat(\Psi)$,

$$Prob^{\mathcal{C}}(s, \Phi\, \mathcal{U}^{\leq t}\, \Psi) = \int_0^t \sum_{s' \in S} \mathbf{P}(s,s')\cdot E(s)\cdot e^{-E(s)\cdot x} \cdot Prob^{\mathcal{C}}(s', \Phi\, \mathcal{U}^{\leq t-x}\, \Psi)\, dx$$

if $s \in Sat(\Phi \wedge \neg\Psi)$, and equals 0 otherwise. Here, the density $E(s)\cdot e^{-E(s)\cdot x}$ denotes the probability of taking some outgoing transition from s at time x. Note the resemblance with Equation (4.1) for the PCTL bounded until operator. For CTMC $\mathcal{C} = (S, \mathbf{R}, L)$ and CSL formula Φ let CTMC $\mathcal{C}[\Phi] = (S, \mathbf{R}', L)$ with $\mathbf{R}'(s,s') = \mathbf{R}(s,s')$ if $s \not\models \Phi$ and 0 otherwise. Note that $emb(\mathcal{C}[\Phi]) = emb(\mathcal{C})[\Phi]$. It has been shown in [13] that for a given CTMC $\mathcal{C}$ and state s in $\mathcal{C}$, the measure $Prob^{\mathcal{C}}(s, \Phi\, \mathcal{U}^{\leq t}\, \Psi)$ can be calculated by means of a transient analysis of the CTMC $\mathcal{C}'$, which can easily be derived from $\mathcal{C}$ using the $[\cdot]$ operator. Let $\pi^{\mathcal{C}}(s,t)(s')$ denote the probability of being in state s' at time t given that the system started in state s, i.e., $\pi^{\mathcal{C}}(s,t)(s') = \Pr_s\left\{\sigma \in Path^{\mathcal{C}} \mid \sigma@t = s'\right\}$. It follows that for any CTMC $\mathcal{C}$:

$$Prob^{\mathcal{C}}(s, \Phi\, \mathcal{U}^{\leq t}\, \Psi) = \sum_{s' \models \Psi} \pi^{\mathcal{C}[\neg\Phi \vee \Psi]}(s,t)(s'). \tag{4.5}$$

Note that this is just the generalization of Equation (4.2) to the continuous setting. Verifying time-bounded until-properties in a CTMC thus amounts to computing transient state probabilities in a derived CTMC. These probabilities are obtained by solving a linear differential equation that can efficiently (and numerically stable) be determined by uniformization [28]. Uniformization is a transformation of a CTMC into a DTMC: For CTMC $\mathcal{C} = (S, \mathbf{R}, L)$ the *uniformized* DTMC is given by $unif(\mathcal{C}) = (S, \mathbf{P}, L)$ where $\mathbf{P} = \mathbf{I} + \mathbf{Q}/q$ for $q \geq \max\{E(s) \,|\, s \in S\}$ and $\mathbf{Q} = \mathbf{R} - \mathrm{diag}(E)$. Here, $\mathbf{I}$ denotes the identity matrix and $\mathrm{diag}(E)$ is the diagonal matrix of E. The *uniformization rate* q is determined by the state with the shortest mean residence time. All (exponential) delays in the CTMC $\mathcal{C}$ are normalized with respect to q. That is, for each state $s \in S$ with $E(s) = q$, one epoch in $unif(\mathcal{C})$ corresponds to a single exponentially distributed delay with rate q, after which one of its successor states is selected probabilistically. As a result, such states have no self-loop in the DTMC. If $E(s) < q$—this state has on average a longer state residence time than $\frac{1}{q}$—one epoch in $unif(\mathcal{C})$ might not be "long enough." Hence, in the next epoch these states might be revisited and, accordingly, are equipped with a self-loop with probability $1 - \frac{E(s)}{q}$. Note the difference between the embedded DTMC $emb(\mathcal{C})$ and the uniformised DTMC $unif(\mathcal{C})$: whereas the epochs in $\mathcal{C}$ and $emb(\mathcal{C})$ coincide and $emb(\mathcal{C})$ can be considered as the time-abstract variant of $\mathcal{C}$, a single epoch in $unif(\mathcal{C})$ corresponds to a single exponentially distributed delay with rate q in $\mathcal{C}$. It now follows that for any CTMC $\mathcal{C}$:

$$Prob^{\mathcal{C}}(s, \Phi\, \mathcal{U}^{\leq t}\, \Psi) = \sum_{k=0}^{\infty} \gamma(k, q \cdot t) \cdot Prob^{unif(\mathcal{C})}(s, \Phi\, \mathcal{U}^{\leq k}\, \Psi) \tag{4.6}$$

where $\gamma(k, q \cdot t)$ denotes the Poisson probability of taking k jumps in the DTMC $unif(\mathcal{C})$ in the interval $[0, t)$, i.e., $\gamma(k, q \cdot t) = e^{-q \cdot t} \cdot (q \cdot t)^k / k!$.

Example. Consider two clusters of workstations that are connected via a backbone connection. Each cluster consists of N workstations, connected in a star topology with a central switch that provides the interface to the backbone. Each of the components of the system (workstations, switches, and backbone) can break down. There is single repair unit that takes care of repairing failed components. The computing power of the cluster is over-dimensioned, in order to be able to accommodate varying levels of traffic volume, as well as to cope with component failures. The system operation is subject to the following informal constraints:

- In order to provide *minimum* quality of service (QoS), at least k ($k < N$) workstations have to be operational, and these workstations have to be connected.

- *Premium* quality of service requires at least N operational workstations, with the same connectivity constraints as mentioned above.

Figure 4.1 indicates the verification times (in seconds) for varying sizes of N (indicated by the absolute number of states of the CTMC). The property that has been checked is: $\mathcal{P}_{\geq .99}(Minimum\, \mathcal{U}^{\leq t}\, Premium)$ for various t. The experiments were conducted on a computer with an Intel P4 3 GHz. processor, 2 GB of RAM running SuSe Linux ver. 9.1. Note that for large values of t, the CTMC may have already reached an equilibrium. This information can be used during the model checking in order to speed up the verification process [45].

4.3.2 Rewards

Costs can be attached to CTMCs to states and to transitions. Cost rates associated with states indicate the cost per unit of time the system stays in that state; rewards associated to edges—these rewards are also called impulse rewards—are fixed and independent of time. For the sake of simplicity, we just consider reward rates. All results and definitions can, however, be extended to incorporate impulse rewards as well (see e.g., [22]).

CMRM. A continuous-time Markov reward model (CMRM) $\mathcal{C}_r$ is a tuple $(\mathcal{C}, r)$ where $\mathcal{C}$ is a CTMC and $r : S \rightarrow \mathbb{R}_{\geq 0}$ is a reward assignment function (as before). The state reward structure is a function r that assigns to each state $s \in S$ a reward rate $r(s)$ such that if t time units are spent in state s, a reward $r(s) \cdot t$ is acquired. A path through a CMRM is a path through its underlying CTMC. Let $\sigma = s_0\, t_0\, s_1\, t_1 \ldots$ be a path. For $t = \sum_{j=0}^{k-1} t_j + t'$ with $t' \leq t_k$ we define $r(\sigma, t) = \sum_{j=0}^{k-1} t_j \cdot r(s_j) + t' \cdot r(s_k)$, the cumulative reward along σ up to time t.

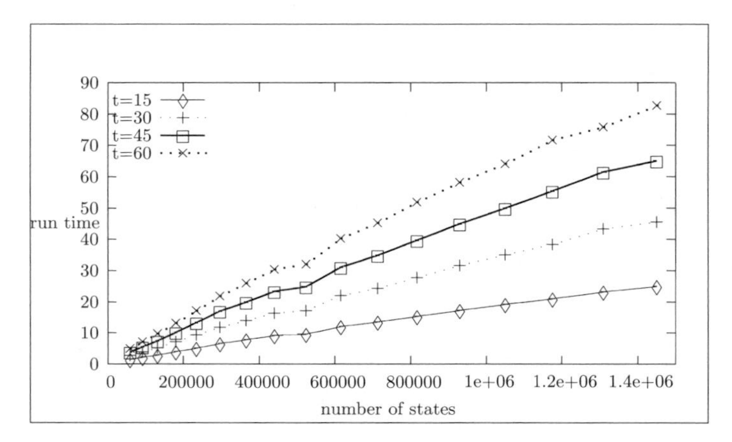

FIGURE 4.1: Verification times for time-bounded until versus the CTMC state space size for various time bounds.

CSRL. Let a, p and $\bowtie$ be as before and $t, r \in \mathbb{R}_{\geq 0}$ (or ∞). The syntax of Continuous Stochastic Reward Logic (CSRL) [13] is:

$$\Phi ::= \mathrm{tt} \;\Big|\; a \;\Big|\; \Phi \wedge \Phi \;\Big|\; \neg \Phi \;\Big|\; \mathcal{P}_{\bowtie p}(\Phi\, \mathcal{U}^{\leq t}_{\leq r}\, \Phi)$$

The semantics of the time- and reward-bounded until operator is given by:

$$\sigma \models \Phi\, \mathcal{U}^{\leq t}_{\leq r}\, \Psi \text{ iff } \exists x \leq t.\, (\sigma@x \models \Psi \wedge \forall y < x.\, \sigma@y \models \Phi \wedge r(\sigma, x) \leq r)\,.$$

Note that the standard until operator is obtained by taking t equal to ∞.

Before continuing with presenting the algorithmic approach for checking cost- and time-bounded until-formulas the following intermezzo is of relevance.

Duality of time and rewards. In the discrete-time setting we have seen that a hop constraint can just be considered as an additional reward constraint. In the continuous-time setting there is also a strong relationship between rewards and time. This is referred to as duality. The basic idea behind this duality, inspired by [14], is that the progress of time can be regarded as the earning of reward and vice versa. First we obtain a duality result for CMRMs where all states have a positive reward. After that we consider the (restricted) applicability of the duality result to CMRMs with zero rewards. Let $\mathcal{C} = ((S, \mathbf{R}, L), r)$ be a CMRM that satisfies $r(s) > 0$ for any state s. Define CMRM $\mathcal{C}^{-1} = ((S, \mathbf{R}', L), r')$ that results from $\mathcal{C}$ by: (i) rescaling the transition rates by the reward of their originating state (as originally proposed in [14]), i.e., $\mathbf{R}'(s, s') = \mathbf{R}(s, s')/r(s)$, and (ii) inverting the reward structure, i.e., $r'(s) = 1/r(s)$. Intuitively, the transformation of $\mathcal{C}$ into $\mathcal{C}^{-1}$ stretches the residence time in state s with a factor that is proportional to the reciprocal of its reward $r(s)$ if $r(s) > 1$, and it compresses the residence time by the same factor if $0 < r(s) < 1$. The reward structure is changed similarly. Note that $\mathcal{C} = (\mathcal{C}^{-1})^{-1}$.

One might interpret the residence of t time units in $\mathcal{C}^{-1}$ as the earning of t reward in state s in $\mathcal{C}$, or (reversely) an earning of a reward r in state s in $\mathcal{C}$ corresponds to a residence of r in $\mathcal{C}^{-1}$. Thus, the notions of time and reward in $\mathcal{C}$ are reversed in $\mathcal{C}^{-1}$. Accordingly [13], for any CMRM $\mathcal{C} = ((S, \mathbf{R}, L), r)$ with $r(s) > 0$ for all $s \in S$ and CSRL state-formula Φ:

$$Sat^{\mathcal{C}}(\Phi) \quad = \quad Sat^{\mathcal{C}^{-1}}(\Phi^{-1}) \tag{4.7}$$

(Recall that $Sat(\Phi) = \{s \in S \mid s \models \Phi\}$). Thus, verifying cost-bounded until formulas on CMRMs with only non-zero rewards can be done in the same way as checking time-bounded until formulas on CTMCs. If CMRM $\mathcal{C}$ contains states equipped with a zero reward, this duality result does not hold, as the reverse of earning a zero reward in $\mathcal{C}$ when considering Φ should correspond to a residence of 0 time units in $\mathcal{C}^{-1}$ for Φ^{-1}, which — as the advance of time in a state cannot be halted — is in general not possible. However, if for each sub-formula of the form $\Phi\, \mathcal{U}^{\leq t}_{\leq r} \Psi$ we have $Sat^{\mathcal{C}}(\Phi) \subseteq \{s \in S \mid r(s) > 0\}$, i.e., all Φ-states are positively rewarded then Equation (4.7) applies. Here, $\mathcal{C}^{-1}$ is defined by setting $\mathbf{R}'(s,s') = \mathbf{R}(s,s')$ and $r'(s) = 0$ in case $r(s) = 0$ and as defined above otherwise.

Verifying time- and cost-bounded until properties. Checking time- and cost-bounded until formulas is based on determining the least solution of the following set of Volterra integral equations: $Prob^{\mathcal{C}}(s, \Phi\, \mathcal{U}^{\leq t}_{\leq k} \Psi)$ equals 1 if $s \in Sat(\Psi)$,

$$Prob^{\mathcal{C}}(s, \Phi\, \mathcal{U}^{\leq t}_{\leq k} \Psi) = \int_{K(s)} \sum_{s' \in S} \mathbf{P}(s,s') \cdot E(s) \cdot e^{-E(s)\cdot x} \cdot Prob^{\mathcal{C}}(s', \Phi\, \mathcal{U}^{\leq t-x}_{\leq r - r(s)\cdot x} \Psi)\, dx$$

if $s \in Sat(\Phi \wedge \neg\Psi)$, and equals 0 otherwise. Here $K(s) = \{x \leq t \mid r(s) \cdot x \leq r\}$ is the subset of $[0,t]$ whose reward lies in $[0,r]$. It is not difficult to see that for $r = \infty$, the above integral equation is exactly the one obtained for time-bounded until properties in CTMCs.

Let now $\pi_r^{\mathcal{C}_r}(s,t)(s') = \Pr_s\{\sigma \in Path^{\mathcal{C}} \mid \sigma@t = s' \wedge r(\sigma,t) \leq r\}$. Then for any CMRM $\mathcal{C}_r$:

$$Prob^{\mathcal{C}_r}(s, \Phi\, \mathcal{U}^{\leq t}_{\leq r} \Psi) = \sum_{s' \models \Psi} \pi_r^{\mathcal{C}_r[\neg\Phi \vee \Psi]}(s,t)(s') \tag{4.8}$$

where for formula Φ, the CMRM $\mathcal{C}_r[\Phi]$ is defined as $(\mathcal{C}[\Phi], r')$ with CTMC $\mathcal{C}[\Phi]$ as before and $r'(s) = r(s)$ if $s \not\models \Phi$ and 0 otherwise. That is, all states that are made absorbing obtain a zero reward (like in the discrete case). The remaining problem, however, is to compute the transient reward probabilities $\pi_r^{\mathcal{C}_r}(s,t)(s')$. We describe an algorithm that is heavily based on that for DMRMs: discretization together with a recursive scheme. A generalization of the path graph generation algorithm for CMRMs can be found in [22].

A discretization approach. This method is based on the algorithm by Tijms and Veldman [58] that discretizes both the time interval and the accumulated reward as multiples of the same step size $d > 0$. d is chosen such that the probability of more than one transition in the CMRM in an interval of length d is negligible. Using this discretization, the probability of accumulating at most r reward at time t is given by:

$$\sum_{i=1}^{i=R} p'_i(s,T)\cdot d \text{ where } R = \frac{r}{d} \text{ and } T = \frac{t}{d}.$$

$p'_i(s,T)$ is the probability of being in state s at discretized time T with cumulated discretized reward i. $p'_i(s,T)$ is defined in a similar way as for the discrete setting, with the exception that the transition probabilities are determined differently. The recursive equation now becomes for $i \geq 0$:

$$\begin{aligned} p'_i(s,k+1) = \quad & p'_{i-r(s)}(s,k)\cdot(1-E(s)\cdot d) \\ & +\sum_{s'} p'_{i-r(s')}(s',k)\cdot\mathbf{R}(s',s)\cdot d. \end{aligned}$$

For the CMRM to be in state s at the $(k+1)$-st time-instant either the CMRM was in state s in the k-th time-instant and remained there for d time-units without traversing a self-loop (the first summand) or it was in state s' and has moved to state s in that period (the second summand). Given that the cumulative reward at the $(k+1)$-st time-instant is i, the cumulative reward in the k-th time-instant is approximated by $i - r(s)$ in the first summand and $i - r(s')$ in the second summand. If this recursive method is implemented by using matrices to store $p(k+1)$ and $p(k)$, then it is necessary to have integer rewards only.

Example. The time complexity of the discretization algorithm is cubic in the number of states in the CMRM and proportional to d^{-2}. To illustrate the complexity of this algorithm on a realistic example, Figure 4.2 depicts the verification times for a CMRM with 276 states. The rewards have been used to model power consumption, and the model has been obtained from a dynamic power management strategy in mobile phones. The property that has been checked is $\mathcal{P}_{>0.5}(\Diamond^{\leq 2000}_{\leq 200}\, done)$. This CSRL-formula asserts that the probability to eventually reach a *done*-state within 2000 ms such that at most 200 mJ is spent, is exceeds $\frac{1}{2}$. Further details on the case study can be found in [22, 1]. Figure 4.3 plots the verification time for an increasing state space size. For the error bound 10^{-3}, time increase is negligible, cf. the plot close to the zero y-axis. Note the significant difference in state space size with the continuous-time setting without rewards. (Expected rewards can be checked in a much faster way as just a system of linear equations needs to be solved where the number of variables is linear in the size of the state space.)

4.3.3 Time-inhomogenity

ICMRM. A time-inhomogeneous CMRM (ICMRM, for short) is a triple $(S,\mathbf{R},r)$ where: S is a finite set of states, $\mathbf{R} : S \times S \times \mathbb{R}_{\geq 0} \to \mathbb{R}_{\geq 0}$ is a *time-indexed* rate matrix, and $r : S \times \mathbb{R}_{\geq 0} \to \mathbb{R}_{\geq 0}$ is a *time-indexed* reward assignment function. The main difference with the continuous-time models treated so far in this survey is that both the transition rates and the rewards are time dependent. The exit rate $E(s,d) = \sum_{s' \in S} \mathbf{R}(s,s',d)$ denotes that the probability of taking a transition from state s within t time units at time d equals $1 - e^{-E(s,d)\cdot t}$. The probability of moving from state s

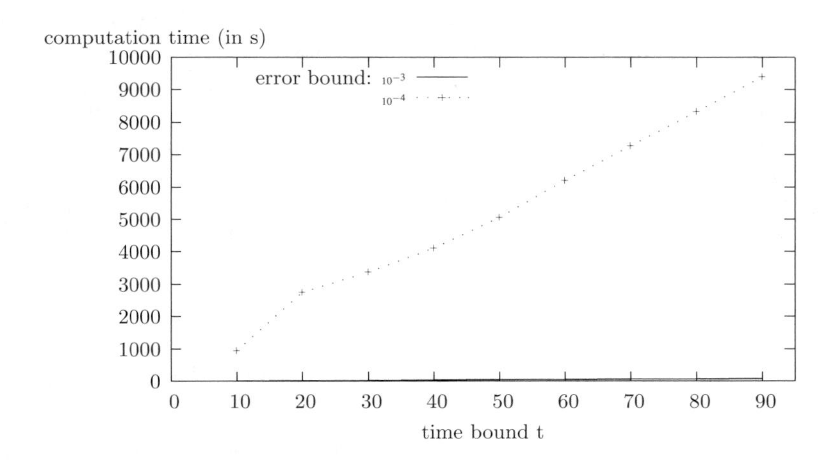

FIGURE 4.2: Computation times versus time bound.

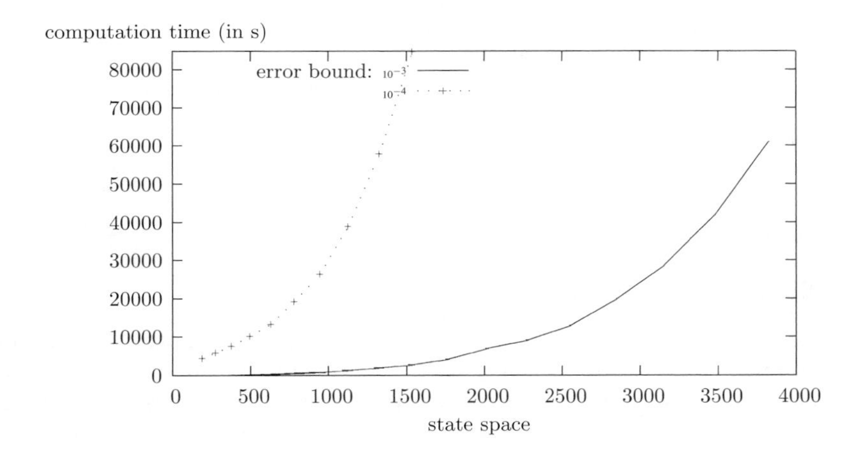

FIGURE 4.3: Computation times versus state space size of CMRM.

to a state s' at time d is given by $\mathbf{P}(s,s',d) = \frac{\mathbf{R}(s,s',d)}{E(s,d)}$. The probability of making a transition from state s to s' at time d within the next t time units is defined by:

$$\frac{\mathbf{R}(s,s',d)}{E(s,d)} \cdot (1 - e^{-E(s,d)\cdot t}).$$

The earned reward when staying for d time units in state s is given by:

$$d \cdot \int_0^d r(s,u)\, du.$$

Note that in case the rate matrix $\mathbf{R}$ and the reward-assignment function r are effectively independent from t, we obtain a time-homogeneous CMRM.

Verifying time- and cost-bounded reachability. Properties of time-inhomogeneous CMRM models can be stated in the logic CSRL. We have seen that the time- and cost-bounded until operator is one of the key ingredients of this logic. The verification of time- and cost-bounded reachability properties boils down to the computation of transient rewards rates in ICMRM. This can be done by generalizing the Tijms-Veldman algorithm for homogeneous CMRMs in the following way [34]. The recursive equation now becomes for $i \geq 0$:

$$\begin{aligned} p'_i(s,k+1) = \ & p'_{i-r(s,k\cdot d)}(s,k)\cdot(1 - E(s,k\cdot d)\cdot d) \\ & + \sum_{s'} p'_{i-r(s',k\cdot d)}(s',k)\cdot\mathbf{R}(s',s,k\cdot d)\cdot d. \end{aligned}$$

This equation is obtained from the recursive scheme for homogeneous CMRMs by replacing transition rates, exit rates, and rewards by their time-dependent counterparts. As time is discretized, at discrete time $k+1$, the transition rate, exit rate, and reward rate at the time instant k are relevant. As transition and reward-rates depend on a continuous-time parameter (and not the discretised notion of time), the real time-instant equals $k\cdot d$. Note that it is straightforward to simplify the above equation in case just the transition rates are time dependent and the reward rates are not, or vice versa.

4.4 Bisimulation and Simulation Relations

The behaviour of Markov chains can be compared by means of equivalence and pre-order relations. Based on the concepts of bisimulation and simulation relations for labeled transitions systems (see e.g., [50]), probabilistic variants thereof have been defined for Markov chains. Three of these notions are treated in more detail in this section, as well as their relationship to the logics CSL and PCTL. For the sake of simplicity, rewards are not considered. The results and definitions in this section can, however, be easily extended toward Markov chains with state rewards.

4.4.1 Strong Bisimulation

One of the most elementary equivalence relations on discrete-time probabilistic systems is probabilistic bisimulation [49]. This variant of strong bisimulation considers two states to be equivalent if the cumulative probability to move to any of the equivalence classes that this relation induces is the same. We consider a slight variant of the original notion in which we require in addition that equivalent states are equally labeled. This is exploited later to establish logical characterizations. For $C \subseteq S$, $\mathbf{P}(s,C) = \sum_{s' \in C} \mathbf{P}(s,s')$ denotes the probability for s to move to a state in C.

Let $\mathcal{D} = (S, \mathbf{P}, L)$ be a DTMC and R an equivalence relation on S. The quotient of S under R is denoted S/R. R is a *strong bisimulation* on $\mathcal{D}$ if for $s_1 \, R \, s_2$:

$$L(s_1) = L(s_2) \quad \text{and} \quad \mathbf{P}(s_1, C) = \mathbf{P}(s_2, C) \text{ for all } C \text{ in } S/R.$$

s_1 and s_2 in $\mathcal{D}$ are strongly bisimilar, denoted $s_1 \sim_d s_2$, if there exists a strong bisimulation R on $\mathcal{D}$ with $s_1 \, R \, s_2$.

Strong bisimulation [17, 36] for CTMCs, also known as ordinary lumpability, is a mild variant of the notion for the discrete-time probabilistic setting where it is required that the cumulative rate (instead of the discrete probability) for two equivalent states to move to any of the induced equivalence classes is equal. Let $\mathcal{C} = (S, \mathbf{R}, L)$ be a CTMC and R an equivalence relation on S. As in the discrete case, for $C \subseteq S$, $\mathbf{R}(s, C) = \sum_{s' \in C} \mathbf{R}(s, s')$ denotes the rate of moving from state s to a state in C via a single transition. Note that $E(s) = \mathbf{R}(s, S)$. R is a *strong bisimulation* on $\mathcal{C}$ if for $s_1 \, R \, s_2$:

$$L(s_1) = L(s_2) \quad \text{and} \quad \mathbf{R}(s_1, C) = \mathbf{R}(s_2, C) \text{ for all } C \text{ in } S/R.$$

s_1 and s_2 in $\mathcal{C}$ are strongly bisimilar, denoted $s_1 \sim_c s_2$, if there exists a strong bisimulation R on $\mathcal{C}$ with $s_1 \, R \, s_2$.

Concerning the relationship between the strong bisimulation notions on discrete-time and continuous-time Markov chains it holds:

$$s \sim_c s' \text{ in CTMC } \mathcal{C} \text{ implies } s \sim_d s' \text{ in DTMC } emb(\mathcal{C}).$$

The reverse does not hold in general, but holds if all states in $\mathcal{C}$ have identical exit rates. A similar strong relationship holds between bisimulation on CTMCs and on their uniformised DTMCs:

$$s \sim_c s' \text{ in CTMC } \mathcal{C} \text{ implies } s \sim_d s' \text{ in DTMC } unif(\mathcal{C}).$$

4.4.2 Weak Bisimulation

Whereas strong bisimulation relates states that mutually mimic all individual steps, *weak bisimulation* only requires this for certain ("observable") transitions and not for other ("silent") transitions. Let $\mathcal{D} = (S, \mathbf{P}, L)$ be a DTMC and $R \subseteq S \times S$ an equivalence relation. Any transition from s to s' (i.e., $\mathbf{P}(s, s') > 0$) where s and s' are R-equivalent is considered an *R-silent* move. Let $Silent_R$ denote the set of states $s \in S$ for which $\mathbf{P}(s, [s]_R) = 1$, i.e., all stochastic states that do not have a successor state outside their R-equivalence class. These states can only perform R-silent moves. Stochastic states outside $Silent_R$ thus may leave their R-equivalence class with positive probability by a single transition. For any state $s \notin Silent_R$, $C \subseteq S$ with $C \cap [s]_R = \emptyset$:

$$\frac{\mathbf{P}(s, C)}{1 - \mathbf{P}(s, [s]_R)}$$

denotes the conditional probability to move from s to some state in C (which is outside $[s]_R$) via a single transition under the condition that from s *no* transition inside

$[s]_R$ is taken. Equivalence R on S is a *weak bisimulation* on $\mathcal{D}$ if for all $s_1 \, R \, s_2$ all following conditions hold:

(i) $L(s_1) = L(s_2)$.

(ii) If $\mathbf{P}(s_i, [s_i]_R) < 1$ for i=1,2 then for all $C \in S/R$, $C \neq [s_1]_R = [s_2]_R$:

$$\frac{\mathbf{P}(s_1, C)}{1 - \mathbf{P}(s_1, [s_1]_R)} = \frac{\mathbf{P}(s_2, C)}{1 - \mathbf{P}(s_2, [s_2]_R)}.$$

(iii) s_1 can reach a state outside $[s_1]_R$ iff s_2 can reach a state outside $[s_2]_R$.

s_1 and s_2 in $\mathcal{D}$ are weakly bisimilar, denoted $s_1 \approx_d s_2$, if and only if there exists a weak bisimulation R on $\mathcal{D}$ such that $s_1 \, R \, s_2$.

Weakly bisimilar states are equally labeled and their conditional probability to move to another equivalence class (given that they do not stay in their own equivalence class) coincides. Furthermore, by the third condition, for any R-equivalence class C, either all states in C are R-silent (i.e., $\mathbf{P}(s, C) = 1$ for all $s \in C$) or for all $s \in C$ there is a sequence of states $s = s_0, s_1, \ldots, s_n$ with $\mathbf{P}(s_i, s_{i+1}) > 0$ that ends in an equivalence class that differs from C (i.e., $s_n \notin C$).

The intuition behind weak bisimulation on CTMCs is that the time-abstract behavior of equivalent states is weakly bisimilar (in the sense of the first two conditions of $\approx_d$), and that the "relative speed" of these states to move to another equivalence class is equal. The following result shows that this formulation can be simplified considerably. Let $\mathcal{C} = (S, \mathbf{R}, L)$ be a CTMC and R an equivalence relation on S with $s_1 \, R \, s_2$. The following statements are equivalent:

(i) If $s_1, s_2 \notin \mathit{Silent}_R$ then for all $C \in S/R$, $C \neq [s_1]_R = [s_2]_R$:

$$\frac{\mathbf{P}(s_1, C)}{1 - \mathbf{P}(s_1, [s_1]_R)} = \frac{\mathbf{P}(s_2, C)}{1 - \mathbf{P}(s_2, [s_2]_R)} \quad \text{and} \quad \mathbf{R}(s_1, S \setminus [s_1]_R) = \mathbf{R}(s_2, S \setminus [s_2]_R).$$

(ii) $\mathbf{R}(s_1, C) = \mathbf{R}(s_2, C)$ for all $C \in S/R$ with $C \neq [s_1]_R = [s_2]_R$.

This result justifies the following definition of weak bisimulation on CTMCs [16]. Let $\mathcal{C} = (S, \mathbf{R}, L)$ be a CTMC and R an equivalence relation on S. R is a *weak bisimulation* on $\mathcal{C}$ if for all $s_1 \, R \, s_2$:

$$L(s_1) = L(s_2) \quad \text{and} \quad \mathbf{R}(s_1, C) = \mathbf{R}(s_2, C) \text{ for all } C \text{ in } S/R \text{ with } C \neq [s_1]_R.$$

s_1 and s_2 in $\mathcal{C}$ are weakly bisimilar, denoted $s_1 \approx_c s_2$, if and only if there exists a weak bisimulation R on $\mathcal{C}$ such that $s_1 \, R \, s_2$. Evidently, any strongly bisimilar pair of states is also weak bisimilar. The reverse, however, does not hold. Thus:

$$\sim_c \subseteq \approx_c \quad \text{and} \quad \sim_d \subseteq \approx_d.$$

Concerning the relationship between the weak bisimulation notions on discrete-time and continuous-time Markov chains we obtain similar results as for strong bisimulation:

$s \approx_c s'$ in CTMC $\mathcal{C}$ implies $s \approx_d s'$ in DTMC $\mathit{emb}(\mathcal{C})$ and $s \approx_d s'$ in DTMC $\mathit{unif}(\mathcal{C})$.

For a CTMC in which all states have the same exit rate, i.e., $E(s)$ equals some constant E for any state s, weak bisimulation $\approx_c$, and strong bisimulation $\sim_c$ coincide.

4.4.3 Strong Simulation

Bisimulation relations are equivalences requiring two bisimilar states to exhibit identical stepwise behavior. On the contrary, simulation relations are preorders on the state space requiring that whenever $s \prec s'$ ("s' simulates s") state s' can mimic all stepwise behaviour of s; the converse, i.e., $s' \prec s$ is not guaranteed, so state s' may perform steps that cannot be matched by s. Thus, if s' simulates s then every successor of s has a corresponding, i.e., related successor of s', but the reverse does not necessarily hold. The use of simulation relies on the preservation of certain classes of formulas, not of all formulas (such as for $\sim$).

For labeled transition systems, state s' simulates state s if for each successor state t of s there is a one-step successor state t' of s' that simulates t. Simulation of two states is thus defined in terms of simulation of their successor states. (It is therefore sometimes called forward simulation.) In the probabilistic setting, the target of a transition is in fact a probability distribution, and thus, the simulation relation $\prec$ needs to be lifted from states to distributions. In fact, strong bisimulation on FPSs was defined as an equivalence on S such that all R-equivalent states s_1 and s_2 are equally labeled and

$$\mathbf{P}(s_1, \cdot) \equiv_R \mathbf{P}(s_2, \cdot)$$

where $\equiv_R$ denotes the lifting of R on $Distr(S)$ defined as:

$$\mu \equiv_R \mu' \quad \text{iff} \quad \mu(C) = \mu'(C) \text{ for all } C \in S/R.$$

(It is easy to see that $\equiv_R$ is an equivalence.) The rough idea behind the definition of simulation relations is to replace the equivalence $\equiv_R$ by a non-symmetric relation $\sqsubseteq_R$ which is obtained using the concept of weight functions [41, 42].

Let S be a set, $R \subseteq S \times S$, and $\mu, \mu' \in Distr(S)$. A *weight function* for μ and μ' with respect to R is a function $\Delta : S \times S \to [0,1]$ such that:

(i) $\Delta(s,s') > 0$ implies $s\,R\,s'$.

(ii) $\mu(s) = \sum_{s' \in S} \Delta(s,s')$ for any $s \in S$.

(iii) $\mu'(s') = \sum_{s \in S} \Delta(s,s')$ for any $s' \in S$.

We write $\mu \sqsubseteq_R \mu'$ (or simply $\sqsubseteq$, if R is clear from the context) if and only if there exists a weight function for μ and μ' with respect to R. $\sqsubseteq_R$ is the lift of R to distributions.

Intuitively, Δ distributes a probability distribution over a set X to a distribution over a set Y such that the total probability assigned by Δ to $y \in Y$ equals the original probability $\mu'(y)$ on Y. In a similar way, the total probability mass of $x \in X$ that is assigned by Δ must coincide with the probability $\mu(x)$ on X. Δ is a probability distribution on $X \times Y$ such that the probability to select (x, y) with xRy is one. In addition, the probability to select an element in R whose first component is x equals $\mu(x)$, and the probability to select an element in R whose second component is y equals $\mu'(y)$.

In the discrete-time setting, simulating states need to be equally labeled, and a weight function must exist that relates their one-step probabilities [42]. Let $\mathcal{D} = (S, \mathbf{P}, L)$ be a DTMC and $R \subseteq S \times S$. R is a *strong simulation* on $\mathcal{D}$ if for all $s_1 \, R \, s_2$:

$$L(s_1) = L(s_2) \quad \text{and} \quad \mathbf{P}(s_1, \cdot) \sqsubseteq_R \mathbf{P}(s_2, \cdot).$$

s_2 strongly simulates s_1 in $\mathcal{D}$, denoted $s_1 \prec_d s_2$, iff there exists a strong simulation R on $\mathcal{D}$ such that $s_1 \, R \, s_2$. For any DTMC $\mathcal{D}$ it holds:

$$\sim_d \text{ coincides with } \prec_d \cap \prec_d^{-1}$$

where $\prec_d^{-1}$ denotes the inverse of the relation $\prec_d$, i.e., $s' \prec_d^{-1} s$ if and only if $s \prec_d s'$.

The intention of a simulation preorder on CTMCs is to ensure that state s_2 simulates s_1 if and only if (i) s_2 is "faster than" s_1 and (ii) the time-abstract behavior of s_2 simulates that of s_1. Note that compared to the discrete-time setting, the only extra requirement is the "faster than" constraint, the other constraints are identical. It therefore directly follows that this notion is a pre-order. Let $\mathcal{C} = (S, \mathbf{R}, L)$ be a CTMC and $R \subseteq S \times S$. R is a *strong simulation* on $\mathcal{C}$ if for all $s_1 \, R \, s_2$:

$$L(s_1) = L(s_2), \quad \mathbf{P}(s_1, \cdot) \sqsubseteq_R \mathbf{P}(s_2, \cdot) \quad \text{and} \quad E(s_1) \leq E(s_2).$$

s_2 strongly simulates s_1 in $\mathcal{C}$, denoted $s_1 \prec_c s_2$, if and only if there exists a strong simulation R on $\mathcal{C}$ such that $s_1 \, R \, s_2$.

Concerning the relationship between the simulation relations on discrete-time and continuous-time Markov chains it holds:

$$s \prec_c s' \text{in CTMC } \mathcal{C} \text{ implies } s \prec_d s' \text{in DTMC } emb(\mathcal{C}) \text{ and } s \prec_d s' \text{in DTMC } unif(\mathcal{C}).$$

The reverse does not hold in general, but holds for the particular case in which all states have the same exit rate; these CTMCs are sometimes refered to as *uniform*.

Similar to the notion of weak bisimulation, weak versions of the simulation preorder relations can be defined that only require that s' mimick the visible steps of s (rather than all possible steps). For the definition of weak simulation relations on DTMCs and CTMCs we refer to [12].

4.4.4 Logical Characterization

Bisimulation. In both the discrete and the continuous setting, strong bisimulation ($\sim_d$ and $\sim_c$) coincide with logical equivalence (for the logics PCTL and CSL, respectively). The latter are denoted $\equiv_{\text{PCTL}}$ and $\equiv_{\text{CSL}}$, respectively. That is, $s_1 \equiv_{\text{PCTL}} s_2$ if and only if s_1 and s_2 satisfy exactly the same PCTL formulas. Similarly, $s_1 \equiv_{\text{CSL}} s_2$ if and only if s_1 and s_2 satisfy exactly the same CSL formulas.

- For any DTMC [5]: $\sim_d$ coincides with $\equiv_{\text{PCTL}}$.
- For any CTMC [8, 27]: $\sim_c$ coincides with $\equiv_{\text{CSL}}$.

Desharnais *et al.* [27] have shown that $\sim_c$ and $\equiv_{\text{CSL}}$ not only coincide for CTMCs with a countable state space but also for continuous-state processes. These results mean that any two bisimilar Markov chains cannot be distinguished by any PCTL (or CSL)-formula, since they satisfy exactly the same formulas. Using efficient algorithms to construct the quotient space under bisimilarity [25], a Markov chain can be lumped prior to the model checking while preserving the results. (The worst case time complexity is logarithmic in the number of states and linear in the number of transition probabilisties.) Another consequence of this result is that in order to disprove that two Markov chains are bisimilar, it suffices to provide a single logical formula that holds in one but not in the other chain. The above definitions and results can easily be lifted to models with rewards, by requiring for $s \sim s'$ that $r(s) = r(s')$. Quotienting reward models with respect to bisimulation has the same time complexity as for DTMCs (and CTMCs).

For weak bisimulation we obtain:

- For any DTMC [5]: $\approx_d$ coincides with $\equiv_{\text{PCTL}}$.
- For any CTMC [12]: $\approx_c$ coincides with $\equiv_{\text{CSL}}$.

A few remarks are in order here as it might be surprising that both strong and weak bisimulation—that have distinguishing expressiveness—coincide with logical equivalence. The main reason for this is that in this chapter we consider the next-less fragment of PCTL (and CSL), i.e., we do not consider the next operator. As the next operator refers to the direct successors of a state, and as these states might be abstracted from in weak (but not in strong) bisimulation, the validity of this operator is not preserved under weak bisimulation whereas it is preserved under strong bisimulation. In absence of the next operator, indeed both weak and strong bisimulation cannot be distinguished from the logical perspective.

Simulation. To consider the relation between simulation relations and the logics PCTL and CSL, we consider the so-called safe fragments of these logics. The syntax of safe PCTL is defined by the grammar:

$$\Phi ::= \text{tt} \;\Big|\; a \;\Big|\; \neg a \;\Big|\; \Phi \wedge \Phi \;\Big|\; \Phi \vee \Phi \;\Big|\; \mathcal{P}_{\geq p}(\Phi\, \mathcal{U}^{\leq k}\, \Phi) \;\Big|\; \mathcal{P}_{\geq p}(\Phi \mathcal{W}^{\leq k} \Phi)$$

where the weak until operator $\mathcal{W}$ (sometimes referred to as unless) is defined by:

$$\sigma \models \Phi \mathcal{W}^{\leq k} \Psi \text{ if and only if } \sigma \models \Phi\, \mathcal{U}^{\leq k}\, \Psi \text{ or } \sigma[i] \models \Phi \text{ for all } i \leq k.$$

In contrast to the (strong) until operator, the weak until operator does not require Ψ to become valid. Note that $\Phi \mathcal{W} \text{ff}$ is identical to $\Box\Phi$. A typical safety property is $\mathcal{P}_{\geq 0.99}(\Box^{\leq 65} \textit{illegal})$ asserting that with probability at least 0.99 the system will not visit an illegal state for the next 65 time units. It is important to realize that we only consider lowerbounds on probabilities in the $\mathcal{P}$-operator, and no upperbounds. In addition, negations only occur adjacent to atomic propositions. (Dually, a live fragment of the logic could be defined that only allows upperbounds. Then for each formula Φ in the safe fragment, there is a liveness formula that is equivalent to $\neg\Phi$.) The safe fragment of CSL is defined in a similar way. The relation between simulation pre-orders $\prec_c$ and $\prec_d$ and the safe fragments of CSL and PCTL, respectively, is as follows. Let $s \leq_{\text{safe PCTL}} s'$ if and only if for any formula Φ in safe PCTL it holds: $s' \models \Phi$ implies $s \models \Phi$. Similarly, the pre-order $\leq_{\text{safe CSL}}$ is defined. Then:

- For any DTMC [26]: $\prec_d$ coincides with $\leq_{\text{safe PCTL}}$.
- For any CTMC [12]: $\prec_c$ coincides with $\leq_{\text{safe CSL}}$.

The definitions and results for DTMCs and CTMCs can easily be adapted to reward extensions thereof by requiring for $s \prec s'$ that $r(s) = r(s')$.

Decision algorithms. For the sake of completeness, we briefly summarize the various decision algorithms that exist for the (bi)simulation relations considered here. Checking strong bisimulation on Markov chains can be done in time $\mathcal{O}(K \cdot \log N)$, where N is the number of states and K is the number of transitions [25]. This algorithm can also be employed for $\approx_c$. In the discrete-time case, checking $\sim_d$ takes $\mathcal{O}(K \cdot \log N)$ time [40], whereas $\approx_d$ takes $\mathcal{O}(N^3)$ time [9]. The computation of $\prec_d$ can be reduced to a maximum flow problem [6] and has a worst case time complexity of $\mathcal{O}((K \cdot N^6 + K^2 \cdot N^3)/\log N)$. The same technique can be applied for computing $\prec_c$. The computation of weak simulation relations can be done in polynomial time by reducing this to linear programming problems.

4.5 Epilogue

4.5.1 Summary of Results

Table 4.1 summarizes the time complexities for the various probabilistic reachability problems discussed in this survey. Note that the indicated complexity figures refer to solve the reachability problem for *all* states in the Markov model at hand. For the sake of completeness, details about the variants of DTMCs and CTMCs that exhibit nondeterminism are included in the table although these models are not further treated in this chapter. Due to the presence of nondeterminism, the problem in these models is to find the maximal (or, dually, the minimal) probability to reach a given goal (set of) state(s) within a given step- or time-bound. We emphasize that all complexity indications refer to model-checking algorithms that are approximate. Stated differently, these algorithms calculate probabilities up to a certain precision that is fixed a priori by the user, and based on these approximate probabilties decide on the validity of the formula under consideration. Interestingly enough, all the approximate algorithms have a polynomial complexity, in contrast to (exact) model-checking algorithms for timed automata that are exponential in, e.g., the number of clock variables.

Table 4.1: Time complexities for verifying bounded probabilistic reachability.

model class	*reachability problem*	*time complexity*	*note (if applicable)*
	the discrete-time setting		
DTMC	step-bounded	$\mathcal{O}(k \cdot N^2)$	
MDP	step-bounded	$\mathcal{O}(poly(N, M))$	extremal probability
DMRM	step- and cost-bounded	$\mathcal{O}(k \cdot r \cdot N^3)$	recursive algorithm
	the continuous-time setting		
CTMC	time-bounded	$\mathcal{O}(E \cdot t \cdot N^2)$	
CTMDP	time-bounded	$\mathcal{O}(E \cdot t \cdot N^2 \cdot M)$	uniform exit rate E
CMRM	time- and cost-bounded	$\mathcal{O}(t \cdot r \cdot N^3 \cdot d^{-2})$	discretization
ICMRM	time- and cost-bounded	$\mathcal{O}(t \cdot r \cdot N^3 \cdot d^{-2})$	discretization

Here, N denotes the state space size, i.e., $N = |S|$, k is the step bound (applicable for discrete-time models only), t is its continuous equivalent, r is the upperbound on the cumulative reward, d is the step size (in case of discretization), E is the largest exit rate in a CTMC, and M is the number of distinct actions in a state (relevant for nondeterministic models only).

Software tools. ETMCC was the first CTMC model checker [35]. It uses a sparse-matrix representation, is based on the algorithms explained in this chapter, and has a simple input format that enables its usage as a back-end to existing performance modeling tools. PRISM [48] is a model checker for (discrete-time and continuous-time) Markov chains as well as Markov decision processes (MDPs). It also contains means for checking expected cumulated reward properties. It uses a mixed representation: a binary-decision diagram for the probability (or rate) matrix and a sparse representation for the solution vector. This tool has been applied to several case studies from different application fields. This includes biological systems, randomized distributed protocols, and security protocols. MRMC [43] is based on the principles of ETMCC and supports besides CSL also the logic CSRL. Due to improved data structures and algorithm implementation this tool is about an order of magnitude faster than ETMCC. This tool contains implementations of the discretization and path graph generation algorithms as well as strong bisimulation minimization.

4.5.2 Further Research Topics

This chapter has surveyed the model-checking approach to discrete and continuous-time Markov (reward) models. We believe that the model-checking approach provides a useful technique for performance and dependability analysis. Logics are useful for specifying performance guarantees, and model-checking algorithms provide effective (and efficient) means for checking these guarantees. This is done in a fully automated way, and provides a *single* framework for checking performance measures as well as functional properties such as absence of deadlocks and reponsiveness. Note that (time-inhomogeneous) continuous-time Markov reward models can be considered as simple stochastic hybrid systems. Further work in this area is needed to consider more expressive models. An interesting topic for future work is to develop model-checking algorithms for (simple variants) of piecewise deterministic Markov processes [24] and to finding logical characterizations for bisimulations on PDMPs. The reader is referred to Chapter 3 in this volume on a compositional specification formalism for PDMPs.

References

[1] A. ACQUAVIVA, A. ALDINI, M. BERNARDO, A. BOGLIOLO, E. BONTA AND E. LATTANZI. Assessing the impact of dynamic power management on the functionality and the performance of battery-powered appliances. In *Dependable Systems and Networks (DSN 2004)*, IEEE CS Press, pp. 731–740, 2004.

[2] S. ANDOVA, H. HERMANNS, J.-P. KATOEN. Discrete-time rewards model-checked. In *Formal Methods for Timed Systems*, LNCS 2791:88-104, 2003.

[3] K.R. APT, N. FRANCEZ AND W.-P. DE ROEVER. A proof system for communicating sequential processes. *ACM Transactions on Programming Languages and Systems*, **2**:359–385, 1980.

[4] A. AZIZ, K. SANWAL, V. SINGHAL AND R. BRAYTON. Model checking continuous time Markov chains. *ACM Transactions on Computational Logic*, **1**(1):162–170, 2000.

[5] A. AZIZ, V. SINGHAL, F. BALARIN, R. BRAYTON AND A. SANGIOVANNI-VINCENTELLI. It usually works: the temporal logic of stochastic systems. In P. Wolper, editor, *Computer-Aided Verification*, LNCS 939:155–165, 1995.

[6] C. BAIER, B. ENGELEN, AND M. MAJSTER-CEDERBAUM. Deciding bisimilarity and similarity for probabilistic processes. *J. of Comp. and System Sc.*, **60**(1):187–231, 2000.

[7] C. BAIER, B. HAVERKORT, H. HERMANNS AND J.-P. KATOEN. On the logical characterisation of performability properties. In: U. Montanari, J.D.P. Rolim, E. Welzl, editors, *Automata, Languages and Programming*, LNCS 1853:780–792, 2000.

[8] C. BAIER, B. HAVERKORT, H. HERMANNS AND J.-P. KATOEN. Model-checking algorithms for continuous-time Markov chains. *IEEE Transactions on Software Engineering*, **29**(6):524–541, 2003.

[9] C. BAIER AND H. HERMANNS. Weak bisimulation for fully probabilistic processes. *Computer-Aided Verification*, LNCS 1254:119-130, 1997.

[10] C. BAIER, J.-P. KATOEN AND H. HERMANNS. Approximate symbolic model checking of continuous-time Markov chains. In *Concurrency Theory*, LNCS 1664:146–162, Springer, 1999.

[11] C. BAIER, H. HERMANNS, J.-P. KATOEN, B. HAVERKORT. Efficient computation of time-bounded reachability probabilities in uniform continuous-time Markov decision processes. *Theoretical Computer Science*, **345**(1):2–26, 2005.

[12] C. BAIER, J.-P. KATOEN, H. HERMANNS, V. WOLF. Comparative branching-time semantics for Markov chains. *Information & Computation* **200**(2):149–214, 2005.

[13] C. BAIER, B.R. HAVERKORT, H. HERMANNS AND J.-P. KATOEN. On the logical characterisation of performability properties. In *Automata, Languages, and Programming*, LNCS 1853:780–792, Springer, 2000.

[14] M.D. BEAUDRY. Performance-related reliability measures for computing systems. *IEEE Trans. on Comp. Sys.*, **27**(6):540–547, 1978.

[15] D. BERTSEKAS. *Dynamic Programming and Optimal Control, volumes 1 and 2.* Athena Scientific, 1995.

[16] M. BRAVETTI. Revisiting interactive Markov chains. In W. Vogler and K.G. Larsen (eds), *Models for Time-Critical Systems*, BRICS Notes Series NS-02-3, pp. 60–80, 2002.

[17] P. BUCHHOLZ. Exact and ordinary lumpability in finite Markov chains. *Journal of Applied Probability*, **31**:59–75, 1994.

[18] P. BUCHHOLZ, J.-P. KATOEN, P. KEMPER AND C. TEPPER. Model-checking large structured Markov chains. *Journal of Logic and Algebraic Programming*, **56**(1-2):69–97, 2003.

[19] E.M. CLARKE AND E.A. EMERSON. Design and synthesis of synchronisation skeletons using branching time temporal logic. In *Logic of Programs*, LNCS 131:52–71, 1981.

[20] E.M. CLARKE AND R. KURSHAN. Computer-aided verification. *IEEE Spectrum*, **33**(6):61–67, 1996.

[21] E.M. CLARKE, O. GRUMBERG AND D. PELED. *Model Checking*. MIT Press, 1999.

[22] L. CLOTH, J.-P. KATOEN, M. KHATTRI, R. PULUNGAN. Model checking Markov reward models with impulse rewards. *Dependable Systems and Networks (DSN 2005)*, pp. 722–731, IEEE CS Press, 2005.

[23] C. COURCOUBETIS AND M. YANNAKAKIS. Verifying temporal properties of finite-state probabilistic programs. In *Found. of Comp. Sc. (FOCS)*, pp. 338–345, 1988.

[24] M.H.A. DAVIS. Piecewise deterministic Markov processes: a general class of non-diffusion stochastic models. *J. Royal Statistical Soc. (B)*, **46**:353–388, 1984.

[25] S. DERISAVI, H. HERMANNS AND W.H. SANDERS. Optimal state-space lumping in Markov chains. *Information Processing Letters*, **87**(6):309–315, 2004.

[26] J. DESHARNAIS. Logical characterisation of simulation for Markov chains. *Workshop on Probabilistic Methods in Verification*, Tech. Rep. CSR-99-8, Univ. of Birmingham, pp. 33–48, 1999.

[27] J. DESHARNAIS AND P. PANANGADEN. Continuous stochastic logic characterizes bisimulation of continuous-time Markov processes. *Journal of Logic and Algebraic Programming*, **56**(1-2):99-115, 2003.

[28] D. GROSS AND D.R. MILLER. The randomization technique as a modeling tool and solution procedure for transient Markov chains. *Op. Res.*, **32**(2):343–361, 1984.

[29] H.A. HANSSON AND B. JONSSON. A logic for reasoning about time and reliability. *Formal Aspects of Comp.*, **6**(5):512–535, 1994.

[30] D. HAREL. Statecharts: a visual formalism for complex systems. *Science of Computer Programming*, **8**(3):231–274, 1987.

[31] S. HART, M. SHARIR AND A. PNUELI. Termination of probabilistic concurrent programs. *ACM Transactions on Programming Languages and Systems*, **5**(3):356–380, 1983.

[32] B. HAVERKORT, L. CLOTH, H. HERMANNS, J.-P. KATOEN AND C. BAIER. Model-checking performability properties. In: *Dependable Systems and Networks*, pp. 103–113, 2002.

[33] B. HAVERKORT, H. HERMANNS, AND J.-P. KATOEN. On the use of model checking techniques for quantitative dependability evaluation. In *IEEE Sym. on Reliable Distributed Systems (SRDS)*, pp. 228–238, IEEE CS Press, 2000.

[34] B. HAVERKORT AND J.-P. KATOEN. The performability distribution for non-homogeneous Markov reward models. In *Performability Workshop*, pp. 32–34, 2005.

[35] H. HERMANNS, J.-P. KATOEN, J. MEYER-KAYSER AND M. SIEGLE. A Markov chain model checker. *J. on Software Tools and Technology Transfer*, **4**(2):153–172, 2003.

[36] J. HILLSTON. *A Compositional Approach to Performance Modelling*. Cambridge Univ. Press, 1996.

[37] C.A.R. HOARE. An axiomatic basis for computer programming. *Communications of the ACM*, **12**:576–580, 583, 1969.

[38] G.J. HOLZMANN. *Design and Validation of Computer Protocols.* (Prentice-Hall, 1991).

[39] R.A. HOWARD. *Dynamic Probabilistic Systems; Vol. I, II.* John Wiley & Sons, 1971.

[40] T. HYUNH AND L. TIAN. On some equivalence relations for probabilistic processes. *Fundamentae Informatica*, **17**:211–234, 1992.

[41] C. JONES AND G. PLOTKIN. A probabilistic powerdomain of evaluations. *IEEE Symp. on Logic in Comp. Sc.*, pp. 186–195, 1989.

[42] B. JONSSON AND K.G. LARSEN. Specification and refinement of probabilistic processes. *IEEE Symp. on Logic in Comp. Sc.*, pp. 266-277, 1991.

[43] J.-P. KATOEN, M. KHATTRI AND I.S. ZAPREEV. A Markov reward model checker. In: *Quantitative Evaluation of Systems (QEST)*. IEEE CS Press, 243–245, 2005.

[44] J.-P. KATOEN, M.Z. KWIATKOWSKA, G. NORMAN AND D. PARKER. Faster and symbolic CTMC model checking. In: L. de Alfaro and S. Gilmore, editors, *Process Algebra and Probabilistic Methods*, LNCS 2165:23–38, 2001.

[45] J.-P. KATOEN AND I.S. ZAPREEV. Safe on-the-fly steady-state detection for time-bounded reachability. In: *Quantitative Evaluation of Systems*, IEEE CS Press, 2006 (to appear).

[46] J.G. KEMENY AND J.L. SNELL. *Finite Markov Chains.* Van Nostrand, 1960.

[47] M.Z. KWIATKOWSKA, G. NORMAN AND R. SEGALA. Automated verification of a randomized distributed consensus protocol using Cadence SMV and PRISM. In: G. Berry et al, editors, *Computer-Aided Verification*, LNCS 2102: 194–206, 2001.

[48] M. KWIATKOWSKA, G. NORMAN AND D. PARKER. Probabilistic symbolic model checking using PRISM: a hybrid approach. *J. on Software Tools for Technology Transfer*, **6**(2):128–142, 2004.

[49] K.G. LARSEN AND A. SKOU. Bisimulation through probabilistic testing. *Inf. and Comput.*, **94**(1):1–28, 1991.

[50] R. MILNER. *Communication and Concurrency*. Prentice-Hall, 1989.

[51] S. OWICKI AND D. GRIES. An axiomatic proof technique for parallel programs. *Acta Informatica*, **6**:319–340, 1976.

[52] M.L. PUTERMAN. *Markov Decision Processes: Discrete Stochastic Dynamic Programming*. John Wiley & Sons, 1994.

[53] J.-P. QUEILLE AND J. SIFAKIS. Specification and verification of concurrent systems in CESAR. In: *Proceedings 5th International Symposium on Programming*, LNCS 137:337–351, 1982.

[54] M.A. QURESHI AND W.H. SANDERS. A new methodology for calculating distributions of reward accumulated during a finite interval. In *Fault-Tolerant Computing Symposium*, IEEE CS Press, pp. 116–125, 1996.

[55] J. STAUNSTRUP, H.R. ANDERSEN, J. LIND-NIELSEN, K.G. LARSEN, G. BEHRMANN, K. KRISTOFFERSEN, H. LEERBERG AND N.B. THEILGAARD. Practical verification of embedded software. *IEEE Computer*, **33**(5):68–75, 2000.

[56] W.J. STEWART. *Introduction to the Numerical Solution of Markov Chains.* Princeton University Press, 1994.

[57] H.C. TIJMS. *A First Course in Stochastic Models.* John Wiley & Sons, 2003.

[58] H.C. TIJMS AND R. VELDMAN. A fast algorithm for the transient reward distribution in continuous-time Markov chains. *Operations Research Letters*, **26**:155–158, 2000.

[59] A.M. TURING. On checking a large routine. In: *Report of a Conference on High Speed Calculating Machines*, pp. 76–69, 1949.

[60] M.Y. VARDI. Automatic verification of probabilistic concurrent finite state programs. In *IEEE Symposium on Foundations of Computer Science*, pp. 327–338, 1985.

Chapter 5

Stochastic Reachability: Theory and Numerical Approximation

Maria Prandini
Politecnico di Milano

Jianghai Hu
Purdue University

5.1 Introduction 107
5.2 Stochastic Hybrid System Model 109
5.3 Reachability Problem Formulation 114
5.4 Numerical Approximation Scheme 116
5.5 Reachability Computations 124
5.6 Possible Extensions 128
5.7 Some Examples 130
5.8 Conclusions 134
References 135

5.1 Introduction

Roughly speaking, hybrid systems are dynamic systems with both continuous and discrete dynamics. The study of hybrid systems has received considerable attention in recent years due to their applications in a diverse range of scientific and engineering problems. Examples include transportation systems such as air traffic management systems ([21], [23]) and automated highway systems [24], robotics [8], computer and communication networks [14], and automotive systems [3]. Besides engineering applications, hybrid systems are also found useful in modeling biological systems [2].

While impressive progress has been made so far in the study of hybrid systems, a majority of the efforts focus exclusively on deterministic hybrid systems, which are not suitable for modeling practical systems with inherent uncertainty. For example, the trajectory of an aircraft is subject to the perturbations of wind [6], the traffic in a computer network may fluctuate and components may break down at random intervals, and stochastic noises exist in the genetic networks regulating the cells [20] as discussed in Chapter 9 of this volume. To model these systems, it is imperative to introduce the notion of *stochastic hybrid systems*, namely, hybrid systems with stochastic continuous dynamics governed by stochastic differential equations and with random discrete mode transitions governed by Markov chains. These are an instance of the stochastic differential equations on hybrid state spaces discussed in Chapter 2 of this volume.

There is a philosophical difference between the study of deterministic and stochastic hybrid systems: In the deterministic case, each system trajectory is treated equally; while in the stochastic case, one weights it according to its likelihood as determined by the probabilistic laws. Due to this philosophical difference, the problems one can study of stochastic hybrid systems are of more variety and "shades" than those of deterministic hybrid systems, and the results obtained are often more robust and less conservative. As an example, a reachability problem in the deterministic case is a yes/no problem, while in the stochastic case one faces a continuous spectrum of "soft" problems with quantitative answers, such as the hitting probability, the expected hitting time, the hitting distribution, etc. Another example is that the asymptotic stability of deterministic hybrid systems has many counterparts in stochastic hybrid systems: recurrence, positive recurrence, ergodicity, mean square and almost sure asymptotic stability, etc.

As a price for their enhanced modeling flexibility and expressiveness, the problems arising in stochastic hybrid systems are in general much more challenging: Analytical solutions are difficult or impossible to obtain; and, compared with the many software packages simulating deterministic hybrid systems, few effective general algorithms exist for the numerical simulation of stochastic hybrid systems. Many problems well studied in the deterministic case remain open for stochastic hybrid systems.

In this chapter, we focus on the reachability analysis of stochastic hybrid systems. In particular, we study the problem of estimating the probability that the system state will enter a certain subset of the state space within a finite or infinite time horizon, starting from an arbitrary initial condition, for a class of stochastic hybrid systems called switching diffusions ([9], [10], [11].) By discretizing the state space and using an interpolated Markov chain to approximate the solutions to the stochastic hybrid systems weakly, we develop a numerical algorithm to compute an estimate of the desired probability. Several immediate extensions of our method are also discussed, including its use in the study of the probabilistic safety problem and the regulation problem. We then demonstrate the efficacy of the proposed algorithm by applying it to two examples referring to manufacturing systems and temperature regulation.

It is worth noting that the proposed methodology for reachability computations was introduced by the authors of the present chapter in [15], and further developed in [16] and [22]. The systems considered in these contributions are described by stochastic differential equations with coefficients changing value at prescribed, a-priori known, time instants. The methodology is extended here to a more complex hybrid setting, where switchings in the dynamics are state dependent.

This chapter is organized as follows. In Section 5.2, a model of the stochastic hybrid systems under study is presented. Then in Section 5.3, a reachability problem for such systems is formulated. Using the numerical approximation scheme discussed in Section 5.4, a numerical algorithm is developed in Section 5.5 to find an approximate solution to the reachability problem. The algorithm can be easily extended to deal with several generalized problems, which are discussed in Section 5.6. To show the efficacy of the developed algorithms, in Section 5.7, simulation results for two examples are presented.

5.2 Stochastic Hybrid System Model

We consider a continuous time stochastic hybrid system, whose state $\mathbf{s}$ is characterized by a continuous component $\mathbf{x}$ and a discrete component $\mathbf{q}$: $\mathbf{s} = (\mathbf{x}, \mathbf{q})$. The discrete state component $\mathbf{q}$ takes values in a finite set $\mathscr{Q} = \{1, 2, \ldots, M\}$, whereas the continuous state component $\mathbf{x}$ takes values in the Euclidean space $\mathbb{R}^n$. Thus the hybrid state space is given by $\mathscr{S} := \mathbb{R}^n \times \mathscr{Q}$.

Starting from some initial value $q_0 \in \mathscr{Q}$ at time $t = 0$, the discrete state component $\mathbf{q}$ evolves following piecewise constant and right continuous trajectories, i.e., for each trajectory, there exists a sequence of consecutive left closed, right open time intervals $\{T_i, i = 0, 1, \ldots\}$, such that $\mathbf{q}(t) = q_i$, $\forall t \in T_i$, with $q_i \in \mathscr{Q}$, $\forall i$, and $q_i \neq q_{i\pm 1}$.

The continuous state component $\mathbf{x}$ evolves starting from some initial value x_0 according to a stochastic differential equation whose coefficients depend on $\mathbf{q}$. More specifically, during each time interval T_i when $\mathbf{q}(t)$ is constant and equal to $q_i \in \mathscr{Q}$, $\mathbf{x}$ is governed by the stochastic differential equation

$$d\mathbf{x}(t) = a(\mathbf{x}(t), q_i)dt + b(\mathbf{x}(t), q_i)\Sigma d\mathbf{w}(t),$$

initialized with $\mathbf{x}(t_i^-) = \lim_{h \to 0^+} \mathbf{x}(t_i - h)$ at time $t_i := \sup\{t : t \in \cup_{k=0}^{i-1} T_k\}$. Functions $a(\cdot, q_i) : \mathbb{R}^n \to \mathbb{R}^n$ and $b(\cdot, q_i) : \mathbb{R}^n \to \mathbb{R}^{n \times n}$ represent the drift and the diffusion terms, respectively, and Σ is a diagonal matrix with positive entries, which modulates the variance of the standard n-dimensional Brownian motion $\mathbf{w}(\cdot)$.

A jump in the discrete state may occur during the continuous state evolution with an intensity that depends on the current value taken by state $\mathbf{s}$. When it actually occurs, $\mathbf{q}$ is reset according to a probabilistic map that depends on the current value taken by $\mathbf{s}$ as well. This is modeled by describing $\mathbf{q}$ as a continuous time stochastic process taking values in the finite state space $\mathscr{Q}$, whose evolution at time t is conditionally independent on the past given $\mathbf{s}(t^-)$, and is governed by the transition probability:

$$\mathrm{P}\left(q(t + \Delta t) = q' \mid \mathbf{q}(t^-) = q, \mathbf{x}(t^-) = x\right) = \lambda_{qq'}(x)\Delta t + o(\Delta t), \; q \neq q',$$

where $\lambda_{qq'} : \mathbb{R}^n \to \mathbb{R}$, $q, q' \in \mathscr{Q}$, $q \neq q'$, are the transition rates satisfying the condition $\lambda_{qq'}(x) \geq 0$, $\forall x \in \mathbb{R}^n$. The transition rates determine both the switching intensity and the reset map for the discrete state component $\mathbf{q}$ of the process $\mathbf{s}$, as is explained below.

Starting from $\mathbf{s}(t^-) = s$, $\mathbf{q}(t)$ will jump during the time interval $[t, t+\Delta t]$ once with probability $\lambda(s)\Delta t + o(\Delta t)$, and two or more times with probability $o(\Delta t)$, where $\lambda : \mathscr{S} \to [0, +\infty)$ is the jump intensity function

$$\lambda(s) = \sum_{q' \in \mathscr{Q}, q' \neq q} \lambda_{qq'}(x), \; s = (x,q) \in \mathscr{S}. \tag{5.1}$$

If $s \in \mathscr{S}$ is such that $\lambda(s) = 0$, then no jump can occur at s.

Let $s = (x,q) \in \mathscr{S}$ be such that $\lambda(s) \neq 0$. Then, when a jump does occur at time t from $\mathbf{s}(t^-) = s$, the distribution of $\mathbf{q}(t)$ over $\mathscr{Q} \setminus \{q\}$ depends on s, and is given by the reset function $R : \mathscr{S} \times \mathscr{Q} \to [0,1]$

$$R(s,q') = \begin{cases} \frac{\lambda_{qq'}(x)}{\lambda(s)}, & q' \neq q \\ 0, & q' = q \end{cases}, \quad s = (x,q) \in \mathscr{S}. \tag{5.2}$$

The stochastic process $\mathbf{s}$ obtained in this way is known in the literature as a switching diffusion (see [9], [10], [11].) This is because, between two consecutive jumps of $\mathbf{q}$, $\mathbf{x}$ behaves as a diffusion process and its dynamics switches as soon as a jump in $\mathbf{q}$ occurs.

Switching diffusion systems are stochastic hybrid system models that arise in a variety of applications involving systems with multiple operating modes, such as in fault tolerant control, multiple target tracking, and flexible manufacturing ([9], [10]), and also applications in finance (see, e.g., [7].)

A formal description of a switching diffusion hybrid system, with the pure jump process $\mathbf{q}$ represented by an integral with respect to a Poisson random measure, is provided next, following [10] and [17]. This construction is closely related to the one used in Chapter 2; here it serves as the reference representation for the Markov chain approximation scheme.

For each $x \in \mathbb{R}^n$, define the consecutive disjoint intervals $\Delta_{ki}(x) \subseteq \mathbb{R}$ of length $\lambda_{ki}(x)$, with $k, i \in \mathscr{Q} = \{1, 2, \ldots, M\}$, $i \neq k$, as follows:

$$\begin{aligned}
\Delta_{12}(x) &= \left[0, \lambda_{12}(x)\right), \\
\Delta_{13}(x) &= \left[\lambda_{12}(x), \lambda_{12}(x) + \lambda_{13}(x)\right), \\
&\vdots \\
\Delta_{1M}(x) &= \left[\sum_{h=2}^{M-1} \lambda_{1h}(x), \sum_{h=2}^{M} \lambda_{1h}(x)\right), \\
\Delta_{21}(x) &= \left[\sum_{h=2}^{M} \lambda_{1h}(x), \sum_{h=2}^{M} \lambda_{1h}(x) + \lambda_{21}(x)\right), \\
\Delta_{23}(x) &= \left[\sum_{h=2}^{M} \lambda_{1h}(x) + \lambda_{21}(x), \sum_{h=2}^{M} \lambda_{1h}(x) + \lambda_{21}(x) + \lambda_{23}(x)\right), \\
&\vdots
\end{aligned}$$

$$\Delta_{2M}(x) = \Big[\sum_{h=2}^{M} \lambda_{1h}(x) + \sum_{h=1,h\neq 2}^{M-1} \lambda_{2h}(x), \sum_{h=2}^{M} \lambda_{1h}(x) + \sum_{h=1,h\neq 2}^{M} \lambda_{2h}(x) \Big),$$

$$\vdots$$

The generic $\Delta_{ki}(x)$ interval of length $\lambda_{ki}(x)$ is given by:

$$\Delta_{ki}(x) = \Big[\sum_{l=1}^{k-1} \sum_{h=1,h\neq l}^{M} \lambda_{lh}(x) + \sum_{h=1,h\neq k}^{i-1} \lambda_{kh}(x), \sum_{l=1}^{k-1} \sum_{h=1,h\neq l}^{M} \lambda_{lh}(x) + \sum_{h=1,h\neq k}^{i} \lambda_{kh}(x) \Big).$$

We associate with each $x \in \mathbb{R}^n$ the interval

$$\Gamma(x) := \cup_{k,i=1,i\neq k}^{M} \Delta_{ki}(x),$$

of length $\sum_{k,i=1,i\neq k}^{M} \lambda_{ki}(x)$. If $\lambda_{qq'}(\cdot)$ is bounded and continuous for all $q, q' \in \mathcal{Q}$, then $x_{\max} := \arg\max_{x\in\mathbb{R}^n} \sum_{k,i=1,i\neq k}^{M} \lambda_{ki}(x)$ is well-defined. Let $\Gamma_{\max} := \Gamma(x_{\max}) = [0, \lambda_{max})$ be the corresponding bounded interval of length $\lambda_{max} := \sum_{k,i=1,i\neq k}^{M} \lambda_{ki}(x_{max})$, and $\mathcal{U}$ be the uniform distribution over $\Gamma_{\max}$.

Define the function $r_q : \mathbb{R}^n \times \mathcal{Q} \times \Gamma_{\max} \to \{0, \pm 1, \pm 2, \cdots \pm M-1\}$ by

$$r_q(x, q, \gamma) = \begin{cases} q' - q, & \text{if } \gamma \in \Delta_{qq'}(x) \\ 0, & \text{otherwise,} \end{cases} \tag{5.3}$$

which describes the entity of the jump in the discrete state starting from (x, q), for each $\gamma \in \Gamma_{\max}$.

Then, the stochastic hybrid system can be represented by the stochastic differential equations

$$\begin{aligned} d\mathbf{x}(t) &= a(\mathbf{x}(t), \mathbf{q}(t))dt + b(\mathbf{x}(t), \mathbf{q}(t)) \Sigma d\mathbf{w}(t) \\ d\mathbf{q}(t) &= \int_{\Gamma_{\max}} r_q(\mathbf{x}(t^-), \mathbf{q}(t^-), \gamma) \mathbf{p}(dt, d\gamma) \end{aligned} \tag{5.4}$$

where $\mathbf{w}(\cdot)$ is a standard n-dimensional Brownian motion, $\mathbf{p}(\cdot, \cdot)$ is a Poisson random measure of intensity $h(dt, d\gamma) = \lambda_{\max} dt \times \mathcal{U}(d\gamma)$, and $\mathbf{w}(\cdot)$ and $\mathbf{p}(\cdot, \cdot)$ are independent.

ASSUMPTION 5.1 $a(\cdot, q)$, $b(\cdot, q)$, and $\lambda_{qq'}(\cdot)$ are bounded and Lipschitz continuous for each $q, q' \in \mathcal{Q}$.

Under this assumption, the system described by Equation (5.4) admits a unique strong solution $\mathbf{s}(t) = (\mathbf{x}(t), \mathbf{q}(t))$, $t \geq 0$, for each initial condition $s_0 = (x_0, q_0) \in \mathcal{S}$. Also, such a solution is a Markov process, and the trajectories of its continuous component $\mathbf{x}(t)$, $t \geq 0$, are continuous since $\mathbf{x}$ is not subject to any reset when a switch in the coefficients of the stochastic differential equation governing its evolution in (5.4) occurs. As observed in [12], the boundedness assumption on the diffusion and the drift terms a and b can be relaxed. As a matter of fact, it has been removed in [4]. In our case, however, this is not a restrictive assumption since the system evolution will be confined to some bounded region for numerical computation purposes.

We now verify that the process defined in (5.4) is indeed the switching diffusion process described at the beginning of this section. The Poisson random measure $\mathbf{p}(\cdot,\cdot)$ generates a sequence $\{(\mathbf{t}_i,\gamma_i), i \geq 1\}$, where $\{\mathbf{t}_i, i \geq 1\}$ is a sequence of increasing nonnegative random variables representing the jump times of a standard Poisson process with intensity $\lambda_{\max}$, and $\{\gamma_i, i \geq 1\}$ is a sequence of independent and identically distributed (i.i.d.) random variables with distribution $\mathcal{U}$, independent of $\{\mathbf{t}_i, i \geq 1\}$.

The random measure $\mathbf{p}(d\tau,d\gamma)$ assigns unit mass to (τ,γ) if there exists $i \geq 1$ such that $\mathbf{t}_i = \tau$ and $\gamma_i = \gamma$. For any measurable subset C of $\Gamma_{\max}$ and any $t > 0$, $\mathbf{p}([0,t] \times C) = \int_0^t \int_C \mathbf{p}(d\tau,d\gamma)$ is in fact given by

$$\mathbf{p}([0,t] \times C) = \sum_{i \geq 1} 1_{\{\mathbf{t}_i \leq t\}} 1_{\{\gamma_i \in C\}},$$

i.e., it is a random variable representing the number of jumps with values in C during the time interval $[0,t]$.

As a consequence of this expression for the Poisson random measure $\mathbf{p}(\cdot,\cdot)$, process $\mathbf{q}$ solving Equation (5.4) with initial condition $q_0 \in \mathcal{Q}$ is given by

$$\begin{aligned}\mathbf{q}(t) &= q_0 + \int_0^t \int_{\Gamma_{\max}} r_q(\mathbf{x}(\tau^-),\mathbf{q}(\tau^-),\gamma)\mathbf{p}(d\tau,d\gamma) \\ &= q_0 + \sum_{i \geq 1: \mathbf{t}_i \leq t} r_q(\mathbf{x}(\mathbf{t}_i^-),\mathbf{q}(\mathbf{t}_i^-),\gamma_i),\end{aligned}$$

whereas between each pair of time instants $\mathbf{t}_i$ and $\mathbf{t}_{i+1}$ the solution $\mathbf{x}$ to (5.4) behaves as a diffusion process with local properties determined by $a(\cdot,\mathbf{q}(\mathbf{t}_i))$ and $b(\cdot,\mathbf{q}(\mathbf{t}_i))$.

Note that a zero jump occurs at time $\mathbf{t}_i$ when $r_q(\mathbf{x}(\mathbf{t}_i^-),\mathbf{q}(\mathbf{t}_i^-),\gamma_i) = 0$. Thus, the actual jump rate is different from $\lambda_{\max}$ and depends on the value taken by $\mathbf{s}(\mathbf{t}_i^-) = (\mathbf{q}(\mathbf{t}_i^-),\mathbf{x}(\mathbf{t}_i^-))$. Consistently with the "informal" definition of switching diffusion systems at the beginning of this section, the jump rate from $s = (x,q) \in \mathcal{S}$ is given by

$$\lambda_{\max} \int_{\{\gamma \in \Gamma_{\max}: r_q(x,q,\gamma) \neq 0\}} \mathcal{U}(d\gamma) = \sum_{q' \in \mathcal{Q}, q' \neq q} \lambda_{qq'}(x) = \lambda(s) \leq \lambda_{max},$$

where $\lambda(\cdot)$ is the jump intensity function defined in (5.1). Also, when a jump occurs from $s = (x,q) \in \mathcal{S}$ such that $\lambda(s) \neq 0$, its distribution over $\mathcal{Q} \setminus \{q\}$ depends on the value taken by $s = (x,q)$ and is given by

$$\frac{\int_{\{\gamma \in \Gamma_{\max}: r_q(x,q,\gamma) = q' - q\}} \mathcal{U}(d\gamma)}{\int_{\{\gamma \in \Gamma_{\max}: r_q(x,q,\gamma) \neq 0\}} \mathcal{U}(d\gamma)} = \frac{\lambda_{qq'}(x)}{\lambda(s)} = R(s,q'),$$

for any $q' \in \mathcal{Q} \setminus \{q\}$, where $R(\cdot,\cdot)$ is the reset function defined in (5.2).

Despite the fact that the function $r_q(x,q,\gamma)$ in (5.3), which determines the jump entity from $(x,q) \in \mathscr{S}$ when γ is the value extracted from $\Gamma_{\max}$, is not continuous as a function of x, the expected value $q + \int_{\mathbb{R}} r_q(x,q,\gamma)\mathfrak{U}(d\gamma)$ of a jump from $s = (x,q)$ is Lipschitz continuous as a function of x. This is shown based on the Lipschitz continuity and boundedness of the transition rates $\lambda_{qq'}(\cdot)$, $q,q' \in \mathscr{Q}$, in the following proposition, whose proof is inspired by [17].

PROPOSITION 5.1 *Assume that $\lambda_{qq'}(\cdot)$ are bounded and Lipschitz continuous for any $q,q' \in \mathscr{Q}$, $q \neq q'$. Then, there exists a constant $C > 0$ such that $\int_{\mathbb{R}} |r_q(x,q,\gamma) - r_q(x',q,\gamma)|\mathfrak{U}(d\gamma) \leq C|x-x'|$, $\forall x,x' \in \mathbb{R}^n$, $q \in \mathscr{Q}$.*

PROOF By the definition of $r_q(x,q,\gamma)$ in (5.3), it is easily derived that

$$\begin{aligned}
&\int_{\mathbb{R}} |r_q(x,q,\gamma) - r_q(x',q,\gamma)|\mathfrak{U}(d\gamma) \\
&\qquad = \frac{1}{\lambda_{\max}} \int_{\Gamma_{\max}} \Big| \sum_{i=1,i\neq q}^{M} (\mathbf{1}_{\Delta_{qi}(x)}(\gamma) - \mathbf{1}_{\Delta_{qi}(x')}(\gamma))(i-q) \Big| d\gamma \\
&\qquad \leq \frac{M}{\lambda_{\max}} \sum_{i=1,i\neq q}^{M} \int_{\Gamma_{\max}} \Big|\mathbf{1}_{\Delta_{qi}(x)}(\gamma) - \mathbf{1}_{\Delta_{qi}(x')}(\gamma)\Big| d\gamma. \qquad (5.5)
\end{aligned}$$

Let C_λ denote the Lipschitz constant of $\lambda_{qq'}(\cdot)$, i.e., $|\lambda_{qq'}(x) - \lambda_{qq'}(x')| \leq C_\lambda |x - x'|$, $\forall x,x' \in \mathbb{R}^n$, $q,q' \in \mathscr{Q}$. We next show that

$$\int_{\Gamma_{\max}} |\mathbf{1}_{\Delta_{qi}(x)}(\gamma) - \mathbf{1}_{\Delta_{qi}(x')}(\gamma)| d\gamma \leq 2(M^2 + M)C_\lambda |x - x'|, \quad q,i \in \mathscr{Q}. \qquad (5.6)$$

Then, the thesis follows by plugging (5.6) into (5.5).

To prove (5.6), we need to distinguish between two cases: (a) $\Delta_{qi}(x) \cap \Delta_{qi}(x') \neq \emptyset$ and (b) $\Delta_{qi}(x) \cap \Delta_{qi}(x') = \emptyset$.

Case (a):

$$\begin{aligned}
&\int_{\Gamma_{\max}} |\mathbf{1}_{\Delta_{qi}(x)}(\gamma) - \mathbf{1}_{\Delta_{qi}(x')}(\gamma)| d\gamma = \int_{\Delta_{qi}(x)\setminus\Delta_{qi}(x')} d\gamma + \int_{\Delta_{qi}(x')\setminus\Delta_{qi}(x)} d\gamma \\
&= \Big| \sum_{l=1}^{q-1} \sum_{h=1,h\neq l}^{M} \lambda_{lh}(x) + \sum_{h=1,h\neq q}^{i-1} \lambda_{qh}(x) - \sum_{l=1}^{q-1} \sum_{h=1,h\neq l}^{M} \lambda_{kl}(x') - \sum_{h=1,h\neq q}^{i-1} \lambda_{qh}(x') \Big| \\
&\quad + \Big| \sum_{l=1}^{q-1} \sum_{h=1,h\neq l}^{M} \lambda_{lh}(x) + \sum_{h=1,h\neq q}^{i} \lambda_{qh}(x) - \sum_{l=1}^{q-1} \sum_{h=1,h\neq l}^{M} \lambda_{lh}(x') - \sum_{h=1,h\neq q}^{i} \lambda_{qh}(x') \Big| \\
&\leq 2(M^2 + M)C_\lambda |x - x'|.
\end{aligned}$$

Case (b):

Let $\Delta_{qi}(x,x')$ denote the interval contiguous to both $\Delta_{qi}(x)$ and $\Delta_{qi}(x')$. Then,

$$\begin{aligned}\int_{\Gamma_{\max}} |\mathbf{1}_{\Delta_{qi}(x)}(\gamma) - \mathbf{1}_{\Delta_{qi}(x')}(\gamma)| d\gamma &= \int_{\Delta_{qi}(x)} d\gamma + \int_{\Delta_{qi}(x')} d\gamma \\ &\leq \int_{\Delta_{qi}(x) \cup \Delta_{qi}(x,x')} d\gamma + \int_{\Delta_{qi}(x') \cup \Delta_{qi}(x,x')} d\gamma.\end{aligned}$$

The sum of the two integrals in the last bound has the same expression as that in case (a), hence it can be bounded by $2(M^2+M)C_\lambda |x-x'|$, which concludes the proof of (5.6). ■

5.3 Reachability Problem Formulation

We now precisely formulate the reachability problem addressed in this chapter. Consider a measurable compact set $\mathscr{D} \subset \mathbb{R}^n$. Our objective is to evaluate the probability that the solution $\mathbf{x}(t)$ to Equation (5.4) initialized with $s_0 = (x_0, q_0)$ reaches $\mathscr{D}$ during some (possibly infinite) look-ahead time horizon $T = [0, t_f]$.
This probability can be expressed as

$$\mathrm{P}_{s_0}\left(\mathbf{x}(t) \in \mathscr{D} \text{ for some } t \in T\right), \tag{5.7}$$

where P_{s_0} is the probability measure induced by the solution $\mathbf{x}$ to (5.4) for the initial condition $s_0 = (x_0, q_0)$.

The set $\mathscr{D}$ can represent some target set or some unsafe/undesirable set for the system. Then, evaluating the probability (5.7) can be of interest for verifying if the system under consideration has been appropriately designed, or if some action has to be taken to appropriately modify it. The discrete component $\mathbf{q}$ of the hybrid state is considered here as instrumental to describing the evolution of the continuous state component $\mathbf{x}$, which is in fact the variable of interest. However, the methodology proposed to estimate (5.7) can be extended straightforwardly to the more general case when both the hybrid state components $\mathbf{x}$ and $\mathbf{q}$ are variables of interest and the reference set is of the form $\cup_{i=1}^{M}(\mathscr{D}_i \times \{i\}) \subset \mathscr{S}$ for some different compact sets $\mathscr{D}_i \subset \mathbb{R}^n$, $i = 1, \ldots, M$.

Note that the reachability event "$\mathbf{x}(t) \in \mathscr{D}$ for some $t \in T$" is well-defined because $\mathscr{D}$ is a Borel set and the process $\mathbf{x}$ has continuous trajectories (actually the less restrictive cadlag property, i.e., all trajectories are right continuous on $[0,\infty)$ with left limit on $(0,\infty)$, would be sufficient [5].)

To evaluate the probability (5.7) numerically, we introduce a bounded open set $\mathscr{U} \subset \mathbb{R}^n$ containing $\mathscr{D}$, $\mathscr{D} \subset \mathscr{U}$. If $\mathscr{D}$ represents an unsafe set, the domain $\mathscr{U}$ should be chosen large enough so that the situation can be declared safe once $\mathbf{x}$ ends up outside $\mathscr{U}$. If $\mathscr{D}$ represents a target set, $\mathscr{U}$ should be chosen large enough so that the objective of reaching the target is failed in practice once $\mathbf{x}$ exits $\mathscr{U}$. This makes sense, for example, in those regulation problems where the system should be driven to operate close to some reference state value $x^\star$ and deviations from the desired operating point are allowed only to some extent.

Let $\mathscr{U}^c$ denote the complement of $\mathscr{U}$ in $\mathbb{R}^n$. Then, with reference to the domain $\mathscr{U}$, the probability of entering $\mathscr{D}$ can be expressed as

$$\mathrm{P}_{s_0} := \mathrm{P}_{s_0}\big(\mathbf{x} \text{ hits } \mathscr{D} \text{ before hitting } \mathscr{U}^c \text{ within the time interval } T\big). \tag{5.8}$$

For the purpose of computing (5.8), we can assume that $\mathbf{x}$ in Equation (5.4) is defined on the open domain $\mathscr{U} \setminus \mathscr{D}$ with initial condition $x_0 \in \mathscr{U} \setminus \mathscr{D}$, and that $\mathbf{x}$ is stopped as soon as it hits the boundary $\partial\mathscr{U} \cup \partial\mathscr{D}$ of $\mathscr{U} \setminus \mathscr{D}$.

Different initial conditions s_0 for the system are characterized by a different probability P_{s_0}. The set of initial conditions such that P_{s_0} does not exceeds ε is given by:

$$S(\varepsilon) = \{s_0 \in \mathscr{S} : \mathrm{P}_{s_0} \leq \varepsilon\}. \tag{5.9}$$

We next describe a methodology for estimating the probability P_{s_0} and the set $S(\varepsilon)$, $\varepsilon \in (0,1)$. The key feature of the proposed methodology is that it is based on the approximation of the solution $\mathbf{x}$ to the switching diffusion Equation (5.4) by an interpolated Markov chain, whose state space is obtained by discretizing the original $\mathbb{R}^n$ space into grids. For properly chosen transition probabilities, the Markov chain interpolated process converges weakly to the solution to the switching diffusion Equation (5.4) as the grid size approaches zero.

Let $D([0,\infty);\mathbb{R}^n)$ denote the space of functions $f : [0,\infty) \to \mathbb{R}^n$ that are continuous from the right and have limit from the left. The Markov chain interpolated process satisfies the cadlag property, hence its trajectories belong to $D([0,\infty);\mathbb{R}^n)$. Given some compact set $\mathscr{A} \subseteq \mathbb{R}^n$, define the first hitting time function $\tau_{\mathscr{A}} : D([0,\infty);\mathbb{R}^n) \to [0,\infty]$ as $\tau_{\mathscr{A}}(f) := \inf\{t \geq 0 : f(t) \in \mathscr{A}\}$ with $\tau_{\mathscr{A}}(f) = \infty$ if $f(t) \in \mathscr{A}^c, \forall t \geq 0$. Then, P_{s_0} can be expressed as $\mathrm{P}_{s_0} = 1 - \mathrm{E}_{s_0}[I_{\mathscr{F}}(\mathbf{x})]$, where E_{s_0} is the expectation taken with respect to P_{s_0}, and $I_{\mathscr{F}}(\cdot)$ is the indicator function of the set $\mathscr{F} := \{f \in D([0,\infty);\mathbb{R}^n) : \tau_{\mathscr{D}}(f) > t_f\} \cup \{f \in D([0,\infty);\mathbb{R}^n) : \tau_{\mathscr{U}^c}(f) < \tau_{\mathscr{D}}(f)\}$.

Suppose that the first hitting time functions $\tau_{\mathscr{D}}$ and $\tau_{\mathscr{U}^c}$ are continuous with probability one relative to the probability measure P_{s_0} induced by the solution $\mathbf{s}$ to (5.4) for the initial conditions of interest $s_0 \in (\mathscr{U} \setminus \mathscr{D}) \times \mathscr{Q}$. Then, by [18, Chapter 9, Theorem 1.5], weak convergence to $\mathbf{s}$ of the approximating Markov chain interpolation implies convergence to the probability of interest P_{s_0} of the corresponding quantity for the approximating Markov chain with probability one. In the sequel, we assume that the appropriate conditions allowing for the estimation of P_{s_0} by weakly approximating $\mathbf{s}$ are satisfied.

REMARK 5.1 The condition above is needed to avoid those pathological situations where the trajectory of the process $\mathbf{x}$ touches the absorbing boundary $\partial\mathscr{D} \cup \partial\mathscr{U}$ without leaving $\mathscr{U} \setminus \mathscr{D}$. If the diffusion matrix $\Sigma(s) := b(x,q)\Sigma^2 b(x,q)^T$, $s = (x,q) \in \mathscr{S}$, is uniformly positive definite over $\mathscr{U} \setminus \mathscr{D}$, then appropriate regularity conditions on $\mathscr{D}$ and $\mathscr{U}^c$ guarantee with probability one that these pathological cases do not occur (see [17] and [18, Chapter 10].)

These pathological situations are known to be critical also for the discrete time approximation schemes used in the simulation of stochastic hybrid systems ([17], [13]), as well as for the detection of guard crossing when simulating non-stochastic hybrid systems with forced transitions. ■

From an algorithmic viewpoint, we introduce an iterative reachability algorithm which computes for each initial state $s_0 \in \mathscr{S}$ an estimate of the probability P_{s_0} of entering $\mathscr{D}$ without exiting the domain $\mathscr{U}$ during the time horizon T of interest, by propagating backwards in time the transition probabilities of the approximating Markov chain starting from the reference set. This iterative procedure directly enables us to determine an estimate of the level set $S(\varepsilon)$ for a specified threshold probability $\varepsilon \in (0,1)$.

5.4 Numerical Approximation Scheme

5.4.1 Markov Chain Approximation

The discrete time Markov chain $\{\mathbf{v}_k, k \geq 0\}$ used for estimating the probability of interest (5.8) is characterized by a two-component state: $\mathbf{v} = (\bar{\mathbf{x}}, \bar{\mathbf{q}})$, where $\bar{\mathbf{x}}$ and $\bar{\mathbf{q}}$ are used to approximate, respectively, the $\mathbf{x}$ and $\mathbf{q}$ components of the switching diffusion $\mathbf{s} = (\mathbf{x}, \mathbf{q})$. $\bar{\mathbf{q}}$ takes values in $\mathscr{Q}$, which is a finite set of cardinality M, whereas $\bar{\mathbf{x}}$ takes values in a finite set $\mathscr{X}_\delta$ obtained by gridding $\mathscr{U} \setminus \mathscr{D}$, where $\delta > 0$ is the gridding scale parameter (see Figure 5.1 for a schematic representation.)

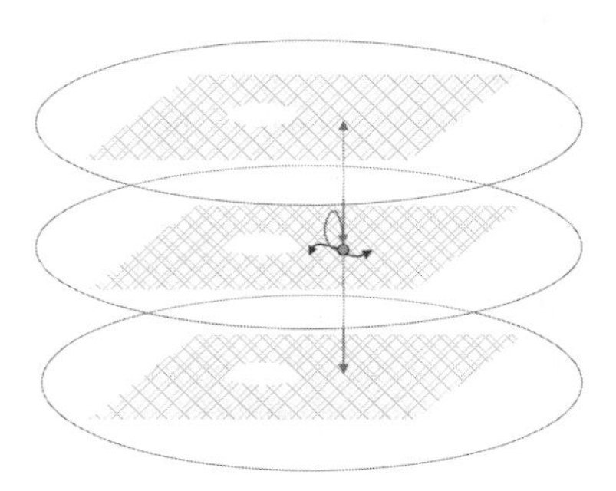

FIGURE 5.1: Schematic representation of the approximating Markov chain ($n = 2$ and $M = 3$).

The switching diffusion $\mathbf{s}$ is approximated by a continuous time process obtained by a piecewise constant interpolation of the discrete time Markov chain $\mathbf{v}$. The interpolation time interval Δt_δ should satisfy the conditions $\Delta t_\delta > 0$, $\forall \delta > 0$, and $\Delta t_\delta = o(\delta)$.

Recall that the $\mathbf{q}$ component of the switching diffusion $\mathbf{s} = (\mathbf{x}, \mathbf{q})$ is a pure jump process. The jump occurrences are governed by a standard Poisson process with intensity $\lambda_{\max}$, independent of the random variables determining the jump entity, and of the Brownian motion affecting the $\mathbf{x}$ component. The distribution of the inter-jump times is exponential with coefficient $\lambda_{\max}$, so that the probability of a single jump within a time interval Δ is $\lambda_{\max}\Delta + o(\Delta)$. The jump entity is state dependent, and jumps of zero entity may occur. When a jump (possibly of zero entity) occurs at time t, then, the $\mathbf{x}$ component is reinitialized with the same value $\mathbf{x}(t^-)$ prior to the jump occurrence.

In order to take this into account when defining the transition probabilities of the approximating Markov chain $\{\mathbf{v}_k, k \geq 0\}$, we start by introducing an enlarged Markov chain process $\{(\mathbf{v}_k, \mathbf{j}_k), k \geq 0\}$, where process $\{\mathbf{j}_k, k \geq 0\}$ represents the jump occurrences. $\{\mathbf{j}_k, k \geq 0\}$ is a sequence of i.i.d. random variables taking values in $\{0,1\}$: If $\mathbf{j}_k = 1$, then a jump, possibly of zero entity, occurs at time k; if $\mathbf{j}_k = 0$, then no jump occurs at time k. For each $k \geq 0$, $\mathbf{j}_k$ is independent of the random variables $\mathbf{v}_i$ up to and including time k, and

$$\mathrm{P}_\delta\left(\mathbf{j}_k = 1\right) = 1 - e^{-\lambda_{\max}\Delta t_\delta} = \lambda_{\max}\Delta t_\delta + o(\Delta t_\delta), \tag{5.10}$$

which tends to the jump rate of the standard Poisson process generating jumps in $\mathbf{q}$, as $\delta \to 0$.

To define the Markov chain process $\{(\mathbf{v}_k, \mathbf{j}_k), k \geq 0\}$, we need to specify how $\mathbf{j}_k$ affects the one-step evolution of $\mathbf{v}_k = (\bar{\mathbf{x}}_k, \bar{\mathbf{q}}_k)$. We distinguish between the two cases when $\mathbf{j}_k = 1$ and $\mathbf{j}_k = 0$. In the former case, we shall define the transitions between the "macro-states" $\bar{q} \in \mathcal{Q}$ of the approximating system, whereas in the latter, we shall define its evolution within a given macro-state $\bar{q} \in \mathcal{Q}$.

Case 1: Inter macro-states transitions

If $\mathbf{j}_k = 1$ (jump occurrence at time k), then, $\bar{\mathbf{x}}_{k+1}$ takes the same value as $\bar{\mathbf{x}}_k$, whereas the value of $\bar{\mathbf{q}}_{k+1}$ is determined based on that of $\mathbf{v}_k$ through appropriate (conditional) transition probabilities:

$$\mathrm{P}_\delta\left((\bar{\mathbf{x}}_{k+1}, \bar{\mathbf{q}}_{k+1}) = (z', \bar{q}') \mid \mathbf{v}_k = (z, \bar{q}), \mathbf{j}_k = 1\right) = \begin{cases} 0, & z' \neq z \\ p_\delta(\bar{q} \to \bar{q}'|z), & z' = z \end{cases} \tag{5.11}$$

where we set

$$p_\delta(\bar{q} \to \bar{q}'|z) := \mathrm{P}_\delta\left(\bar{\mathbf{q}}_{k+1} = \bar{q}' \mid \mathbf{v}_k = (z, \bar{q}), \mathbf{j}_k = 1\right).$$

The transition probability function $p_\delta(\bar{q} \to \bar{q}'|z)$ describes the evolution of $\bar{\mathbf{q}}$ when a jump (possibly of zero entity) occurs from $(z, \bar{q})$ (Figure 5.2.) In order for $\bar{\mathbf{q}}$ to

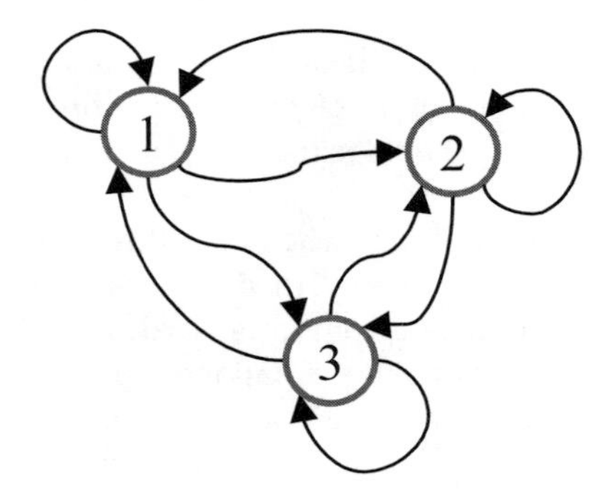

FIGURE 5.2: Schematic representation of inter macro-state transitions ($M = 3$).

reproduce the behavior of $\mathbf{q}$ when a jump occurs, we set

$$
\begin{aligned}
p_\delta(\bar{q} \to \bar{q}'|z) &:= \int_{\{\gamma \in \Gamma_{\max}: r_q(z,\bar{q},\gamma)=\bar{q}'-\bar{q}\}} \mathcal{U}(d\gamma) \\
&= \begin{cases} \dfrac{\lambda_{\bar{q}\bar{q}'}(z)}{\lambda_{\max}}, & \bar{q}' \neq \bar{q} \\ 1 - \dfrac{1}{\lambda_{\max}} \displaystyle\sum_{\bar{q}^* \in \mathcal{Q}, \bar{q}^* \neq \bar{q}} \lambda_{\bar{q}\bar{q}^*}(z), & \bar{q}' = \bar{q}. \end{cases}
\end{aligned}
\tag{5.12}
$$

In this way, the probability distribution of $\bar{\mathbf{q}}_{k+1}$ when a jump of non-zero entity occurs at time k from $(z, \bar{q})$ is given by

$$
\mathrm{P}_\delta\left(\bar{\mathbf{q}}_{k+1} = \bar{q}' \mid \mathbf{v}_k = (z,\bar{q}), \mathbf{j}_k = 1, \bar{\mathbf{q}}_{k+1} \neq \bar{\mathbf{q}}_k\right) = \frac{\lambda_{\bar{q}\bar{q}'}(z)}{\sum_{\bar{q}^* \in \mathcal{Q}, \bar{q}^* \neq \bar{q}} \lambda_{\bar{q}\bar{q}^*}(z)} = R((z,\bar{q}),\bar{q}')
$$

where $R(\cdot,\cdot)$ is the reset function in (5.2). Also, the probability that a jump of non zero entity occurs at time k from $(z, \bar{q})$ is given by

$$
\begin{aligned}
&\mathrm{P}_\delta\left(\mathbf{j}_k = 1, \bar{\mathbf{q}}_{k+1} \neq \bar{q} \mid \mathbf{v}_k = (z,\bar{q})\right) \\
&\quad = \sum_{\bar{q}' \in \mathcal{Q}, \bar{q}' \neq \bar{q}} \mathrm{P}_\delta\left(\bar{\mathbf{q}}_{k+1} = \bar{q}' \mid \mathbf{v}_k = (z,\bar{q}), \mathbf{j}_k = 1\right) \mathrm{P}_\delta\left(\mathbf{j}_k = 1\right) \\
&\quad = \left(1 - e^{-\lambda_{\max}\Delta t_\delta}\right) \frac{\sum_{\bar{q}' \in \mathcal{Q}, \bar{q}' \neq \bar{q}} \lambda_{\bar{q}\bar{q}'}(z)}{\lambda_{\max}} \\
&\quad = \lambda(z)\Delta t_\delta + o(\Delta t_\delta),
\end{aligned}
$$

where $\lambda(\cdot)$ is the jump intensity function defined in (5.1).

Case 2: Intra macro-state transitions

If $\mathbf{j}_k = 0$ (no jump occurrence at time k), then, $\bar{\mathbf{q}}_{k+1}$ takes the same value as $\bar{\mathbf{q}}_k$, whereas the value of $\bar{\mathbf{x}}_{k+1}$ is determined based on that of $\mathbf{v}_k$, through appropriate (conditional) transition probabilities:

$$\mathrm{P}_\delta\left((\bar{\mathbf{x}}_{k+1},\bar{\mathbf{q}}_{k+1}) = (z',\bar{q}') \mid \mathbf{v}_k = (z,\bar{q}), \mathbf{j}_k = 0\right) = \begin{cases} 0, & \bar{q}' \neq \bar{q} \\ p_\delta(z \to z'|\bar{q}), & \bar{q}' = \bar{q} \end{cases} \tag{5.13}$$

where we set

$$p_\delta(z \to z'|\bar{q}) := \mathrm{P}_\delta\left(\bar{\mathbf{x}}_{k+1} = z' \mid \mathbf{v}_k = (z,\bar{q}), \mathbf{j}_k = 0\right). \tag{5.14}$$

The transition probability function $p_\delta(z \to z'|\bar{q})$ describes the evolution of $\bar{\mathbf{x}}$ within the "macro-state" $\bar{q} \in \mathscr{Q}$. For the weak convergence result to hold, this function should be suitably selected so as to approximate "locally" the evolution of the $\mathbf{x}$ component of the switching diffusion $\mathbf{s} = (\mathbf{x}, \mathbf{q})$ with absorption on the boundary $\partial\mathscr{U} \cup \partial\mathscr{D}$ when no jump occurs in $\mathbf{q}$.

To formally define this "local consistency" notion, we need first to introduce some notation and definitions. Let Σ be a diagonal matrix, i.e., $\Sigma = \mathrm{diag}(\sigma_1, \sigma_2, \ldots, \sigma_n)$, with $\sigma_1, \sigma_2, \ldots, \sigma_n > 0$. Fix a grid parameter $\delta > 0$. Denote by $\mathbb{Z}^n_\delta$ the integer grids of $\mathbb{R}^n$ scaled according to the grid parameter δ and the positive diagonal entries of matrix Σ as follows

$$\mathbb{Z}^n_\delta = \{(m_1\eta_1\delta, m_2\eta_2\delta, \ldots, m_n\eta_n\delta) | \ (m_1, m_2, \ldots, m_n) \in \mathbb{Z}^n\},$$

where $\eta_i := \frac{\sigma_i}{\sigma_{\max}}$, $i = 1, \ldots, n$, with $\sigma_{\max} = \max_i \sigma_i$. For each grid point $z \in \mathbb{Z}^n_\delta$, define the immediate neighbors set as a subset of all the points in $\mathbb{Z}^n_\delta$ whose distance from z along the coordinate axis x_i is at most $\eta_i\delta$, $i = 1, \ldots, n$. Formally:

$$\mathscr{N}_\delta(z) = \{z + (i_1\eta_1\delta, i_2\eta_2\delta, \ldots, i_n\eta_n\delta) \in \mathbb{Z}^n_\delta | \ (i_1, i_2, \ldots, i_n) \in \mathscr{I}\}, \tag{5.15}$$

where $\mathscr{I} \subseteq \{0, 1, -1\}^n \setminus \{(0, 0, \ldots, 0)\}$.

The immediate neighbors set $\mathscr{N}_\delta(z)$ represents the set of states to which $\bar{\mathbf{x}}$ can evolve in one time step within a macro-state, starting from z. We remark that, depending on the diffusion matrix $b(s)$, $s \in \mathscr{S}$, in (5.4), different choices for $\mathscr{I}$ appearing in $\mathscr{N}_\delta(z)$ can be adopted for the convergence result to hold. The one with $\mathscr{I} = \{0, 1, -1\}^n \setminus \{(0, 0, \ldots, 0)\}$ is a typical choice. As another example, in Section 5.4.2, a set $\mathscr{N}_\delta(z)$ with smaller cardinality is chosen for the purpose of reducing computation time in the case when b is the identity matrix multiplied by a scalar function. For the time being, we assume that a proper immediate neighbors set is adopted and fixed.

The finite set $\mathscr{Z}_\delta$ where $\bar{\mathbf{x}}$ takes value is defined as the set of all those grid points in $\mathbb{Z}^n_\delta$ that lie inside $\mathscr{U}$ but outside $\mathscr{D}$: $\mathscr{Z}_\delta = (\mathscr{U} \setminus \mathscr{D}) \cap \mathbb{Z}^n_\delta$. The interior $\mathscr{Z}^\circ_\delta$ of $\mathscr{Z}_\delta$ consists of all those points in $\mathscr{Z}_\delta$ which have all their neighbors in $\mathscr{Z}_\delta$. The boundary of $\mathscr{Z}_\delta$ is given by $\partial\mathscr{Z}_\delta = \mathscr{Z}_\delta \setminus \mathscr{Z}^\circ_\delta$. $\partial\mathscr{Z}_\delta$ is the union of two disjoint sets: $\partial\mathscr{Z}_\delta = \partial\mathscr{Z}_{\delta\mathscr{U}} \cup \partial\mathscr{Z}_{\delta\mathscr{D}}$. $\partial\mathscr{Z}_{\delta\mathscr{U}}$ is the set of points with at least one neighbor outside $\mathscr{U}$, and $\partial\mathscr{Z}_{\delta\mathscr{D}}$ is the set of points with at least one neighbor inside $\mathscr{D}$. The points that satisfy both the conditions are assigned all to either $\partial\mathscr{Z}_{\delta\mathscr{D}}$ or $\partial\mathscr{Z}_{\delta\mathscr{U}}$, so as to make these two sets disjoint. Assigning them to $\partial\mathscr{Z}_{\delta\mathscr{D}}$ ($\partial\mathscr{Z}_{\delta\mathscr{U}}$) will eventually lead to an over-estimation (under-estimation) of the probability of interest. However, by choosing $\mathscr{U}$ sufficiently large, the over-estimation (under-estimation) error becomes negligible.

For each $\bar{q} \in \mathscr{Q}$, we now define the transition probability function $p_\delta(z \to z'|\bar{q})$ in (5.14) so that:

- each state z in $\partial \mathscr{Z}_\delta$ is an absorbing state:

$$p_\delta(z \to z'|\bar{q}) = \begin{cases} 1, & z' = z \\ 0, & \text{otherwise,} \end{cases} \qquad z \in \partial \mathscr{Z}_\delta$$

- starting from a state z in $\mathscr{Z}_\delta^\circ$, $\bar{\mathbf{x}}$ moves to one of its neighbors in $\mathscr{N}_\delta(z)$ or stays at the same state according to probabilities determined by its current location:

$$p_\delta(z \to z'|\bar{q}) = \begin{cases} \pi_\delta(z'|(z,\bar{q})), & z' \in \mathscr{N}_\delta(z) \cup \{z\} \\ 0, & \text{otherwise,} \end{cases} \qquad z \in \mathscr{Z}_\delta^\circ \tag{5.16}$$

where $\pi_\delta(z'|(z,\bar{q}))$ are appropriate functions of the drift and diffusion terms in (5.4) evaluated at $(z,\bar{q})$.

Figure 5.3 shows a possible choice for the immediate neighbors set $\mathscr{N}_\delta(z)$ in the two-dimensional case ($n = 2$), where $\mathbb{Z}_\delta^2$ is obtained by uniformly gridding $\mathbb{R}^2$ (i.e., $\eta_1 = \eta_2$.) The ellipsoidal region in the plot is $\mathscr{D}$, whereas $\mathscr{U}$ is the rectangular area containing $\mathscr{D}$. Examples of z states internal to the resulting set $\mathscr{Z}_\delta$ and on the boundaries $\partial \mathscr{Z}_{\delta\mathscr{D}}$ and $\partial \mathscr{Z}_{\delta\mathscr{U}}$ are shown, with the corresponding transitions to the immediate neighbors set.

Let the Markov chain be at state $v = (z,\bar{q}) \in \mathscr{Z}_\delta^\circ \times \mathscr{Q}$ at some time step k. Define

$$\begin{aligned} m_\delta(z,\bar{q}) &= \frac{1}{\Delta t_\delta} \mathrm{E}_\delta \left[\bar{\mathbf{x}}_{k+1} - \bar{\mathbf{x}}_k \mid \mathbf{v}_k = (z,\bar{q}), \mathbf{j}_k = 0 \right] \\ &= \frac{1}{\Delta t_\delta} \sum_{z' \in \mathscr{N}_\delta(z)} (z' - z) \pi_\delta(z'|(z,\bar{q})), \\ V_\delta(z,\bar{q}) &= \frac{1}{\Delta t_\delta} \mathrm{E}_\delta \left[(\bar{\mathbf{x}}_{k+1} - \bar{\mathbf{x}}_k)(\bar{\mathbf{x}}_{k+1} - \bar{\mathbf{x}}_k)^T \mid \mathbf{v}_k = (z,\bar{q}), \mathbf{j}_k = 0 \right] \\ &= \frac{1}{\Delta t_\delta} \sum_{z' \in \mathscr{N}_\delta(z)} (z' - z)(z' - z)^T \pi_\delta(z'|(z,\bar{q})). \end{aligned}$$

The immediate neighbors set $\mathscr{N}_\delta(z)$ and the family of distribution functions $\{\pi_\delta(\cdot|(z,\bar{q})) : \mathscr{N}_\delta(z) \cup \{z\} \to [0,1], z \in \mathscr{Z}_\delta^\circ\}$ should be selected so that as $\delta \to 0$,

$$\begin{aligned} m_\delta(z,\bar{q}) &\to a(x,\bar{q}), \\ V_\delta(z,\bar{q}) &\to b(x,\bar{q})\Sigma^2 b(x,\bar{q})^T, \end{aligned} \tag{5.17}$$

$\forall x \in \mathscr{U} \setminus \mathscr{D}$, where, for each $\delta > 0$, z is a point in $\mathscr{Z}_\delta^\circ$ closest to x. Different choices are possible so as to satisfy these "local consistency" properties for a Markov chain

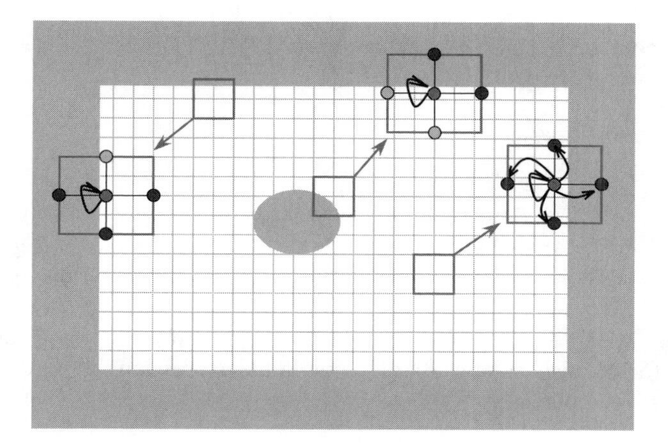

FIGURE 5.3: Example of immediate neighbors set and intra macro-state transitions in the two-dimensional case.

to approximate a diffusion process, and they affect the computational complexity of the approximation. The reader is referred to [18] for more details on this. In Section 5.4.2, we shall present possible choices for them in some case of interest.

Now that we have defined the transition probabilities of the enlarged Markov chain process $\{(\mathbf{v}_k,\mathbf{j}_k),k\geq 0\}$ (see Equations (5.10), (5.11), and (5.13)), we can characterize process $\{\mathbf{v}_k,k\geq 0\}$. It is easily shown that $\{\mathbf{v}_k,k\geq 0\}$ is a Markov chain since

$$\begin{aligned}
&\mathrm{P}_\delta\left(\mathbf{v}_{k+1}=v' \mid \mathbf{v}_k=v,\mathbf{v}_i=v_i,\, i<k\right)\\
&\quad=\sum_{j,j'\in\{0,1\}}\mathrm{P}_\delta\left(\mathbf{v}_{k+1}=v',\mathbf{j}_{k+1}=j',\mathbf{j}_k=j \mid \mathbf{v}_k=v,\mathbf{v}_i=v_i,\, i<k\right)\\
&\quad=\sum_{j,j'\in\{0,1\}}\mathrm{P}_\delta\left(\mathbf{v}_{k+1}=v',\mathbf{j}_{k+1}=j' \mid \mathbf{v}_k=v,\mathbf{j}_k=j\right)\mathrm{P}_\delta\left(\mathbf{j}_k=j \mid \mathbf{v}_k=v\right)\\
&\quad=\mathrm{P}_\delta\left(\mathbf{v}_{k+1}=v' \mid \mathbf{v}_k=v\right),
\end{aligned}$$

where the second equation follows from the fact that $\{(\mathbf{v}_k,\mathbf{j}_k),k\geq 0\}$ is a Markov chain, and $\mathbf{j}_k$ is independent of $\mathbf{v}_i$, $i\leq k$.

Moreover, the transition probabilities of $\{\mathbf{v}_k,k\geq 0\}$ can be expressed as follows

$$\mathrm{P}_\delta\left(\mathbf{v}_{k+1}=v' \mid \mathbf{v}_k=v\right)=\sum_{j\in\{0,1\}}\mathrm{P}_\delta\left(\mathbf{v}_{k+1}=v' \mid \mathbf{v}_k=v,\mathbf{j}_k=j\right)\mathrm{P}_\delta\left(\mathbf{j}_k=j\right).$$

Let us define

$$p_\delta(v\to v'):=\mathrm{P}_\delta\left(\mathbf{v}_{k+1}=v' \mid \mathbf{v}_k=v\right)$$

for ease of notation. By plugging Equations (5.10), (5.11), and (5.13) in the expression above, we finally get

$$p_\delta((z,\bar{q}) \to (z',\bar{q}')) =$$
$$\begin{cases} (1-e^{-\lambda_{\max}\Delta t_\delta})p_\delta(\bar{q}\to\bar{q}'|z)+e^{-\lambda_{\max}\Delta t_\delta}p_\delta(z\to z'|\bar{q}), & z'=z,\bar{q}'=\bar{q} \\ (1-e^{-\lambda_{\max}\Delta t_\delta})p_\delta(\bar{q}\to\bar{q}'|z), & z'=z,\bar{q}'\neq\bar{q} \\ e^{-\lambda_{\max}\Delta t_\delta}p_\delta(z\to z'|\bar{q}), & z'\neq z,\bar{q}'=\bar{q} \\ 0, & z'\neq z,\bar{q}'\neq\bar{q}, \end{cases}$$

for all $(z,\bar{q}),(z',\bar{q}') \in \mathscr{Z}_\delta \times \mathscr{Q}$, where $p_\delta(\bar{q}\to\bar{q}'|z)$ and $p_\delta(z\to z'|\bar{q})$ are given in (5.12) and (5.16), respectively. By plugging in the expressions of $p_\delta(\bar{q}\to\bar{q}'|z)$ and $p_\delta(z\to z'|\bar{q})$, we obtain:

$$p_\delta((z,\bar{q}) \to (z',\bar{q}')) =$$
$$\begin{cases} (1-e^{-\lambda_{\max}\Delta t_\delta})\big(1-\frac{\sum_{\bar{q}^*\neq\bar{q}}\lambda_{\bar{q}\bar{q}^*}(z)}{\lambda_{\max}}\big)+e^{-\lambda_{\max}\Delta t_\delta}, & z'=z\in\partial\mathscr{Z}_\delta,\bar{q}'=\bar{q} \\ (1-e^{-\lambda_{\max}\Delta t_\delta})\big(1-\frac{\sum_{\bar{q}^*\neq\bar{q}}\lambda_{\bar{q}\bar{q}^*}(z)}{\lambda_{\max}}\big)+e^{-\lambda_{\max}\Delta t_\delta}\pi_\delta(z|(z,\bar{q})), & z'=z\in\mathscr{Z}_\delta^\circ,\bar{q}'=\bar{q} \\ (1-e^{-\lambda_{\max}\Delta t_\delta})\frac{\lambda_{\bar{q}\bar{q}'}(z)}{\lambda_{\max}}, & z'=z,\bar{q}'\neq\bar{q} \\ e^{-\lambda_{\max}\Delta t_\delta}\pi_\delta(z'|(z,\bar{q})), & z'\in\mathscr{N}_\delta(z),\bar{q}'=\bar{q} \\ 0, & \text{otherwise.} \end{cases} \tag{5.18}$$

Fix $\delta > 0$ and consider the corresponding discrete time Markov chain $\{\mathbf{v}_k, k \geq 0\}$ with state space $\mathscr{Z}_\delta \times \mathscr{Q}$ and transition probabilities defined in (5.18), where $\mathscr{N}_\delta(z)$, $\pi_\delta(z'|(z,\bar{q}))$, and Δt_δ are such that the local consistency properties (5.17) are satisfied. Let $\{\overline{\Delta \mathbf{t}}_k, k \geq 0\}$ be an i.i.d. sequence of random variables independent of $\{\mathbf{v}_k, k \geq 0\}$, exponentially distributed with mean value Δt_δ satisfying $\Delta t_\delta > 0$ and $\Delta t_\delta = o(\delta)$. Denote by $\{\mathbf{v}(t), t \geq 0\}$ the continuous time stochastic process that is equal to $\mathbf{v}_k$ on the time interval $[\bar{\mathbf{t}}_k, \bar{\mathbf{t}}_{k+1})$ for all k, where $\bar{\mathbf{t}}_0 = 0$ and $\bar{\mathbf{t}}_{k+1} = \bar{\mathbf{t}}_k + \overline{\Delta \mathbf{t}}_k$, $k \geq 0$. If the chain $\{\mathbf{v}_k, k \geq 0\}$ starts from a point $v_0 \in \mathscr{Z}_\delta^\circ \times \mathscr{Q}$ closest to $s_0 \in (\mathscr{U} \setminus \mathscr{D}) \times \mathscr{Q}$, then, we conclude that the following proposition holds.

PROPOSITION 5.2 *Under Assumption 5.1, as $\delta \to 0$, the process $\{\mathbf{v}(t), t \geq 0\}$, obtained by interpolation of the approximating Markov chain $\{\mathbf{v}_k, k \geq 0\}$, converges weakly to the solution $\{\mathbf{s}(t) = (\mathbf{x}(t), \mathbf{q}(t)), t \geq 0\}$ to Equation* (5.4) *initialized with s_0, with $\mathbf{x}(t)$ defined on $\mathscr{U} \setminus \mathscr{D}$ and absorption on the boundary $\partial\mathscr{U} \cup \partial\mathscr{D}$.*

Weak convergence of Markov chain approximations is proven in [18, Theorem 4.1, Chapter 10] for jump diffusion processes. This theorem does not directly imply Proposition 5.2, because it would require $r_q(x,q,\gamma)$ in (5.3) to be continuous as a function of x, which is not the case. However, the continuity property shown in Proposition 5.1 for the integral $\int_{\mathbb{R}} r_q(\cdot,q,\gamma)\mathfrak{U}(d\gamma)$ can be used in the proof of [18,

Theorem 4.1, Chapter 10] in place of the continuity of $r_q(\cdot,q,\gamma)$ to assess the weak convergence result. Intuitively, this is because, when a jump occurs from (x,q), the new value of the discrete state component is determined, for both the approximating Markov chain as well as the switching diffusion, as $q+r_q(x,q,\gamma)$ where γ is the value extracted from the uniform distribution $\mathcal{U}$ over $\Gamma_{\max}$, independently of all the other random variables up to the time of the jump. Thus, what really matters is the continuity in x of the expected value of $r_q(x,q,\gamma)$ with respect to γ for each $q \in \mathcal{Q}$, which is obviously implied by that of $r_q(\cdot,q,\gamma)$, for any $q \in \mathcal{Q}$ and $\gamma \in \Gamma_{\max}$.

5.4.2 Locally Consistent Transition Probability Functions

In this section, we describe a possible choice for the immediate neighbors set $\mathcal{N}_\delta(z)$, the transition probability function $\pi_\delta(\cdot|v)$ in (5.16) from $v=(z,\bar{q}) \in \mathcal{Z}_\delta^\circ \times \mathcal{Q}$ to $\mathcal{N}_\delta(z) \cup \{z\}$, and the interpolation interval Δt_δ, that is effective in guaranteeing that the local consistency properties (5.17) hold. This is to complete the definition of the transitions probabilities (5.18) of the approximating Markov chain $\{\mathbf{v}_k, k \geq 0\}$ in Proposition 5.2.

We consider the case when the diffusion term in (5.4) is of the form $b(s)=\beta(s)I$, where $\beta : \mathcal{S} \to \mathbb{R}$ is a scalar function and I is the identity matrix of size n. In this case, each of the n components of the n-dimensional Brownian motion in (5.4) affects directly the corresponding single component of $\mathbf{x}$. This is the reason why the immediate neighbors set $\mathcal{N}_\delta(z)$, $z \in \mathcal{Z}_\delta$, can be confined to the set of points along each one of the x_i, $i=1,\dots,n$, directions whose distance from q is $\eta_i\delta$, $i=1,\dots,n$, respectively (see Figure 5.3 for an example in the case when $n=2$ and $\eta_1=\eta_2$.) For each $z \in \mathcal{Z}_\delta$, $\mathcal{N}_\delta(z)$ is then composed of the following $2n$ elements:

$$
\begin{aligned}
z_{1_+} &= z+(+\eta_1\delta,0,\dots,0) & z_{1_-} &= z+(-\eta_1\delta,0,\dots,0)\\
z_{2_+} &= z+(0,+\eta_2\delta,\dots,0) & z_{2_-} &= z+(0,-\eta_2\delta,\dots,0)\\
&\vdots & & \\
z_{n_+} &= z+(0,0,\dots,+\eta_n\delta) & z_{n_-} &= z+(0,0,\dots,-\eta_n\delta).
\end{aligned}
$$

The transition probability function $\pi_\delta(\cdot|v)$ over $\mathcal{N}_\delta(z) \cup \{z\}$ from $v=(z,\bar{q}) \in \mathcal{Z}_\delta^\circ \times \mathcal{Q}$ can be defined as follows:

$$
\pi_\delta(z'|v) = \begin{cases} c(v)\,\xi_0(v), & z'=z \\ c(v)\,e^{+\delta\xi_i(v)}, & z'=z_{i_+},\ i=1,\dots,n \\ c(v)\,e^{-\delta\xi_i(v)}, & z'=z_{i_-},\ i=1,\dots,n \end{cases} \tag{5.19}
$$

with

$$\xi_0(v) = \frac{2}{\rho\sigma_{\max}^2\beta(v)^2} - 2n \quad \xi_i(v) = \frac{[a(v)]_i}{\eta_i\sigma_{\max}^2\beta(v)^2}, \quad i = 1,\dots,n$$

$$c(v) = \big(2\sum_{i=1}^{n}\mathrm{csh}(\delta\xi_i(v)) + \xi_0(v)\big)^{-1},$$

where for any $y \in \mathbb{R}^n$, $[y]_i$ denotes the component of y along the x_i direction, $i = 1,2,\dots,n$. ρ is a positive constant that has to be chosen small enough such that $\xi_0(v)$ defined above is positive for all $v \in \mathscr{Z}_\delta^\circ \times \mathscr{Q}$. In particular, this is guaranteed if

$$0 < \rho \leq (n\sigma_{\max}^2 \max_{s\in(\mathscr{U}\setminus\mathscr{D})\times\mathscr{Q}} \beta(s)^2)^{-1}. \tag{5.20}$$

As for Δt_δ, we set $\Delta t_\delta = \rho\delta^2$.

A direct computation shows that for this choice of the neighbors set, the transition probabilities, and the interpolation interval, for each $v \in \mathscr{Z}_\delta^\circ \times \mathscr{Q}$,

$$m_\delta(v) = \tfrac{2c(v)}{\rho\delta}\begin{bmatrix}\eta_1\,\mathrm{sh}(\delta\xi_1(v))\\ \eta_2\,\mathrm{sh}(\delta\xi_2(v))\\ \vdots \\ \eta_n\,\mathrm{sh}(\delta\xi_n(v))\end{bmatrix},$$

$$V_\delta(v) = \tfrac{2c(v)}{\rho}\mathrm{diag}(\eta_1^2\mathrm{csh}\left(\delta\xi_1(v)\right), \eta_2^2\mathrm{csh}(\delta\xi_2(v)),\dots,\eta_n^2\mathrm{csh}(\delta\xi_n(v)))\,.$$

It is then easily verified that the equations in (5.17) are satisfied, which in turn leads to the weak convergence result in Proposition 5.2.

5.5 Reachability Computations

Consider the look-ahead time horizon $T = [0,t_f]$. Fix $\delta > 0$ and set $k_f := \frac{t_f}{\Delta t_\delta}$ (if $t_f = \infty$, then $k_f = \infty$; if t_f is finite, then δ should be chosen so that k_f is an integer.) As a result of Proposition 5.2, an estimate of the probability of interest P_{s_0} in (5.8) is provided by the corresponding quantity for the Markov chain $\{\mathbf{v}_k = (\bar{\mathbf{x}}_k, \bar{\mathbf{q}}_k), k \geq 0\}$ starting from a point $v_0 = (z_0, \bar{q}_0) \in \mathscr{Z}_\delta^\circ \times \mathscr{Q}$ closest to s_0:

$$\begin{aligned}\hat{\mathrm{P}}_{s_0} &:= \mathrm{P}_\delta\left(\bar{\mathbf{x}}_k \text{ hits } \partial\mathscr{Z}_{\delta\mathscr{D}} \text{ before hitting } \partial\mathscr{Z}_{\delta\mathscr{U}} \text{ within } 0 \leq k \leq k_f\right) \\ &= \mathrm{P}_\delta\left(\bar{\mathbf{x}}_k \in \partial\mathscr{Z}_{\delta\mathscr{D}} \text{ for some } 0 \leq k \leq k_f\right) \\ &= \mathrm{P}_\delta\left(\bar{\mathbf{x}}_{k_f} \in \partial\mathscr{Z}_{\delta\mathscr{D}}\right),\end{aligned} \tag{5.21}$$

where the second equality follows from the fact that the boundary $\partial\mathscr{Z}_{\delta\mathscr{U}}$ is absorbing, and the third one from the fact that the boundary $\partial\mathscr{Z}_{\delta\mathscr{D}}$ is absorbing as well. This estimate asymptotically converges to P_{s_0} as δ tends to zero.

We now describe an iterative algorithm to compute (5.21). This algorithm was first introduced by the authors of the present chapter in [15], and further developed in [16] and [22]. For the sake of self-containedness, we recall it here. We also point out that it can be used to determine an estimate of the set $S(\varepsilon)$ defined in (5.9).

Let $\hat{\mathrm{p}}^{(k)} : \mathscr{Z}_\delta \times \mathscr{Q} \to [0,1]$ with

$$\hat{\mathrm{p}}^{(k)}(v) := \mathrm{P}_\delta\left(\bar{\mathbf{x}}_{k_f} \in \partial\mathscr{Z}_{\delta\mathscr{D}} \mid \mathbf{v}_{k_f-k} = v\right) \tag{5.22}$$

be a set of probability maps defined on $\mathscr{Z}_\delta \times \mathscr{Q}$ and indexed by $k = 0,1,\ldots,k_f$. Then, the desired quantity $\hat{\mathrm{P}}_{s_0}$ can be expressed as

$$\hat{\mathrm{P}}_{s_0} = \hat{\mathrm{p}}^{(k_f)}(v_0).$$

Also, the set $S(\varepsilon)$ of initial conditions s_0 for the system such that the probability P_{s_0} does not exceeds some $\varepsilon \in (0,1)$ in (5.9) can be approximated by the level set of the probability map $\hat{\mathrm{p}}^{(k_f)}$ corresponding to ε

$$\hat{S}(\varepsilon) := \mathrm{P}_\delta\left(v \in \mathscr{Z}_\delta \times \mathscr{Q} : \hat{\mathrm{p}}^{(k_f)}(v) \leq \varepsilon\right).$$

The proposed iterative algorithm to compute $\hat{\mathrm{p}}^{(k_f)}$ determines the whole set of maps $\hat{\mathrm{p}}^{(k)} : \mathscr{Z}_\delta \times \mathscr{Q} \to [0,1]$ for $k = 0,1,\ldots,k_f$. Despite the increased computation burden, this has the advantage that, at any $t \in (0,t_f)$, an estimate of the probability of interest over the residual time horizon $[t_f - t, t_f]$ of length t is readily available, and is given by the map $\hat{\mathrm{p}}^{(\lfloor (t_f-t)/\Delta t_\delta \rfloor)} : \mathscr{Z}_\delta \times \mathscr{Q} \to [0,1]$ evaluated at the value taken by the state at time $t_f - t$; in other words, one does not need to recompute the probability map.

Fix a k such that $0 \leq k < k_f$. It is easily seen then that the map $\hat{\mathrm{p}}_\delta^{(k)} : \mathscr{Z}_\delta \times \mathscr{Q} \to [0,1]$ satisfies the following recursive equation

$$\hat{\mathrm{p}}^{(k+1)}(v) = \sum_{v' \in \mathscr{Z}_\delta \times \mathscr{Q}} p_\delta(v \to v')\hat{\mathrm{p}}^{(k)}(v'), \; v \in \mathscr{Z}_\delta \times \mathscr{Q}.$$

Recalling that any $v \in \partial\mathscr{Z}_\delta \times \mathscr{Q}$ is an absorbing state and that, for each $k \in [0,k_f]$, $\hat{\mathrm{p}}^{(k)}(v) = 1$ if $v \in \partial\mathscr{Z}_{\delta\mathscr{D}} \times \mathscr{Q}$, and $\hat{\mathrm{p}}^{(k)}(v) = 0$ if $v \in \partial\mathscr{Z}_{\delta\mathscr{U}} \times \mathscr{Q}$, we get

$$\hat{\mathrm{p}}^{(k+1)}(v) = \begin{cases} \sum_{v' \in \mathscr{Z}_\delta \times \mathscr{Q}} p_\delta(v \to v')\hat{\mathrm{p}}^{(k)}(v'), & v \in \mathscr{Z}_\delta^\circ \times \mathscr{Q} \\ 1, & v \in \partial\mathscr{Z}_{\delta\mathscr{D}} \times \mathscr{Q} \\ 0, & v \in \partial\mathscr{Z}_{\delta\mathscr{U}} \times \mathscr{Q}. \end{cases} \tag{5.23}$$

Let $v = (z,\bar{q}) \in \mathscr{Z}_\delta^\circ \times \mathscr{Q}$ and $v' = (z',\bar{q}') \in \mathscr{Z}_\delta \times \mathscr{Q}$. By distinguishing between the cases when (i) $v' = v$ (inter or intra macro-state transition), (ii) $z' = z$ and $\bar{q}' \neq \bar{q}$ (inter macro-state transition), and (iii) $z' \neq z$ and $\bar{q}' = \bar{q}$ (intra macro-state transition), the summation in (5.23) can be expanded as follows

$$\sum_{v' \in \mathscr{Z}_\delta \times \mathscr{Q}} p_\delta(v \to v')\hat{\mathrm{p}}^{(k)}(v') = p_\delta(v \to v)\hat{\mathrm{p}}^{(k)}(v) + \sum_{v'=(z,\bar{q}'):\bar{q}' \in \mathscr{Q}\setminus\{\bar{q}\}} p_\delta(v \to v')\hat{\mathrm{p}}^{(k)}(v') + \sum_{v'=(z',\bar{q}):z' \in \mathscr{N}_\delta(z)} p_\delta(v \to v')\hat{\mathrm{p}}^{(k)}(v'),$$

with the transition probabilities appearing in (5.18).

Finite horizon case: In the finite horizon case ($k_f < \infty$), the probability map $\hat{\mathrm{p}}^{(k_f)}$ can be computed by iterating equation (5.23) k_f times starting from $k = 0$ with the initialization

$$\hat{\mathrm{p}}^{(0)}(v) = \begin{cases} 1, & \text{if } v \in \partial \mathscr{Z}_{\delta\mathscr{D}} \times \mathscr{Q} \\ 0, & \text{otherwise,} \end{cases} \tag{5.24}$$

which is easily obtained from the definition (5.22) of $\hat{\mathrm{p}}^{(k)}$.

We remark that the grid size δ should be chosen properly to balance the following two conflicting considerations:

(i) Small δ is required to approximate fast diffusion processes. To see this, observe that since the intra macro-state transitions are limited to the immediate neighbors set, the maximal distance that the $\bar{\mathbf{x}}$ component of the Markov chain can travel in each single time interval of average value Δt_δ is $\eta_i\delta$ along the direction x_i, $i = 1,\dots,n$. Thus, for the continuous state component $\mathbf{x}$ of the stochastic hybrid system to be approximated by $\bar{\mathbf{x}}$, the component along the x_i axis $|[a(\cdot,q)]_i|$ of $a(\cdot,q)$ has to be upper bounded roughly by $\frac{\eta_i\delta}{\Delta t_\delta}$ over $\mathscr{U} \setminus \mathscr{D}$, for any $i = 1,\dots,n$, uniformly over the macro-state set $\mathscr{Q}$. Since $\Delta t_\delta = o(\delta)$, this condition imposes an upper bound on the admissible values for δ. For the choice $\Delta t_\delta = \rho\delta^2$ in Section 5.4.2, for example, $\delta \leq \min_{i=1,2,\dots,n, x \in \mathscr{U}\setminus\mathscr{D}, q \in \mathscr{Q}} \frac{\eta_i}{\rho|[a(x,q)]_i|}$.

(ii) Computation complexity of the algorithm grows with δ. Specifically, the number of iterations to determine $\hat{\mathrm{p}}^{(k_f)}$ is given by $k_f = t_f/\Delta t_\delta$, which is of the order $1/\delta^2$ if Δt_δ is chosen to be proportional to δ^2 as in Section 5.4.2. On the other hand, the computation time of each iteration is of the order $1/\delta^n$ as the state space $\mathscr{Z}_\delta$ is of the order of $1/\delta^n$. Thus the computation time of the algorithm grows with δ in the order of $1/\delta^{n+2}$, which increases rapidly as the dimension n increases.

This discussion shows that large δ's may not allow for the simulation of fast moving processes (point i), but for small δ's the running time may be too long (point ii).

Infinite horizon case: In the infinite horizon case, the iterative algorithm adopted in the finite horizon case would require infinite iterations.

Note that the iterative Equation (5.23) relating $\hat{\mathrm{p}}^{(k+1)}$ to $\hat{\mathrm{p}}^{(k)}$ is linear, hence can be written in matrix form as

$$\vec{P}^{(k+1)} = \vec{A}_\delta \vec{P}^{(k)} + \vec{b}_\delta, \tag{5.25}$$

where the sequence $\{\hat{\mathrm{p}}^{(k)}(v), v \in \mathscr{X}_\delta{}^\circ \times \mathscr{Q}\}$ has been arranged into a column vector $\vec{P}^{(k)} \in \mathbb{R}^{|\mathscr{X}_\delta^\circ \times \mathscr{Q}|}$ according to some fixed ordering of the points in $\mathscr{X}_\delta^\circ \times \mathscr{Q}$, and the square matrix $\vec{A}_\delta$ and column vector $\vec{b}_\delta$ of size $|\mathscr{X}_\delta^\circ \times \mathscr{Q}|$ are chosen properly. Let $(\mathscr{X}_\delta^\circ)^\circ$ denote the interior of $\mathscr{X}_\delta^\circ$ consisting of all those points in $\mathscr{X}_\delta^\circ$ whose immediate neighbors all belong to $\mathscr{X}_\delta^\circ$. Matrix $\vec{A}_\delta$ is a sparse positive matrix with the property that the sum of its elements on each row is smaller than or equal to 1, where equality holds if and only if that row corresponds to a point in $(\mathscr{X}_\delta^\circ)^\circ \times \mathscr{Q}$. As for $\vec{b}_\delta$, it is a positive vector with nonzero elements on exactly those rows corresponding to points on the boundary $\partial(\mathscr{X}_\delta^\circ) \times \mathscr{Q} = (\mathscr{X}_\delta^\circ \setminus (\mathscr{X}_\delta^\circ)^\circ) \times \mathscr{Q}$ of $\mathscr{X}_\delta^\circ \times \mathscr{Q}$.

Equation (5.25) can be interpreted as the dynamic equation of a discrete time system with state $\vec{P}^{(k)}$, dynamic matrix $\vec{A}_\delta$ and constant input equal to $\vec{b}_\delta$, evolving over an infinite time horizon starting from $k = 0$. The following propositions show that this system is asymptotically stable, and hence the solution to (5.25) converges to some (unique) $\vec{P}$ value, irrespectively of the initialization. $\vec{P}$ is exactly the probability map $\hat{\mathrm{p}}^{(k_f)}$ of interest, which can then be determined by solving the fixed point equation associated with (5.25).

PROPOSITION 5.3 *The eigenvalues of $\vec{A}_\delta$ are all in the interior of the unit disk of the complex plane.*

As a result of Proposition 5.3, we have that the following proposition holds.

PROPOSITION 5.4 *Consider equation*

$$\vec{P}^{(k+1)} = \vec{A}_\delta \vec{P}^{(k)} + \vec{b}_\delta. \tag{5.26}$$

(i) There is a unique $\vec{P} \in \mathbb{R}^{|\mathscr{X}_\delta{}^\circ \times \mathscr{Q}|}$ satisfying

$$\vec{P} = \vec{A}_\delta \vec{P} + \vec{b}_\delta. \tag{5.27}$$

(ii) Starting from any initial value $\vec{P}^{(0)}$ at $k = 0$, the solution $\vec{P}^{(k)}$ to Equation (5.26) converges to the fix point $\vec{P}$ as $k \to \infty$. Moreover, if $\vec{P}^{(0)} \geq \vec{P}$, then $\vec{P}^{(k)} \geq \vec{P}$ for all $k \geq 0$. Conversely, if $\vec{P}^{(0)} \leq \vec{P}$, then $\vec{P}^{(k)} \leq \vec{P}$ for all $k \geq 0$. Here the symbols $\geq$ and $\leq$ denote component-wise comparison between vectors.

In [22], results similar to the above two propositions are proved for the non-hybrid case. Since their proofs can be easily extended to the hybrid case studied in this chapter, we omit them here.

The desired quantity $\hat{p}^{(k_f)}$ can be obtained in several ways. For example, one can solve the linear equation $(I - \vec{A}_\delta)\vec{P} = \vec{b}_\delta$ directly, by aid of sparse matrix computation tools. Alternatively, one can iterate Equation (5.25) starting at $k = 0$ from two initial conditions, one an upper bound and the other a lower bound of $\vec{P}$. Proposition 5.4 implies that the iteration results for the two cases will remain an upper bound and a lower bound of $\vec{P}$, respectively, for all k, and will converge toward each other and hence to $\vec{P}$ as $k \longrightarrow \infty$, enabling one to approximate $\vec{P}$ within arbitrary precision.

REMARK 5.2 The convergence rate of the iteration (5.25) is determined by the largest eigenvalue of the stochastic matrix $\vec{A}_\delta$, which tends to 1 as $\delta \to 0$ with a corresponding eigenvector tending to $(1, \ldots, 1)$. Thus for small δ, convergence of the iteration (5.25) is slow. To alleviate this difficulty, techniques such as adaptive gridding can be adopted. ∎

5.6 Possible Extensions

The approach proposed in the previous section can be easily extended to address different stochastic reachability problems.

5.6.1 Probabilistic Safety

Given an open set $\mathscr{W} \subset \mathbb{R}^n$ with compact support, consider the problem of determining the probability that the continuous state $\mathbf{x}$ of the switching diffusion system (5.4) initialized at $s_0 = (x_0, q_0) \in \mathscr{W} \times \mathscr{Q}$ will remain within the safe set $\mathscr{W}$ during some finite time horizon $T = [0, t_f]$:

$$\mathrm{P}_{s_0}\left(\mathbf{x}(t) \in \mathscr{W} \text{ for all } t \in T\right). \tag{5.28}$$

Note that this quantity can be rewritten as follows

$$\mathrm{P}_{s_0}\left(\mathbf{x}(t) \in \mathscr{W} \text{ for all } t \in T\right) = 1 - \mathrm{P}_{s_0}\left(\mathbf{x}(t) \in \mathscr{W}^c \text{ for some } t \in T\right),$$

thus the problem can be reduced to that of estimating

$$\mathrm{P}_{s_0}(\mathscr{W}^c) := \mathrm{P}_{s_0}\left(\mathbf{x}(t) \in \mathscr{W}^c \text{ for some } t \in T\right). \tag{5.29}$$

$\mathrm{P}_{s_0}(\mathscr{W}^c)$ has the same expression as the probability introduced in (5.7). Despite the fact that the set $\mathscr{D}$ appearing in (5.7) is bounded, whereas $\mathscr{W}^c$ in (5.29) is unbounded, it is easily seen that the numerical approximation scheme proposed for estimating (5.7) can still be applied to estimate (5.29).

In this setting, there is no need to introduce the set $\mathscr{U}$ so as to reduce the state space of the system to a bounded set, since the component $\mathbf{x}$ of the switching diffusion (5.4) can be confined to the bounded set $\mathscr{W}$ with absorption on its boundary $\partial\mathscr{W}$ for the purpose of computing (5.29).

The approximating Markov chain transition probabilities can be defined as described in Section 5.4.1 with $\mathscr{W}$ and $\mathscr{W}^c$ respectively replacing $\mathscr{U} \setminus \mathscr{D}$ and $\mathscr{D}$. The finite set where the component $\bar{\mathbf{x}}$ of the approximating Markov chain takes values is then given by $\mathscr{Z}_\delta = \mathscr{W} \cap \mathbb{Z}^n_\delta$. Its boundary $\partial\mathscr{Z}_\delta$ is only composed of the set of points in $\mathscr{Z}_\delta$ with at least one neighbor inside $\mathscr{W}^c$. Consequently, the iterative Equation (5.23) to compute $\hat{\mathrm{p}}^{(k_f)}(v) = \mathrm{P}_\delta\left(\bar{\mathbf{x}}_{k_f} \in \partial\mathscr{Z}_\delta \,|\, \mathbf{v}_0 = v\right)$ can still be applied with $\partial\mathscr{Z}_\delta$ replacing $\partial\mathscr{Z}_{\delta\mathscr{D}}$, and no absorbing $\partial\mathscr{Z}_{\delta\mathscr{U}}$ boundary.

One can also determine an estimate of the set with safety level $1-\varepsilon$, where $\varepsilon \in (0,1)$, i.e., the set $S(\varepsilon;\mathscr{W}^c) = \{s_0 \in \mathscr{S} : \mathrm{P}_{s_0}(\mathscr{W}^c) \leq \varepsilon\}$ of initial conditions s_0 such that the probability that the system will remain within $\mathscr{W}$ during the time horizon T is at least $1-\varepsilon$.

5.6.2 Regulation

Reachability analysis can be useful in the framework of regulation theory where the aim is to drive the state of the system close to some desired operating condition. The effectiveness of the designed controlled system can be assessed by considering a small region around the reference point and considering a time-varying set shrinking toward this region.

Let $\mathscr{W}^\star \subset \mathbb{R}^n$ be an open set with compact support representing some small region around a reference point $x^\star$. Consider a time varying open set $\mathscr{W}(t)$ with compact support that progressively shrinks toward $\mathscr{W}^\star$ during some finite time horizon T and the probability

$$\mathrm{P}_{s_0}\left(\mathbf{x}(t) \in \mathscr{W}(t) \text{ for all } t \in T\right) = 1 - \mathrm{P}_{s_0}\left(\mathbf{x}(t) \in \mathscr{W}^c(t) \text{ for some } t \in T\right).$$

For the purpose of computing

$$\mathrm{P}_{s_0}(\mathscr{W}^c(\cdot)) := \mathrm{P}_{s_0}\left(\mathbf{x}(t) \in \mathscr{W}^c(t) \text{ for some } t \in T\right), \tag{5.30}$$

we can confine the component $\mathbf{x}$ of the switching diffusion (5.4) to the largest $\mathscr{W}(0)$ with absorption on the boundary $\partial\mathscr{W}(0)$.

The approximating Markov chain transition probabilities can be defined as described in Section 5.4.1 with $\mathscr{W}(0)$ and $\mathscr{W}^c(0)$ respectively replacing $\mathscr{U} \setminus \mathscr{D}$ and $\mathscr{D}$. The finite set where the component $\bar{\mathbf{x}}$ of the Markov chain takes values is then given by $\mathscr{Z}_\delta = \mathscr{W}(0) \cap \mathbb{Z}^n_\delta$, whereas its boundary $\partial\mathscr{Z}_\delta$ is composed of the set of points in $\mathscr{Z}_\delta$ with at least one neighbor inside $\mathscr{W}^c(0)$. According to Proposition 5.2, the interpolated Markov chain converges weakly to the solution $\{\mathbf{s}(t) = (\mathbf{x}(t), \mathbf{q}(t)), t \geq 0\}$ to Equation (5.4) initialized with s_0, with $\mathbf{x}(t)$ defined on $\mathscr{W}(0)$ with absorption on the boundary $\partial\mathscr{W}(0)$.

Note that $\mathrm{P}_{s_0}(\mathscr{W}^c(\cdot))$ in (5.30) can be expressed as the probability of the process $(t, \mathbf{x}(t))$ hitting the set $\{(\tau, \mathscr{W}^c(\tau)) : \tau \in T\}$ within the time interval T. Then, as discussed in Section 5.3, under appropriate regularity conditions on the enlarged time-space process, weak convergence to $\mathbf{s}$ of the approximating Markov chain interpolation implies convergence to the probability $P_{s_0}(\mathscr{W}^c(\cdot))$ of the corresponding quantity for the approximating Markov chain with probability one.

Let $\mathscr{X}_{\delta,t} = \mathscr{W}(t) \cap \mathbb{Z}^n_\delta$. Denote by $\mathscr{X}^\circ_{\delta,t}$ the interior of $\mathscr{X}_{\delta,t}$, i.e., the set of all those points in $\mathscr{X}_{\delta,t}$ which have all their neighbors in $\mathscr{X}_{\delta,t}$. Clearly, $\mathscr{X}_{\delta,0} = \mathscr{X}_\delta$ and $\partial\mathscr{X}_\delta = \mathscr{X}_\delta \setminus \mathscr{X}^\circ_{\delta,0}$. Then, with reference to the Markov chain approximation, an estimate of the probability (5.30) is provided by

$$\hat{P}_{s_0}(\mathscr{W}^c(\cdot)) := P_\delta\left(\bar{\mathbf{x}}_k \in \mathscr{X}_\delta \setminus \mathscr{X}^\circ_{\delta,k} \text{ for some } 0 \le k \le k_f\right), \tag{5.31}$$

with the approximating Markov chain starting from a point v_0 closest to s_0. This expression is different from (5.21) in the time-invariant case. To derive a recursive equation similar to (5.23), we need to redefine the probabilistic maps $\hat{p}^{(k)} : \mathscr{X}_\delta \times \mathscr{Q} \to [0,1]$ in (5.22) as follows:

$$\hat{p}^{(k)}(v) := P_\delta\left(\max_{h\in[k_f-k,k_f]} I_{\mathscr{X}_\delta\setminus\mathscr{X}^\circ_{\delta,h}}(\bar{\mathbf{x}}_h) = 1 \mid \mathbf{v}_{k_f-k} = v\right),$$

where $I_{\mathscr{X}_\delta\setminus\mathscr{X}^\circ_{\delta,h}}(\cdot)$ is the indicator function of the set $\mathscr{X}_\delta \setminus \mathscr{X}^\circ_{\delta,h}$. It is then easily seen that

$$\begin{aligned}
\hat{p}^{(k+1)}(v) &= E_\delta\left[\max_{h\in[k_f-k-1,k_f]} I_{(\mathscr{X}_\delta\setminus\mathscr{X}^\circ_{\delta,h})\times\mathscr{Q}}(\mathbf{v}_h) \mid \mathbf{v}_{k_f-k-1} = v\right] \\
&= I_{(\mathscr{X}_\delta\setminus\mathscr{X}^\circ_{\delta,k_f-k-1})\times\mathscr{Q}}(v) + \left(1 - I_{(\mathscr{X}_\delta\setminus\mathscr{X}^\circ_{\delta,k_f-k-1})\times\mathscr{Q}}(v)\right) \\
&\quad E_\delta\left[\max_{h\in[k_f-k,k_f]} I_{(\mathscr{X}_\delta\setminus\mathscr{X}^\circ_{\delta,h})\times\mathscr{Q}}(\mathbf{v}_h) \mid \mathbf{v}_{k_f-k-1} = v\right] \\
&= \begin{cases} \sum_{v'\in\mathscr{X}_\delta\times\mathscr{Q}} p_\delta(v\to v')\hat{p}^{(k)}(v'), & v \in \mathscr{X}^\circ_{\delta,k_f-k-1}\times\mathscr{Q} \\ 1, & v \in (\mathscr{X}_\delta\setminus\mathscr{X}^\circ_{\delta,k_f-k-1})\times\mathscr{Q}. \end{cases}
\end{aligned}$$

We can then compute $\hat{p}^{(k_f)}(v_0) = \hat{P}_{s_0}(\mathscr{W}^c(\cdot))$ by iterating the equation above starting from the initialization

$$\hat{p}^{(0)}(v) = \begin{cases} 1, & v \in \partial\mathscr{X}_\delta \times \mathscr{Q}, \\ 0, & \text{otherwise.} \end{cases}$$

Note that in the time invariant case the expression for $\hat{p}^{(k)}$ and the recursive scheme to compute $\hat{p}^{(k_f)}$ reduce to (5.22) and (5.23).

5.7 Some Examples

In this section we present some examples of application of the methodology for reachability analysis discussed in the previous sections. In the first example, motivated by a manufacturing system, a probabilistic safety problem is discussed, where the efficacy of a control strategy in maintaining the inventory level within a desired "safe" range is addressed. A single machine is considered, which is modeled by a switching diffusion as described in [11]. In the second example a regulation problem is discussed, where the efficacy of a threshold-based strategy in driving the average temperature of a room within a desired range is verified. This example is inspired by [19] and [1]. In these two example, the continuous state space $\mathbb{R}^n$ is one-dimensional ($n = 1$). Two and three-dimensional examples can be found in [16] and [22].

5.7.1 Manufacturing System

We consider a machine that produces some commodity. When the machine is operating, then, the inventory level $\mathbf{x}$ is governed by

$$d\mathbf{x}(t) = (u - \alpha)\,dt + \sigma\,d\mathbf{w}(t)$$

where $\alpha > 0$ is the demand rate (assumed to be constant), u is the production rate taking values in $[0, r]$ with $r > \alpha$, and $\mathbf{w}$ is a one-dimensional Brownian motion with variance modulated by σ, modeling demand fluctuations, sales returns, etc. Note that $\mathbf{x}$ can possibly take negative values. A negative value for $\mathbf{x}$ has to be understood as a backlog in demand.

Some failure may occur while the machine is operating. If a failure occurs, then the dynamics governing $\mathbf{x}$ switches to

$$d\mathbf{x}(t) = -\alpha\,dt + \sigma\,d\mathbf{w}(t),$$

and some intervention has to be taken on the machine so as to drive it back to the operating condition. This is modeled by a continuous time Markov chain process $\mathbf{q}$, independent of $\mathbf{w}$, which takes values in $\mathscr{Q} = \{1, 2\}$, where 1 stands for the operating condition and 2 for the down condition. The transition rates $\lambda_{12} > 0$ and $\lambda_{21} > 0$, respectively representing the infinitesimal rates of failure and repair, are assumed to be constant.

Let $x^\star > 0$ be some upper bound on the admissible inventory levels. In the results reported below, u is assumed to be a sigmoidal function $f : \mathbb{R} \to [0, r]$ of the inventory level x, satisfying $f(x^\star/2) = \alpha$ (production rate equal to the demand rate at half the maximum inventory level $x^\star$) and decreasing from r to 0 in a neighborhood of $x^\star/2$:

$$u = f(x) = \frac{r}{1 + (\frac{r}{\alpha} - 1)\exp(-100(1 - 2\frac{x}{x^\star}))}$$

Our objective is verifying whether the production strategy u is effective in maintaining the inventory level within the set $\mathscr{W} = (0, x^\star)$, during some finite time interval $[0, t_f]$ with high probability, with the machine that is in the operating condition at time $t = 0$.

By applying the approach described in Section 5.6.1, we determine an estimate of the safety probability

$$\mathrm{P}_{s_0}\left(\mathbf{x}(t) \in (0, x^\star) \text{ for all } t \in [0, t_f]\right),$$

with $s_0 = (x_0, q_0)$, as a function of the initial inventory level $x_0 \in (0, x^\star)$ when the machine is initially operating ($q_0 = 1$.) We set $r = 8$, $\alpha = 5$, $\lambda_{12} = 0.01$, $\sigma = 1$, $x^\star = 50$, $t_f = 100$, and $\delta = (10\rho \max\{r-\alpha, \alpha\})^{-1}$, with $\rho = 1/\sigma^2$, and consider different values for the repair rate λ_{21}.

Figure 5.4 shows the dependence of the safety probability map on the repair rate λ_{21}. Plots corresponding to decreasing values of the repair rate λ_{21} ($\lambda_{21} = 1, 0.7$, and 0.4) are reported from left to right in this figure. As expected, when the repair rate decreases, then the probability that the inventory level $\mathbf{x}$ will remain within the safe set $\mathscr{W}$ during the time horizon $[0, t_f]$ decreases as well. This decrease is particularly evident when the initial inventory level is close to the boundary of $\mathscr{W}$.

Note that there is some asymmetry in the reported plots of the safety probability map, especially close to the boundaries of $\mathscr{W}$. This is due to some "asymmetry" in the drift term: For inventory levels close to the lower bound of $\mathscr{W}$, the drift term is equal to $u - \alpha \simeq 3$ if the machine is operating and to $-\alpha = -5$ if the machine is down, whereas for inventory levels close to the upper bound, the drift term is equal to $u - \alpha \simeq -5$ if the machine is operating and to $-\alpha = -5$ if the machine is down. Thus, the drift term is more effective in maintaining the inventory level within $\mathscr{W}$ when $\mathbf{x}$ is close to the upper bound than to the lower bound of $\mathscr{W}$.

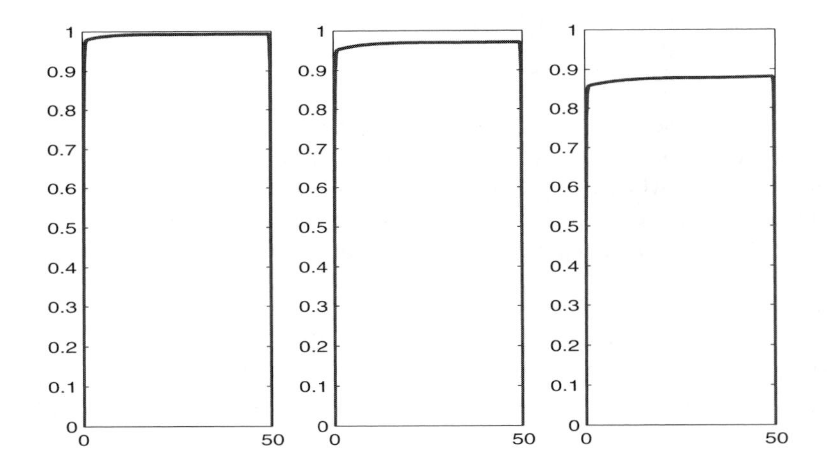

FIGURE 5.4: Safety probability as a function of the initial inventory level $x_0 \in (0, x^\star)$, when the machine is initially operating ($q_0 = 1$), for repair rate λ_{21} equal to 1, 0.7, and 0.4 from left to right.

5.7.2 Temperature Regulation

We consider the problem of progressively driving the temperature of a room within some desired range $\mathscr{W}^\star = (x_-^\star, x_+^\star)$, with $x_-^\star < x_+^\star$, by turning on and off a heater. Starting from a temperature value in a set $\mathscr{W}(0) \supset \mathscr{W}^\star$, the desired set $\mathscr{W}^\star$ should be reached within a certain time $t_f > 0$.

The average temperature of the room $\mathbf{x}$ evolves according to the following stochastic differential equation:

$$d\mathbf{x}(t) = \begin{cases} (-\frac{l}{C}(\mathbf{x}(t) - x_a) + \frac{r}{C})dt + \frac{\sigma}{C}d\mathbf{w}(t), & \text{if the heater is on,} \\ -\frac{l}{C}(\mathbf{x}(t) - x_a)dt + \frac{\sigma}{C}d\mathbf{w}(t), & \text{if the heater is off,} \end{cases} \tag{5.32}$$

where l is the average heat loss rate, C is the average thermal capacity of the room, x_a is the ambient temperature (assumed to be constant), r is the rate of heat gain supplied by the heater, and $\mathbf{w}$ is a standard Brownian motion modeling the uncertainty and disturbances affecting the temperature evolution with variance modulated by $\sigma > 0$.

Consider a discrete state $\mathbf{q}$ taking values in $\mathscr{Q} = \{1,2\}$ representing the two conditions when the heater is either "on" ($\mathbf{q} = 1$) or "off" ($\mathbf{q} = 2$). We assume that the heater is turned on or off with a rate that depends on the average room temperature. More specifically, $\mathbf{q}$ is taken to be a continuous time process with state space $\mathscr{Q}$ and transition rates $\lambda_{12} : \mathbb{R} \to \mathbb{R}$ and $\lambda_{21} : \mathbb{R} \to \mathbb{R}$ that depends on the continuous state component as follows

$$\lambda_{12}(x) = \frac{\bar{\lambda}_{12}}{1 + \exp(-100(\frac{x}{x_{\text{high}}} - 0.9))} \qquad \lambda_{21}(x) = \frac{\bar{\lambda}_{21}}{1 + \exp(100(\frac{x}{x_{\text{low}}} - 1.1))}$$

so that the largest rate values $\bar{\lambda}_{12} > 0$ and $\bar{\lambda}_{21} > 0$ are reached as soon as x gets respectively higher than x_{high} and smaller than x_{low} with these threshold values satisfying $x_-^\star < x_{\text{low}} < x_{\text{high}} < x_+^\star$. This can model the fact that a command of switching on (off) is issued to the heater when the temperature gets close to the threshold values, but it takes some time for the heater to actually commute.

The temperature is measured in Fahrenheit degrees and the time in minutes. The parameters in Equation (5.32) are assigned the following values: $x_a = 28$, $l/C = 0.1$, $r/C = 10$, and $\sigma/C = 1$. The infinitesimal switching rates are both chosen to be equal to 10. As for the gridding parameter, we set $\delta = (\rho \max_{x \in \mathscr{W}(0)} \{a(x,1), a(x,2)\})^{-1}$, where $a(x,1) = -\frac{l}{C}(x - x_a) + \frac{r}{C}$ and $a(x,2) = -\frac{l}{C}(x - x_a)$, and $\rho = (C/\sigma)^2$.

As illustrated in Section 5.6.2, this problem can be formulated as a stochastic reachability analysis problem by introducing a time-varying set $\mathscr{W}(t)$ shrinking from $\mathscr{W}(0)$ to $\mathscr{W}^\star$ during the time interval $[0, t_f]$. The results reported below refer to the case when $\mathscr{W}^\star = (66,76)$, $\mathscr{W}(0) = (20,80)$, $\mathscr{W}(t) = (20 + (66-20)\frac{t}{t_f}, 80 + (76-80)\frac{t}{t_f})$, $t \in [0, t_f]$, and $t_f = 120$.

In Figure 5.5, we plot the probability that the temperature is progressively conveyed to the desired range $(66,76)$ along the time horizon $[t, 120]$, as a function of the temperature value at time t, for t taking values in $[0, 120]$. The top row refers to the case when the heater is initially on, whereas the bottom row refers to the case when the heater is initially off. In each plot we represent a three dimensional surface. The value taken by a point of this surface at $(x,t) \in [20,80] \times [0,120]$ represents the probability that the temperature is progressively conveyed within $(66,76)$ during the time horizon $[t,120]$, starting from $\mathbf{x}(t) = x$.

The effect of two different pair of threshold values x_{low} and x_{high} is evaluated. The probability maps corresponding to $x_{\text{low}} = 69$ and $x_{\text{high}} = 73$ are represented on the left side of Figure 5.5, whereas the ones corresponding to $x_{\text{low}} = 67$ and $x_{\text{high}} = 75$ are represented on the right side. Not surprisingly, these plots show that the probability that the temperature is progressively conveyed to the desired range $(66,76)$ is smaller in the latter case, and this is because the heater is switched on/off when the temperature is closer to the boundaries than in the former case.

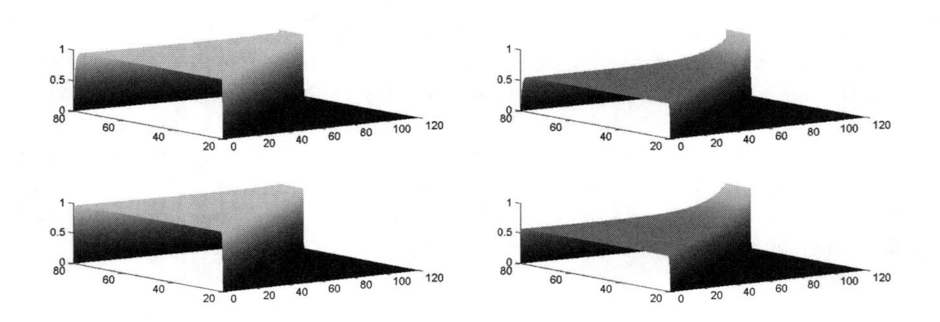

FIGURE 5.5: Maps of the probability that the temperature is progressively conveyed to the desired range $\mathscr{W}^\star = (66,76)$ in the prescribed time horizon $[t,120]$, as a function of the temperature value at time t, for $t \in [0,120]$. The top (bottom) row refers to the case when the heater is initially on (off.) The plots corresponding to the threshold temperatures $x_{\text{low}} = 69$ and $x_{\text{high}} = 73$ ($x_{\text{low}} = 67$ and $x_{\text{high}} = 75$) are reported on the left (right).

5.8 Conclusions

The reachability problem for a class of stochastic hybrid systems called switching diffusions is studied in this chapter. For such systems, the objective is to compute the probability that the system state will enter a certain subset of the state space within finite or infinite time horizon.

By discretizing the state space into a grid and constructing an interpolated Markov chain on the grid that approximates the stochastic hybrid system solution weakly as the discretization step size goes to zero, a numerical algorithm is proposed to compute an estimate of the reachability probability. Extensions of the proposed algorithm are presented with reference to other related problems such as the probabilistic safety and regulation problems.

Simulation results obtained by applying the developed algorithms to some application examples show that the method is effective in finding an approximate solution to the reachability problem. In these examples the computational cost is still reasonable because of the low-dimensionality of the considered stochastic hybrid system. Reachability computations in fact become more intensive as the dimension of the continuous state space grows. This, however, is a well-known problem also in the context of deterministic hybrid systems.

Further work is needed to extend the weak approximation result to a more general class of stochastic hybrid systems as those described in [4] and [5], and in Chapter 2 of the present volume, the main issue being that of coping with jumps in the hybrid state due to boundary hitting. Theoretically, such "forced transitions" can be modeled in the current framework by making the transition rates $\lambda_{qq'}(x)$, from the discrete state q to the discrete state q', $q' \neq q$, go to infinity as the continuous state component x approaches the switching boundary. However, this procedure may incur some intricacy for the proof of convergence results in this chapter. From an algorithmic viewpoint, we are investigating different possibilities for constructing the Markov chain approximation and implementing the reachability algorithm, so as to exploit the structure of the sparse Markov chain transition probability matrix.

Acknowledgments. The authors would like to thank John Lygeros and Henk Blom for their valuable comments on this chapter. The work of the first author was partially supported by Ministero dell'Università e della Ricerca (MIUR) under the project "New techniques for the identification and adaptive control of industrial systems." The work of the second author was supported by the Purdue Research Foundation.

References

[1] S. Amin, A. Abate, M. Prandini, J. Lygeros, and S. Sastry. Reachability analysis for controlled discrete time stochastic hybrid systems. In J. Hespanha and A. Tiwari, editors, *Hybrid Systems: Computation and Control*, volume 3927 of *Lecture Notes in Computer Science*, pages 49–63. Springer Verlag, Berlin, 2006.

[2] K. Amonlirdviman, N. A. Khare, D. R. P. Tree, W.-S. Chen, J. D. Axelrod, and C. J. Tomlin. Mathematical modeling of planar cell polarity to understand domineering nonautonomy. *Science*, 307(5708):423–426, Jan. 2005.

[3] A. Balluchi, L. Benvenuti, M. D. D. Benedetto, G. M. Miconi, U. Pozzi,

T. Villa, H. Wong-Toi, and A. L. Sangiovanni-Vincentelli. Maximal safe set computation for idle speed control of an automotive engine. In N. Lynch and B. H. Krogh, editors, *Hybrid Systems: Computation and Control*, volume 1790 of *Lecture Notes in Computer Science*, pages 32–44. Springer Verlag, Berlin, 2000.

[4] H.A.P. Blom. Stochastic hybrid processes with hybrid jumps. In *IFAC Conference Analysis and Design of Hybrid Systems*, Saint-Malo, Brittany, France, June 2003.

[5] M.L. Bujorianu. Extended stochastic hybrid systems and their reachability problem. In R. Alur and G. Pappas, editors, *Hybrid Systems: Computation and Control*, volume 2993 of *Lecture Notes in Computer Science*, pages 234–249. Springer Verlag, Berlin, 2004.

[6] R. E. Cole, C. Richard, S. Kim, and D. Bailey. An assessment of the 60 km rapid update cycle (RUC) with near real-time aircraft reports. Technical Report NASA/A-1, MIT Lincoln Laboratory, July 1998.

[7] R. Cont and P. Tankov. *Financial modelling with Jump Processes*. Chapman & Hall/CRC Financial Mathematics Series. Chapman & Hall/CRC, Boca Raton, FL, 2004.

[8] M. Egerstedt. Behavior based robotics using hybrid automata. In N. Lynch and B. H. Krogh, editors, *Hybrid Systems: Computation and Control*, volume 1790 of *Lecture Notes in Computer Science*, pages 103–116. Springer Verlag, Berlin, 2000.

[9] M.K. Ghosh, A. Arapostathis, and S.I. Marcus. An optimal control problem arising in flexible manufacturing systems. In *IEEE Conference on Decision and Control*, Dec. 1991.

[10] M.K. Ghosh, A. Arapostathis, and S.I. Marcus. Optimal control of switching diffusions with application to flexible manufacturing systems. *SIAM J. Control Optim.*, 31:1183–1204, 1993.

[11] M.K. Ghosh, A. Arapostathis, and S.I. Marcus. Ergodic control of switching diffusions. *SIAM J. Control Optim.*, 35(6):1952–1988, 1997.

[12] M.K. Ghosh and A. Bagchi. Modeling stochastic hybrid systems. In J. Cagnol and J.P. Zolesio, editors, *System Modeling and Optimization*, pages 269–280. Kluwer Academic Publishers, Boston, MA, 2005.

[13] E. Gobet. Weak approximation of killed diffusion using Euler schemes. *Stochastic Processes and their Applications*, 87:167–197, 2000.

[14] J. Hespanha. Polynomial stochastic hybrid systems. In M. Morari, L. Thiele, and F. Rossi, editors, *Hybrid Systems: Computation and Control*, volume 3414 of *Lecture Notes in Computer Science*, pages 322–338. Springer Verlag, Berlin, 2005.

[15] J. Hu and M. Prandini. Aircraft conflict detection: A method for computing the probability of conflict based on Markov chain approximation. In *European Control Conference*, Cambridge, UK, September 2003.

[16] J. Hu, M. Prandini, and S. Sastry. Aircraft conflict prediction in presence of a spatially correlated wind field. *IEEE Trans. on Intelligent Transportation Systems*, 6(3):326–340, 2005.

[17] J. Krystul and A. Bagchi. Approximation of first passage times of switching diffusion. In *Intern. Symposium on Mathematical Theory of Networks and Systems*, Leuven, Belgium, July 2004.

[18] H.J. Kushner and P.G. Dupuis. *Numerical Methods for Stochastic Control Problems in Continuous Time*. Springer-Verlag, New York, 2001.

[19] R. Malhame and C-Y Chong. Electric load model synthesis by diffusion approximation of a high-order hybrid-state stochastic system. *IEEE Trans. on Automatic Control*, 30(9):854–860, 1985.

[20] H. H. McAdams and A. Arkin. Stochastic mechanisms in gene expression. *Proc. Natl. Acad. Sci.*, 94:814–819, 1997.

[21] G. Pola, M. L. Bujorianu, J. Lygeros, and M.D. Di Benedetto. Stochastic hybrid models: An overview. In *IFAC Conference on Analysis and Design of Hybrid Systems*, Saint-Malo, France, 2003.

[22] M. Prandini and J. Hu. A stochastic approximation method for reachability computations. In H.A.P. Blom and J. Lygeros, editors, *Stochastic hybrid systems: theory and safety applications*, volume 337 of *Lecture Notes in Control and Informations Sciences*, pages 107–139. Springer, Berlin, 2006.

[23] C. Tomlin, G.J. Pappas, and S. Sastry. Conflict resolution for air traffic management: A study in multi-agent hybrid systems. *IEEE Trans. on Automatic Control*, 43:509–521, 1998.

[24] P. Varaiya. Smart cars on smart roads: problems of control. *IEEE Trans. on Automatic Control*, 38(2):195–207, 1993.

Chapter 6

Stochastic Flow Systems: Modeling and Sensitivity Analysis

Christos G. Cassandras
Boston University

6.1 Introduction 139
6.2 Modeling Stochastic Flow Systems 142
6.3 Sample Paths of Stochastic Flow Systems 146
6.4 Optimization Problems in Stochastic Flow Systems 148
6.5 Infinitesimal Perturbation Analysis (IPA) 150
6.6 Conclusions 164
References 165

6.1 Introduction

In this chapter, we consider a class of stochastic hybrid systems referred to as *stochastic flow systems* (or *stochastic fluid systems*). The dynamics of a basic flow (or fluid) system are given by

$$\dot{x}(t) = \alpha(t) - \beta(t)$$

where $x(t)$ describes the state, $\alpha(t)$ is the incoming flow, and $\beta(t)$ is the outgoing flow. Thus, $x(t)$ represents the content of some "tank" which changes as incoming and outgoing flows vary over time. As such, the content can never be negative, so the dynamics above must be modified to reflect this fact. Similarly, the content may not exceed a given capacity $C < \infty$. Thus, we rewrite the dynamics as

$$\dot{x}(t) = \begin{cases} 0 & \text{if } x(t) = 0 \text{ and } \alpha(t) - \beta(t) \leq 0 \\ 0 & \text{if } x(t) = C \text{ and } \alpha(t) - \beta(t) \geq 0 \\ \alpha(t) - \beta(t) & \text{otherwise.} \end{cases} \tag{6.1}$$

Such systems are particularly interesting when they consist of interconnected components forming a flow network. The output of a node in such a network can become the input of one or more other nodes. Moreover, controls may be applied (e.g., valves) to regulate the amount of flow allowed in/out of various nodes so as to achieve desired specifications. The hybrid nature of such systems is seen in (6.1) where the events "$x(t)$ reaches/leaves 0" or "$x(t)$ reaches/leaves C" cause a switch in the operating mode of the system. Similar switches may occur as a result of *controllable* events such as "shut down the outgoing flow" or *uncontrollable* ones such as "incoming flow changes from one constant value to another." The dynamics become stochastic when $\alpha(t)$ and $\beta(t)$ are random processes, normally assumed to be independent of each other.

Systems of this type naturally arise in settings such as the management of water resources or chemical processes. In addition, however, they are extremely useful as models of complex discrete event systems where the movement of discrete entities (such as parts in a manufacturing system, packets in a network, or vehicles in a transportation system) are abstracted into "flows." As an example, consider the Internet, where the natural modeling framework (similar to any packet-based communication network) is provided by queueing systems. However, on one hand the enormous traffic volume involved makes packet-by-packet analysis infeasible, and on the other, many of the standard assumptions under which queueing theory gives useful analytical results no longer apply to such a setting. For instance, traffic processes in the Internet rarely conform to Poisson characteristics; they are typically bursty and largely time-varying. In addition, the need to explicitly model buffer overflow phenomena defies tractable analytical derivations. Finally, various flow control mechanisms are feedback-based, an area where queueing theory has had limited success.

The main argument for fluid models as abstractions of inherently discrete event behavior lies on the observation that random phenomena may play different roles at different time scales. When the variations on the faster time scale have less impact than those on the slower time scale, the use of fluid models is justified. The efficiency of such a model rests on its ability to aggregate multiple events. By ignoring the micro-dynamics of each discrete entity and focusing on the change of the aggregated flow rate instead, a fluid model allows the aggregation of events associated with the movement of multiple entities within a time period of a constant flow rate into a single rate change event. In the context of communication networks, fluid models were introduced in [1] and later proposed in [2],[3] for the analysis of multiplexed data streams and network performance. They have also been shown to be useful in simulating various kinds of high speed networks [4],[5],[6],[7],[8]. A *Stochastic* Fluid Model (SFM), as introduced in [9], has the extra feature of treating flow rates as general stochastic processes, as opposed to deterministic quantities. We should also point out that fluid models have been used in other settings as well, such as manufacturing systems [10],[11],[12],[13].

If one is interested in studying the performance of a system modeled through a SFM, then the accuracy of the model depends on traffic conditions, the structure of the underlying system, and the nature of the performance metrics of interest. Moreover, some metrics may depend on higher-order statistics of the distributions of the underlying random variables involved, which a fluid model may not be able to accurately capture. Our main interest, however, is in using SFMs for the purpose of *control and optimization* rather than performance analysis. In this case, the value of an SFM lies in capturing only those features of the underlying "real" system that are needed to design an effective controller that can potentially optimize performance without actually estimating the corresponding optimal performance value with accuracy. Even if the exact solution to an optimization problem cannot be obtained by such "lower-resolution" models, one can still identify near-optimal points with useful robustness properties. Such observations have been made in several contexts (e.g., [14]), including results related to SFMs reported in [15] where a connection between the SFM and queueing-system-based solution is established for various optimization problems in queueing systems.

There is another attractive feature of SFMs that motivates their study. Specifically, sensitivity analysis of SFMs can be carried out and provide simple and efficient performance gradient estimators that, in turn, greatly facilitate system design as well as on-line control and optimization tasks. In the case of Discrete Event Systems (DES), *Infinitesimal Perturbation Analysis* (IPA) is a gradient estimation technique developed in the 1980s which allows the evaluation of state and performance metric sensitivities with respect to controllable parameters based only on observable sample path data (such as counting events and recording their occurrence times). The resulting gradient estimators are not only very simple, but they can also be shown to be unbiased in many cases of interest under well-defined conditions [16],[17],[18]. However, the scope of IPA does not include buffer overflow phenomena, multiclass networks, and systems with feedback control. The reason is that in such systems IPA gradient estimates are statistically biased, hence unreliable for control purposes. Enhanced estimators can be derived, but the appealing simplicity of the IPA approach is subsequently lost. In contrast, fluid models have been shown to circumvent these limitations, thus extending the application domain of IPA by providing unbiased estimators for many interesting types of stochastic flow systems.

Once sensitivity estimates of a performance measure of interest (e.g., packet loss rate in a network) with respect to control parameters of interest (e.g., thresholds on buffer contents for admission control) are obtained, they can be used in standard gradient-based algorithms to steer the system toward improved performance and ultimately optimize it. This approach has some very important advantages:

- The gradient estimation is done *on-line*, so that it can be implemented on the real system. As operating conditions change, it will aim at *continuously* seeking to optimize a generally time-varying performance metric (since such systems often fail to achieve steady state.)

- The gradient estimation process is driven by observed system data and does not require distributional knowledge of any underlying stochastic processes in the system; in other words, it is model-free. Since obtaining actual distributions for flow processes is extremely hard, this property implies that such a task becomes unnecessary.

- The estimators are shown to be unbiased when evaluated based on SFM sample paths (this property allows us to reliably use them with stochastic optimization algorithms e.g., [19].)

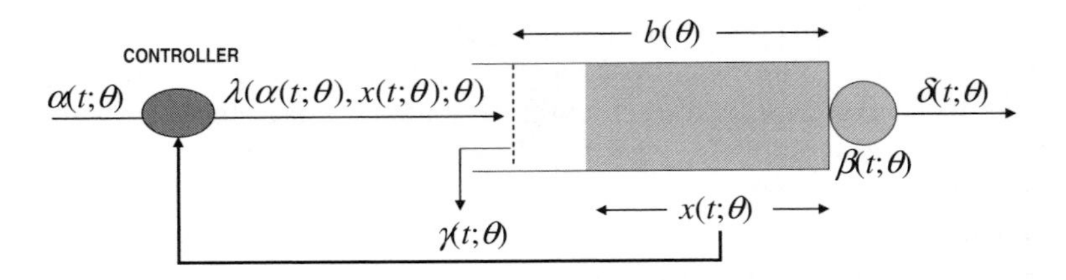

FIGURE 6.1: A stochastic flow system with feedback.

- It turns out that the estimators consist only of accumulators and timers and are generally easy to implement.

It is also worth pointing out that, even though the estimators are derived based on a SFM, their simplicity allows us to evaluate them along sample paths of the underlying DES. In other words, the functional form of an estimator is derived based on a SFM, but the data required to compute the estimates of interest are directly obtainable from the underlying system.

In this chapter, we will first describe models of stochastic flow systems, drawing mostly from applications to communication networks. We will then address the issue of sensitivity analysis for such systems: given a system parameter $\theta \in \mathbb{R}$ and some performance metric $J(\theta)$ which cannot be analytically evaluated, we are interested in estimating the derivative $dJ/d\theta$. The IPA approach is based on sample path analysis: we evaluate state trajectory perturbations and, hence, the derivative $dx/d\theta$ through which a sample performance sensitivity of the form

$d\mathcal{L}(\theta)/d\theta$ can be derived, where $J(\theta) = E[\mathcal{L}(\theta)]$. In Section 6.5.1 we will analyze a single node with a single class of fluid. Section 6.5.2 extends the analysis to multiple nodes in tandem. Throughout this chapter, we limit ourselves to IPA for systems with no feedback control. The case of IPA for systems where feedback mechanisms are incorporated will be treated in the next chapter. In addition, Chapter 8 considers more elaborate stochastic hybrid models for the analysis of the Transmission Control Protocol (TCP) widely used for congestion control in the Internet.

6.2 Modeling Stochastic Flow Systems

A basic stochastic flow system is shown in Figure 6.1. Associated with such a system are several random processes which are all defined on a common probability space $(\Omega, \mathscr{F}, P)$. The *arrival* flow process $\{\alpha(t;\theta)\}$ and the *service* flow process $\{\beta(t;\theta)\}$, along with a controllable parameter (generally vector) $\theta \in \Theta \subseteq \mathbb{R}^n$, are externally defined and referred to as the *defining processes* of the system. The *derived processes* are those determined by $\{\alpha(t;\theta)\}$ and $\{\beta(t;\theta)\}$, i.e., the state (content)

$\{x(t;\theta)\}$, the outflow $\{\delta(t;\theta)\}$, and the overflow $\{\gamma(t;\theta)\}$. The latter depends on $b(\theta)$, which defines the capacity of the system or a threshold beyond which the overflow process is triggered by choice (even if the capacity is greater than this value.) In addition, this system incorporates a controller which modifies the arrival flow based on state information, resulting in the *inflow* process $\lambda(\alpha(t;\theta),x(t;\theta);\theta)$.

A simple instance of this system arises when the two defining processes are independent of θ and we set $b(\theta)=\theta$, i.e., the only controllable parameter is a threshold (or the actual buffer capacity.) Moreover, let $\lambda(\alpha(t;\theta),x(t;\theta);\theta)=\alpha(t)$, i.e., no control is applied. In this case, the state dynamics are

$$\frac{dx(t;\theta)}{dt^+}=\begin{cases}0 & \text{if } x(t;\theta)=0 \text{ and } \alpha(t)-\beta(t)\leq 0\\ 0 & \text{if } x(t;\theta)=\theta \text{ and } \alpha(t)-\beta(t)\geq 0\\ \alpha(t)-\beta(t) & \text{otherwise}\end{cases} \tag{6.2}$$

and we can see that in this hybrid system there are two modes: whenever the state reaches the values 0 or θ, the system operates with $\frac{dx(t;\theta)}{dt^+}=0$, otherwise it operates with $\frac{dx(t;\theta)}{dt^+}=\alpha(t)-\beta(t)$. We use the explicit derivative notation $\frac{dx(t;\theta)}{dt^+}$ in order to differentiate it from the state sensitivity with respect to θ, $\frac{dx(t;\theta)}{d\theta}$, which we will also use in the sequel.

Exogenous and endogenous events. In order to specify the way in which transitions from one mode to another can occur in (6.2), we define *exogenous* and *endogenous* events. An exogenous event refers to any change in the defining processes that causes the sign of $\alpha(t)-\beta(t)$ to change either in a continuous fashion or due to a discontinuity in $\alpha(t)$ or $\beta(t)$. An endogenous event occurs when the state reaches either one of the two critical values 0 and θ. For more complex systems, the precise definition of exogenous and endogenous events may be adjusted. However, the former always refers to a change in the defining processes, whereas the latter refers to points where the state *enters* a certain region of the state space. In view of this discussion, the system described by (6.2) can also be modeled through the simple hybrid automaton of Figure 6.2 (where function arguments are omitted for simplicity.) We use the notation "$x\downarrow a$" and "$x\uparrow a$" to indicate an event that causes $x(t;\theta)$ to reach a value a from above or from below respectively and observe that these are both endogenous. We also define σ^+ to be the event "$\alpha(t)-\beta(t)$ switches sign from negative (or zero) to strictly positive," which is exogenous and results from a change in one or both of the defining processes. Similarly, σ^- is the event "$\alpha(t)-\beta(t)$ switches sign from positive to negative (or zero)." In Figure 6.2, the mode with $\frac{dx(t;\theta)}{dt^+}=0$ corresponds to either $x(t;\theta)=0$ or $x(t;\theta)=\theta$. A transition from that mode is the result of an event σ^+ in the former case or σ^- in the latter.

Introducing state feedback generally changes the hybrid automaton model to include additional modes. For example, consider a form of multiplicative feedback such as the one used in [20]:

$$\lambda(\alpha(t),x(t);\theta,\phi,c)=\begin{cases}c\alpha(t) & \text{if } \phi<x(t)\leq\theta\\ \alpha(t) & \text{if } 0\leq x(t)<\phi\end{cases}$$

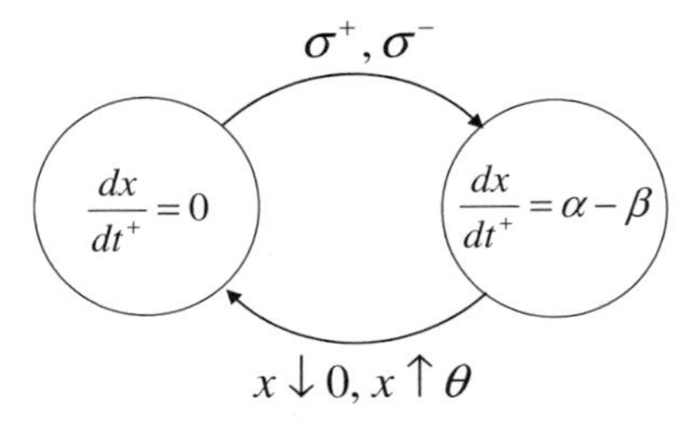

FIGURE 6.2: Hybrid automaton for a simple stochastic flow system. The event "$x \downarrow 0$" means "state reaches 0 from above" and "$x \uparrow \theta$" means "state reaches θ from below." The events σ^+, σ^- denote "$\alpha(t) - \beta(t)$ switches sign from negative (or zero) to strictly positive" and "$\alpha(t) - \beta(t)$ switches sign from positive to negative (or zero)."

where $0 < c \leq 1$ is a gain parameter and and $\phi < \theta$ is a threshold parameter. The controller above is unspecified for $x(t) = \phi$ because chattering may arise and some thought is required in order to avoid it. In particular, consider three possible cases that may arise when $x(t) = \phi$: (i) $\beta(t) < c\alpha(t)$. Then, the state at t^+ becomes $x(t^+) > \phi$; (ii) $\alpha(t) < \beta(t)$. Then, that state at t^+ becomes $x(t^+) < \phi$; and (iii) $c\alpha(\tau) \leq \beta(\tau) \leq \alpha(\tau)$ for all $\tau \in [t, t+\varepsilon)$ for some $\varepsilon > 0$. Assuming strict inequalities apply, there are two further cases to consider: (a) If we set $\lambda(\tau) = c\alpha(\tau)$, it follows that $\frac{dx}{dt}\big|_{t=\tau^+} = c\alpha(\tau) - \beta(\tau) < 0$, which implies that the state immediately starts decreasing. Therefore $x(\tau^+) < \phi$ and the actual inflow becomes $\lambda(\tau^+) = \alpha(\tau^+)$. Thus, $\frac{dx}{dt}\big|_{t=\tau^+} = \alpha(\tau^+) - \beta(\tau^+) > 0$ and the state starts increasing again. This process repeats, resulting in a chattering behavior. (b) If, on the other hand, we set $\lambda(\tau) = \alpha(\tau)$, it follows that $\frac{dx}{dt}\big|_{t=\tau^+} = \alpha(\tau^+) - \beta(\tau^+) > 0$. Then, upon crossing ϕ, the actual inflow must switch to $c\alpha(\tau^+)$ which gives $c\alpha(\tau^+) - \beta(\tau^+) < 0$. This implies that the state immediately decreases below ϕ and a similar chattering phenomenon occurs. In order to prevent the chattering arising in case (iii), we set $\lambda(\tau) = \beta(\tau)$ so that $\frac{dx}{d\tau^+} = 0$ for all $\tau \geq t$, i.e., the state is maintained at ϕ (in the case where $c\alpha(\tau) = \beta(\tau)$ or $\alpha(\tau) = \beta(\tau)$, it is obvious that $\frac{dx}{d\tau^+} = 0$). We then complete the specification of the controller as follows:

$$\lambda(\alpha(t), x(t); \theta, \phi, c) = \begin{cases} \alpha(t) & \text{if } 0 < x < \phi \\ c\alpha(t) & \text{if } x(t) = \phi \text{ and } \beta(t) < c\alpha(t) \\ \beta(t) & \text{if } x(t) = \phi \text{ and } c\alpha(t) \leq \beta(t) \leq \alpha(t) \\ \alpha(t) & \text{if } x(t) = \phi \text{ and } \alpha(t) < \beta(t) \\ c\alpha(t) & \text{if } \phi < x \leq \theta. \end{cases}$$

The corresponding hybrid automaton model is shown in Figure 6.3 where σ^+ and σ^- are the same exogenous events as before and ρ^+ and ρ^- are defined as "$c\alpha(t) - \beta(t)$ switches sign from negative (or zero) to strictly positive" and "$c\alpha(t) - \beta(t)$ switches sign from positive to negative (or zero)" respectively. In addition, a term in brackets

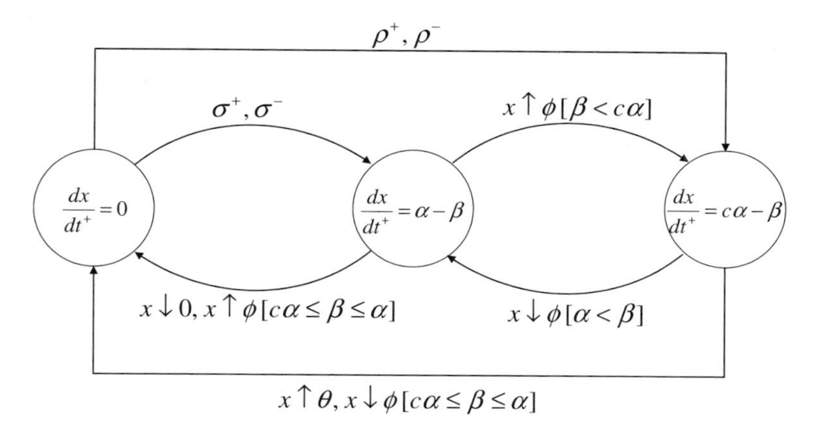

FIGURE 6.3: Hybrid automaton for a stochastic flow system with feedback.

denotes a condition which must hold at the time the accompanying event takes place. For example, "$x \uparrow \phi\ [\beta < c\alpha]$" means that when the state reaches the value ϕ from below, the condition $[\beta < c\alpha]$ must be true in order for the transition shown to occur. We can see that the conditions describing the transitions are substantially more complicated than those of Figure 6.2.

As a last example, let us consider a stochastic flow system consisting of two coupled nodes as shown in Figure 6.4 where we assume that the buffer capacities are both infinite. The dynamics of the two nodes are given by

$$\frac{dx_m(t;\theta)}{dt^+} = \begin{cases} 0 & \text{if } x_m(t;\theta) = 0 \text{ and } \alpha_m(t) - \beta_m(t) \leq 0 \\ \alpha_m(t) - \beta_m(t) & \text{otherwise} \end{cases}$$

where $m = 1,2$ and

$$\alpha_2(\theta;t) \equiv \begin{cases} \beta_1(t), & \text{if } x_1(\theta;t) > 0 \\ \alpha_1(t;\theta), & \text{if } x_1(\theta;t) = 0. \end{cases}$$

The corresponding hybrid automaton model is shown in Figure 6.5 where we have defined the following exogenous events: σ_1^+ means "$\alpha_1(t) - \beta_1(t)$ switches sign from negative (or zero) to strictly positive," σ_2^+ means "$\beta_1(t) - \beta_1(t)$ switches sign from negative (or zero) to strictly positive," and ρ^+ means "$\alpha_1(t) - \beta_2(t)$ switches sign from negative (or zero) to strictly positive."

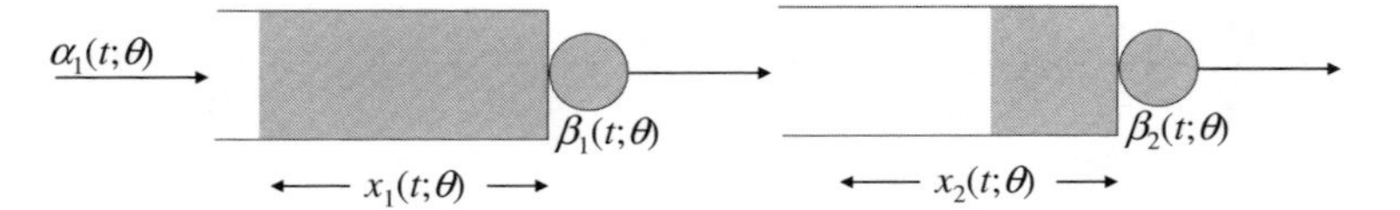

FIGURE 6.4: A two-node stochastic flow system.

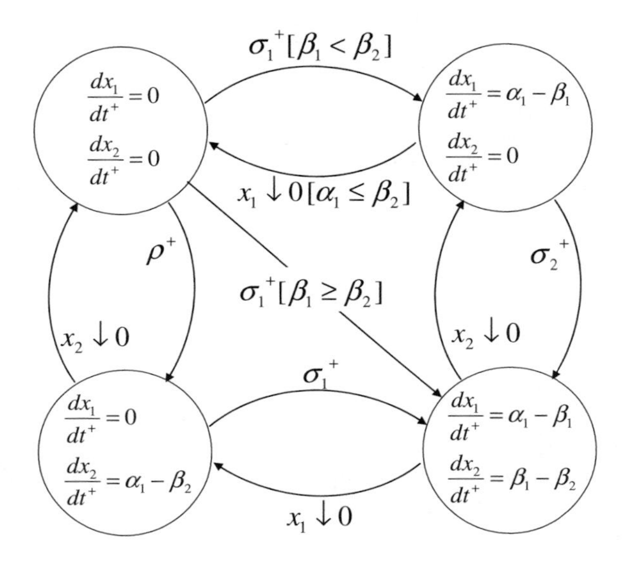

FIGURE 6.5: Hybrid automaton for a two-node stochastic flow system.

6.3 Sample Paths of Stochastic Flow Systems

A sample path of a stochastic flow system is characterized by a particular structure reflecting the mode switches in the dynamics, e.g., in (6.2). Let us consider one of the simplest possible versions of the system in Figure 6.1, i.e., the two defining processes are independent of θ, $b(\theta)=\theta$, and $\lambda(\alpha(t;\theta),x(t;\theta);\theta)=\alpha(t)$ (no control is applied). Figure 6.6 shows a typical sample path in terms of the state trajectory $x(t;\theta)$. The sequence $\{v_i : i=1,2,\ldots\}$ denotes the occurrence times of all mode switching events. For example, at time v_{i-2} an endogenous event "$x\uparrow\theta$" occurs, while at time v_{i+3} an exogenous event σ^+ occurs, i.e., $\alpha(t)-\beta(t)$ switches sign from negative (or zero) to strictly positive.

Boundary and Non-Boundary Periods. A *Boundary Period* (BP) is a maximal interval where the state $x(t;\theta)$ is constant and equal to one of the critical values in

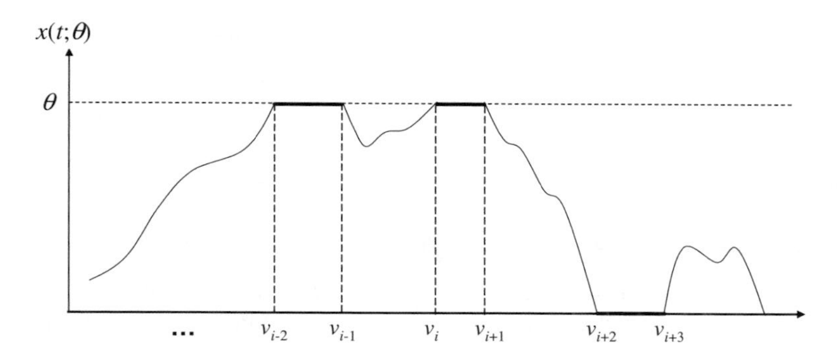

FIGURE 6.6: A sample path of a simple stochastic flow system.

our model, i.e., 0 or θ. A *Non-Boundary Period* (NBP) is a supremal interval such that $x(t;\theta)$ is not equal to a critical value. Thus, any sample path can be partioned into BPs and NBPs. In Figure 6.6, $[v_{i-2}, v_{i-1}]$, $[v_i, v_{i+1}]$, and $[v_{i+2}, v_{i+3}]$ are BPs, while (v_{i-1}, v_i) and (v_{i+1}, v_{i+2}) are NBPs. If a sample path is defined over a finite interval $[0,T]$, let N_B and $N_{\bar{B}}$ denote the random number of BPs and NBPs respectively observed over $[0,T]$.

Resetting Cycles. A NBP and its ensuing BP define a *resetting cycle*. The term is motivated by the fact that in carrying out IPA we find that all event time derivatives with respect to the parameter θ evaluated over such an interval are independent of all past history. However, we caution the reader that a *resetting cycle* should not be confused with a *regenerating cycle* often used in random process theory, because the evolution of the stochastic process $\{x(t;\theta)\}$ itself is generally not independent of its past history in a resetting cycle. The kth resetting cycle is denoted by $\mathscr{C}_k = (v_{k,0}, v_{k,R_k}]$, where $v_{k,j}$ corresponds to some event time v_i defined above by re-indexing to identify events in this resetting cycle. Thus, $\mathscr{C}_k$ includes $R_k + 1$ events. In Figure 6.6, the intervals $(v_{i-1}, v_{i+1}]$ and $(v_{i+1}, v_{i+3}]$ correspond to resetting cycles each of which includes two events. For a sample path defined over $[0,T]$, let $N_{\mathscr{C}}$ denote the random number of resetting cycles contained in $[0,T]$.

Empty and Non-Empty Periods. An *empty period* is an interval over which $x(t;\theta) = 0$, i.e., the buffer in Figure 6.1 is empty. Any other interval is referred to as non-empty. Further, a BP $[v_i, v_{i+1}]$ such that $x(t;\theta) = \theta$ for all $t \in [v_i, v_{i+1}]$ is referred to as a *full period*. Note that it is possible for $x(t;\theta) = 0$ with some positive outgoing flow present, as long as $\alpha(t) > 0$ and $\alpha(t) \leq \beta(t)$. This is in contrast to a queueing system where an empty period is equivalent to an interval over which the server is idling (i.e., it is not busy). In Figure 6.6, the interval (v_{i-3}, v_{i+2}) corresponds to a non-empty period, while the interval $[v_{i+2}, v_{i+3}]$ corresponds to an empty period. The kth non-empty period is denoted by $\bar{\mathscr{E}}_k = (v_{k,0}, v_{k,S_k})$, where $v_{k,j}$ correspond to some event time v_i by re-indexing and $\bar{\mathscr{E}}_k$ includes $S_k + 1$ events. For a sample path defined over $[0,T]$, let $N_{\mathscr{E}}$ and $N_{\bar{\mathscr{E}}}$ denote the random number of empty and non-empty periods respectively contained in $[0,T]$.

6.4 Optimization Problems in Stochastic Flow Systems

A large class of optimization problems is defined by viewing θ as a controllable parameter (scalar or vector) and seeking to optimize cost functions of the form

$$J(\theta;x(0),T) = E\left[\mathscr{L}(\theta;x(0),T)\right]$$

where $\mathscr{L}(\theta;x(0),T)$ is a sample function of interest evaluated in the interval $[0,T]$ with initial conditions $x(0)$. Note that $x(t;\theta)$ is generally a vector representing buffer contents over a network of interconnected nodes. Typical cost functions of interest in stochastic flow systems are the loss volume (due to overflow processes), the loss probability, the average workload (i.e., the buffer contents), and the system throughput. In addition, delay metrics can be incorporated through fluid versions of Little's law (see [21].) In this chapter, we shall limit ourselves to the loss volume, $L(\theta;x(0),T)$, and the average workload, $Q(\theta;x(0),T)$, which will be explicitly defined in the sequel.

Given that we do not wish to impose any limitations on the defining processes $\{\alpha(t;\theta)\}$ and $\{\beta(t;\theta)\}$ (other than mild technical conditions), it is infeasible to obtain closed-form expressions for $J(\theta;x(0),T)$. Therefore, we resort to iterative methods such as stochastic approximation algorithms (e.g., [19]) which are driven by estimates of the cost function gradient with respect to the parameter vector of interest. For the cost minimization problem above, we are interested in estimating $\partial J/\partial\theta$ based on sample path data, where a sample pathof the system may be directly observed or obtained through simulation. We then seek to obtain θ^* minimizing $J(\theta;x(0),T)$ through an iterative scheme of the form

$$\theta_{n+1} = \theta_n - \eta_n H_n(\theta_n;x(0),T,\omega_n), \quad n = 0,1,\ldots \tag{6.3}$$

where $H_n(\theta_n;x(0),T,\omega_n)$ is an estimate of $dJ/d\theta$ evaluated at $\theta = \theta_n$ and based on information obtained from a sample path denoted by ω_n. Furthermore, $\{\eta_n\}$ is an appropriate sequence of step sizes; in the case where $T \to \infty$ and stationarity conditions apply to the system, there are standard conditions that $\{\eta_n\}$ must satisfy [19] to guarantee convergence (w.p. 1) to θ^*. Obviously, the existence of a global minimum depends on the nature of $J(\theta;x(0),T)$. Although $J(\theta;x(0),T)$ is often a convex function of θ, in general we can only attain local minima. More critical, however, is the fact that stationarity may not be generally assumed and the optimization process above is largely intended to track a moving optimum that varies from one interval $[0,T]$ to the next. For our purposes, we shall consider T as a fixed time horizon and evaluate performance over $[0,T]$. To simplify the analysis that follows, we will assume that $x(0) = 0$ (in practice, it is possible to avoid this issue as explained, for example, in [22].) In addition, we will omit the initial condition, the observation interval T and the sample path ω_n unless it is necessary to stress such dependence. We will also assume that θ is a scalar parameter.

In order to execute an algorithm such as (6.3), we need to estimate $H_n(\theta_n)$, i.e., the derivative $dJ/d\theta$. The IPA approach is based on using the sample derivative $d\mathcal{L}/d\theta$ as an estimate of $dJ/d\theta$. The strength of this approach is that $d\mathcal{L}/d\theta$ can be obtained from observable sample path data alone and, usually, in a very simple manner that can be readily imlemented on line. Moreover, it is often the case that $d\mathcal{L}/d\theta$ is an *unbiased* estimate of $dJ/d\theta$, a property that allows us to use (6.3) in obtaining θ^*. An IPA estimator is unbiased if

$$\frac{dJ(\theta)}{d\theta} \equiv \frac{dE\left[\mathcal{L}(\theta)\right]}{d\theta} = E\left[\frac{d\mathcal{L}(\theta)}{d\theta}\right] \equiv E\left[\mathcal{L}'(\theta)\right].$$

The unbiasedness of an IPA derivative $\mathcal{L}'(\theta)$ has been shown to be ensured by the following two conditions (see [23], Lemma A2, p. 70):

C1. For every $\theta \in \Theta$ (where Θ is a closed bounded set), the sample derivative $\mathcal{L}'(\theta)$ exists w.p.1.

C2. W.p.1, the random function $\mathcal{L}(\theta)$ is Lipschitz continuous throughout Θ, and the (generally random) Lipschitz constant has a finite first moment.

Regarding **C1**, the existence of $\mathcal{L}'(\theta)$ for all problems considered in this chapter is guaranteed by the following assumption.

ASSUMPTION 6.1

a. W.p.1, all *defining processes*, i.e., flow rate functions $\alpha(t) \geq 0$ and $\beta(t) \geq 0$, are piecewise analytic in the interval $[0,T]$.

b. For every $\theta \in \Theta$, w.p. 1, two events cannot occur at exactly the same time, unless the occurrence of an event triggers the occurrence of the other at the same time.

c. W.p.1, no two processes $\{\alpha(t)\}$ or $\{\beta(t)\}$, have identical values during any open subinterval of $[0,T]$.

All three parts of Assumption 6.1 are mild technical conditions. Regarding parts b and c, we point out that even if they do not hold, it is possible to use one-sided derivatives and still carry out similar analysis, as in [9]. However, in order to keep the analysis and notation manageable we impose these conditions.

Consequently, establishing the unbiasedness of $\mathcal{L}'(\theta)$ reduces to verifying the Lipschitz continuity of the sample function $\mathcal{L}(\theta)$ with appropriate Lipschitz constants.

6.5 Infinitesimal Perturbation Analysis (IPA)

In the remainder of this chapter, we describe the IPA approach for stochastic flow systems. This entails the evaluation of derivatives of the form $\frac{dx(t;\theta)}{d\theta}$ which we shall henceforth express as $x'(t;\theta)$. These derivatives, in turn, critically depend on the event time derivatives $\frac{dv_i(\theta)}{d\theta}$ which we will express as $v_i'(\theta)$; we will normally omit the argument and simply write v_i' unless it is essential to indicate the dependence on θ. Given a specific sample function $\mathcal{L}(\theta)$, it is a relatively simple matter to evaluate $\mathcal{L}'(\theta)$ from $x'(t;\theta)$.

6.5.1 Single-Class Single-Node System

We consider the stochastic flow system of Figure 6.1 with the two defining processes independent of θ, $b(\theta) = \theta$, and $\lambda(\alpha(t;\theta), x(t;\theta);\theta) = \alpha(t)$, i.e., no control is applied. This system was originally studied in [9],[24]. A typical sample path was shown in Figure 6.6 and the dynamics of this system were given in (6.2).

The performance measures of interest are the average workload $Q(\theta)$ and the loss volume $L(\theta)$ defined as follows:

$$Q(\theta) = \int_0^T x(t;\theta)dt \tag{6.4}$$

$$L(\theta) = \int_0^T \gamma(t;\theta)dt. \tag{6.5}$$

Partitioning a sample path into resetting periods indexed by $k = 1, 2, \ldots$, let $\mathscr{C}_k = (v_{k,0}, v_{k,1}) \cup [v_{k,1}, v_{k,2}]$ where $(v_{k,0}, v_{k,1})$ is a NBP and $[v_{k,1}, v_{k,2}]$ is a BP. Note that in this case $R_k = 2$ for all $k = 1, 2, \ldots$ Let

$$q_k(\theta) = \int_{v_{k,0}}^{v_{k,2}} x(t;\theta)dt, \quad k = 1, \ldots, N_{\mathscr{C}} \tag{6.6}$$

where $N_{\mathscr{C}}$ is the number of resetting cycles in the interval $[0, T]$. Then, we can rewrite (6.4)–(6.5) as follows:

$$Q(\theta) = \sum_{k=1}^{N_{\mathscr{C}}} q_k(\theta) = \sum_{k=1}^{N_{\mathscr{C}}} \int_{v_{k,0}}^{v_{k,2}} x(t;\theta)dt \tag{6.7}$$

$$L(\theta) = \sum_{k=1}^{N_{\mathscr{C}}} \int_{v_{k,1}}^{v_{k,2}} \gamma(t;\theta)dt. \tag{6.8}$$

Observe that the events at $v_{k,0}$ and $v_{k,2}$ end BPs and are, therefore, both exogenous, i.e., $v_{k,0}$ and $v_{k,2}$ are independent of θ. Therefore, differentiating (6.7) with respect to θ gives

$$Q'(\theta) = \sum_{k=1}^{N_{\mathscr{C}}} q_k'(\theta) = \sum_{k=1}^{N_{\mathscr{C}}} \int_{v_{k,0}}^{v_{k,2}} x'(t;\theta)dt. \tag{6.9}$$

On the other hand, the event at $v_{k,1}$ is endogenous (either "$x \downarrow 0$" and "$x \uparrow \theta$"), so that $v'_{k,1} \neq 0$ in general. Thus, (6.8) gives

$$L'(\theta) = \sum_{k=1}^{N_{\mathscr{C}}} \left[-\gamma(v_{k,1};\theta)v'_{k,1} + \int_{v_{k,1}}^{v_{k,2}} \gamma'(t;\theta)dt \right].$$

Moreover, the loss rate during any BP is

$$\gamma(t;\theta) = \begin{cases} 0 & \text{if } x(t;\theta) = 0 \\ \alpha(t) - \beta(t) & \text{if } x(t;\theta) = \theta \end{cases}, \quad t \in [v_{k,1}, v_{k,2}),\ k = 1, \ldots, N_{\mathscr{C}}$$

so that $\gamma'(t;\theta) = 0$ in either case. It follows that

$$L'(\theta) = -\sum_{k=1}^{N_{\mathscr{C}}} \gamma(v_{k,1};\theta)v'_{k,1}. \tag{6.10}$$

We also make the following observation: During any NBP, the state starts with $x(v_{k,0}) = 0$ or θ and ends with $x(v_{k,1}) = 0$ or θ. Therefore,

$$\theta \mathbf{1}[x(v_{k,0}) = \theta] + \int_{v_{k,0}}^{v_{k,1}} [\alpha(t) - \beta(t)]dt = \theta \mathbf{1}[x(v_{k,1}) = \theta]$$

where $\mathbf{1}[\cdot]$ is the usual indicator function. Differentiating with respect to θ gives

$$[\alpha(v_{k,1}) - \beta(v_{k,1})]v'_{k,1} = \mathbf{1}[x(v_{k,1}) = \theta] - \mathbf{1}[x(v_{k,0}) = \theta]. \tag{6.11}$$

Observe that the right hand side above is 0 whenever the NBP starts and ends with $x(v_{k,0}) = x(v_{k,1}) = 0$ or $x(v_{k,0}) = x(v_{k,1}) = \theta$. Thus, recalling that $\alpha(v_{k,1}) - \beta(v_{k,1}) \neq 0$ by Assumption 6.1, $v'_{k,1} \neq 0$ only when a NBP is preceded by an empty period and followed by a full period or vice versa. In order to differentiate between different types of resetting cycles, we will use the following notation: EE denotes a cycle that starts just after an empty period and ends with an empty period; EF denotes a cycle that starts just after an empty period and ends with a full period; FE denotes a cycle that starts just after a full period and ends with an empty period; and FF denotes a cycle that starts just after a full period and ends with a full period.

With these observations in mind, we can now obtain explicit expressions for $Q'(\theta)$ and $L'(\theta)$ as shown in the next theorem.

THEOREM 6.1 *The sample derivatives $Q'(\theta)$ and $L'(\theta)$ with respect to θ are*

$$Q'(\theta) = \sum_{j=1}^{N_{\bar{\mathscr{E}}}} [v_{j,S_j} - v_{j,1}] \tag{6.12}$$

$$L'(\theta) = -N_{\mathscr{C}_{EF}} \tag{6.13}$$

where j counts the number of non-empty periods, $v_{j,1}$ is the time of the first overflow point in the jth non-empty period and v_{j,S_j} is the time when it ends (if there is no overflow in this period, then $v_{j,1} = v_{j,S_j}$).

PROOF For the kth resetting cycle $\mathscr{C}_k$, consider the following four possible cases:

Case 1 (EE): The cycle starts just after an empty period and ends with an empty period. In this case,

$$x(t;\theta) = \begin{cases} \int_{v_{k,0}}^{t} (\alpha(t) - \beta(t))\,dt & \text{if } t \in (v_{k,0}, v_{k,1}) \\ 0 & \text{if } t \in [v_{k,1}, v_{k,2}] \end{cases}$$

and $x'(t;\theta) = 0$ for all $t \in \mathscr{C}_k$. Moreover, $\gamma(t;\theta) = 0$ for all $t \in \mathscr{C}_k$, since no overflow event is included in such a cycle; hence, $\gamma(v_{k,1};\theta) = 0$. Recalling (6.6) and the fact that $v_{k,0}$, $v_{k,2}$ are independent of θ, it follows that

$$q_k'(\theta) = 0 \quad \text{and} \quad \gamma(v_{k,1};\theta)v_{k,1}' = 0. \tag{6.14}$$

Case 2 (EF): The cycle starts just after an empty period and ends with a full period. In this case,

$$x(t;\theta) = \begin{cases} \int_{v_{k,0}}^{t} (\alpha(t) - \beta(t))\,dt & \text{if } t \in (v_{k,0}, v_{k,1}) \\ \theta & \text{if } t \in [v_{k,1}, v_{k,2}]. \end{cases}$$

Differentiating with respect to θ we get

$$x'(t;\theta) = \begin{cases} 0 & \text{if } t \in (v_{k,0}, v_{k,1}) \\ 1 & \text{if } t \in [v_{k,1}, v_{k,2}]. \end{cases}$$

Using (6.9), we get $q_k'(\theta) = v_{k,2} - v_{k,1}$. Moreover, since $\gamma(v_{k,1},\theta) = \alpha(v_{k,1}) - \beta(v_{k,1})$, (6.11) gives $(\alpha(v_{k,1}) - \beta(v_{k,1}))v_{k,1}' = 1$. Therefore,

$$q_k'(\theta) = v_{k,2} - v_{k,1} \quad \text{and} \quad \gamma(v_{k,1};\theta)v_{k,1}' = 1. \tag{6.15}$$

Case 3 (FF): The cycle starts starts just after a full period and ends with a full period. In this case,

$$x(t;\theta) = \begin{cases} \theta + \int_{v_{k,0}}^{t} (\alpha(t) - \beta(t))\,dt & \text{if } t \in (v_{k,0}, v_{k,1}) \\ \theta & \text{if } t \in [v_{k,1}, v_{k,2}]. \end{cases}$$

Therefore, $x'(t;\theta) = 1$ for all $t \in \mathscr{C}_k$. As a result, by (6.9), $q_k'(\theta) = v_{k,2} - v_{k,0}$. In addition, from (6.11) we get that $v_{k,1}' = 0$. Thus,

$$q_k'(\theta) = v_{k,2} - v_{k,0} \quad \text{and} \quad \gamma(v_{k,1};\theta)v_{k,1}' = 0. \tag{6.16}$$

Case 4 (FE): The cycle starts just after a full period and ends with an empty period. In this case,

$$x(t;\theta) = \begin{cases} \theta + \int_{v_{k,0}}^{t} (\alpha(t) - \beta(t))\,dt & \text{if } t \in (v_{k,0}, v_{k,1}) \\ 0 & \text{if } t \in [v_{k,1}, v_{k,2}]. \end{cases}$$

Therefore,

$$x'(t;\theta) = \begin{cases} 1 & \text{if } t \in (v_{k,0}, v_{k,1}) \\ 0 & \text{if } t \in [v_{k,1}, v_{k,2}]. \end{cases}$$

It follows from (6.9) that $q_k'(\theta) = v_{k,1} - v_{k,0}$. In addition, $x(v_{k,1};\theta) = 0$, therefore $\gamma(v_{k,1};\theta) = 0$. Thus,

$$q_k'(\theta) = v_{k,1} - v_{k,0} \quad \text{and} \quad \gamma(v_{k,1};\theta)v_{k,1}' = 0. \tag{6.17}$$

Combining the results above for $q_k'(\theta)$ and using (6.9) we get

$$Q'(\theta) = \sum_{k=1}^{N_{\mathscr{C}}} \left[(v_{k,2} - v_{k,1})\mathbf{1}_{EF} + (v_{k,2} - v_{k,0})\mathbf{1}_{FF} + (v_{k,1} - v_{k,0})\mathbf{1}_{FE} \right]$$

where $\mathbf{1}_{EF}$, $\mathbf{1}_{FF}$, and $\mathbf{1}_{FE}$ are indicator functions associated with resetting cycles of type EF, FF. and FE respectively. Observe that the union of a non-empty period and the empty period following it defines either (i) a single EE cycle or (ii) an EF cycle followed by m FF cycles, $m = 0, 1, 2, \cdots$, and ending with an FE cycle. Therefore, the sum above is identical to $\sum_{j=1}^{N_{\bar{\mathscr{E}}}} [v_{j,S_j} - v_{j,1}]$ and (6.12) is obtained.

Finally, combining the results above for $\gamma(v_{k,1};\theta)v_{k,1}'$ we get a nonzero contribution in (6.10) from the EF case only. Observe that the number of non-empty periods with at least some overflow is equal to the number of EF cycles, $N_{\mathscr{C}_{EF}}$, thus yielding (6.13) and completing the proof. ■

It is important to observe that the two IPA estimators above are *model-free*. That is, not only do they not depend on any distributional information characterizing the defining processes, but they are also independent of all model parameters. Moreover, the implementation of the estimators is extremely simple. In the case of $L'(\theta)$, it suffices to count the number of nonempty periods within which an event "$x \uparrow \theta$" occurs (i.e., some overflow is observed). In the case of $Q'(\theta)$, we accumulate time intervals defined by the first "$x \uparrow \theta$" event (if one is observed) in a nonempty period and the end of this period. Note that if the stochastic flow system we have analyzed is a SFM of an underlying DES, then overflow events can be directly observed on a sample path of the actual DES itself. This implies that the estimators can be implemented on line and evaluated using real time data; the SFM is only implicitly constructed to generate the IPA estimators.

Unbiasedness. As mentioned earlier, the unbiasedness of the IPA derivatives is established under condition **C1** and **C2** given in Section 6.4. Condition **C2** rests on the following two lemmas which we will state without proof (their proof, which is tedious but straightforward, may be found in [9] with more general versions in [24].) Let $\Delta\theta > 0$ be a perturbation in the parameter θ and let $\Delta x(t;\theta,\Delta\theta)$ be the resulting state perturbation. The first lemma asserts that a perturbation $\Delta x(t;\theta,\Delta\theta)$ is bounded by the change in buffer capacity $\Delta\theta$, which is to be expected.

LEMMA 6.1 $0 \leq \Delta x(t;\theta,\Delta\theta) \leq \Delta\theta$ *for all* $t \in [0,T]$.

The second lemma considers the change in loss volume, $\Delta L(t;\theta,\Delta\theta)$, resulting from a perturbation $\Delta\theta > 0$ and asserts that its magnitude cannot exceed the change in buffer capacity $\Delta\theta$.

LEMMA 6.2 $-\Delta\theta \leq \Delta L(t;\theta,\Delta\theta) \leq 0$ *for all* $t \in [0,T]$.

We can then establish unbiasedness as follows.

THEOREM 6.2 *Let* $N(t)$ *be the random number of exogenous events in* $[0,T]$. *Under Assumption 6.1, 1. If* $E[N(T)] < \infty$, *then the IPA derivative* $L_T^{'}(\theta)$ *is an unbiased estimator of* $\frac{dE[L_T(\theta)]}{d\theta}$. *2. The IPA derivative* $Q_T^{'}(\theta)$ *is an unbiased estimator of* $\frac{dE[Q_T(\theta)]}{d\theta}$.

PROOF Under Assumption 6.1, **C1** holds for $L_T(\theta)$ and $Q_T(\theta)$. Therefore, we only need to establish **C2**. Let τ_i denote the occurrence of the ith exogenous event, so we can partition $[0,T]$ into intervals $[\tau_{i-1},\tau_i)$. Given a perturbation $\Delta\theta > 0$, we can then write

$$\Delta L_T(\theta) = \sum_{i=1}^{N(T)} \Delta L_i(\theta,\Delta\theta)$$

where $\Delta L_i(\theta,\Delta\theta)$ is the loss volume perturbation after an exogenous event takes place. By Lemma 6.2, $-\Delta\theta \leq \Delta L_i \leq 0$, so that

$$|\Delta L_T(\theta)| \leq N(T)\,|\Delta\theta|,$$

i.e., $L_T(\theta)$ is Lipschitz continuous with Lipschitz constant $N(T)$. Since $E[N(T)] < \infty$, this establishes unbiasedness.

Next, consider $Q_T(\theta)$ and fix θ and $\Delta\theta > 0$. Using Lemma 6.1 and recalling (6.4),

$$|\Delta Q_T(\theta)| = \left| \int_0^T \Delta x(t;\theta,\Delta\theta)dt \right| \leq T\,|\Delta\theta|,$$

that is, $Q_T(\theta)$ is Lipschitz continuous with constant T. This completes the proof. ■

We conclude this section by noting that when the parameter θ affects the system through one or both of the defining processes, proceeding along the same lines provides IPA estimators with similiar characteristics to those of (6.12) and (6.13) that can also be shown to be unbiased [24]. We also add that extensions to multiclass stochastic flow systems are possible [25]. In this case, there are multiple incoming flow processes, each with a different priority and associated with a different controllable threshold parameter. The IPA estimators provide sensitivities of the loss volume and workload metrics with respect to each of these thresholds.

6.5.2 Multi-node Tandem System

In this section, we consider a network setting where all nodes are in series and the parameter of interest is a buffer threshold. What complicates matters in this setting is the fact that a state perturbation at one node will generally propagate to other nodes. Therefore, the state derivatives $\frac{dx_m(t;\theta)}{dt^+}$, $m = 1,2,\ldots$ are coupled to each other. Understanding the form that this coupling can take is the key to deriving IPA estimators for performance metrics of interest in such a system.

Consider a stochastic flow system with M nodes in series indexed by $m = 1,\ldots,M$. The outflow of node m is the inflow to node $m+1$, and we assume there is no feedback in the system. Let $b_m > 0$ denote the buffer size of node m. At the first node, we assume that there is a threshold parameter θ limiting any incoming flow to $x_1(t;\theta) \leq \theta$. For notational simplicity, we will also write $b_1 = \theta$. Extending the notation used in the single-node case, the incoming flow at each node $m = 2,\ldots,M$ is denoted by $\alpha_m(t;\theta)$, to indicate the fact that it generally depends on θ, whereas $\alpha_1(t)$ is an external process independent of θ. The rate with which node $m = 1,\ldots,M$ outputs flow at time t is denoted by $\beta_m(t)$ and is independent of θ. The overflow rate is denoted by $\gamma_m(t;\theta)$. The state dynamics at node $m = 1,\ldots,M$, are given by

$$\frac{dx_m(t;\theta)}{dt^+} = \begin{cases} 0 & \text{if } x_m(t;\theta) = 0 \text{ and} \\ & \quad \alpha_m(t;\theta) - \beta_m(t) \leq 0, \\ 0 & \text{if } x_m(t;\theta) = b_m \text{ and} \\ & \quad \alpha_m(t;\theta) - \beta_m(t) \geq 0, \\ \alpha_m(t;\theta) - \beta_m(t) & \text{otherwise} \end{cases} \tag{6.18}$$

where, to maintain uniformity in the notation, it is understood that $\alpha_1(t;\theta) = \alpha_1(t)$. With this convention in mind, the outflow rate from node $m = 1,\ldots,M-1$ is the inflow rate to the downstream node $m+1$, so that for all $m = 2,\ldots,M$ we have

$$\alpha_m(t;\theta) = \begin{cases} \beta_{m-1}(t) & \text{if } x_{m-1}(t;\theta) > 0 \\ \alpha_{m-1}(t;\theta) & \text{if } x_{m-1}(\theta;t) = 0. \end{cases} \tag{6.19}$$

Finally, the overflow rate $\gamma_m(t;\theta)$ at node m due to a full buffer is defined by

$$\gamma_m(t;\theta) = \begin{cases} \alpha_m(t;\theta) - \beta_m(t) & \text{if } x_m(t;\theta) = b_m \text{ and} \\ & \quad \alpha_m(t;\theta) - \beta_m(t) \geq 0, \\ 0 & \text{otherwise.} \end{cases} \tag{6.20}$$

For convenience, we define

$$A_m(t;\theta) \equiv \alpha_m(t;\theta) - \beta_m(t) \tag{6.21}$$

and remind the reader that the defining processes in this system, $\{\alpha_1(t)\}$ and $\{\beta_m(t)\}$, $m = 1,\ldots,M$, are stochastic processes representing the random instantaneous rates of the inflows and of node processing rates.

Taking a closer look at (6.19), note that the value of $\alpha_m(t;\theta)$, $m > 1$, is given by either $\beta_{m-1}(t)$, which is independent of θ, or by $\alpha_{m-1}(t;\theta)$. In turn, the value of $\alpha_{m-1}(t;\theta)$ is given by either $\beta_{m-2}(t)$ or by $\alpha_{m-2}(t;\theta)$. Proceeding recursively, we see that the value of $\alpha_m(t;\theta)$ is ultimately given by one of the processes $\{\alpha_1(t)\}$ and $\{\beta_i(t),\, i = 1,\ldots,m\}$ which are all independent of θ. The way in which $\alpha_m(t;\theta)$ switches among them depends on θ through the states $x_i(t;\theta)$, $i = 1,\ldots,m-1$ and the points in time when this switching occurs defines *switchover points* which are crucial in our analysis.

For the purpose of our analysis, we define an *event* of node $m = 1,...,M$ to be one of the following:

(i) e_1 corresponds to a jump (discontinuity) in either $\alpha_m(t;\theta)$ or $\beta_m(t)$.

(ii) e_2 occurs when $A_m(t;\theta)$ becomes 0 with no discontinuity in $A_m(t;\theta)$ at t.

(iii) e_3 occurs when the state $x_m(t;\theta)$ reaches the value b_m or θ (i.e., the buffer becomes full or empty.)

Similar to the single-node case, a typical sample path of the process $\{x_m(t;\theta)\}$ can be decomposed into Boundary Periods (BPs) and Non-Boundary Periods (NBPs). A BP is further classified as either an Empty Period (EP) during which $x_m(t;\theta) = 0$ or as a Full Period (FP) during which $x_m(t;\theta) = b_m$. Since the function $x_m(t;\theta)$ is generally continuous in t for a fixed θ, we will consider EPs and FPs to be closed intervals and NBPs to be open intervals in the relative topology induced by $[0,T]$. Let

$$B_{m,n} = [\tau_{m,n}(\theta), \sigma_{m,n}(\theta)]$$

denote the nth BP, $n = 1,\ldots,N_m$, where N_m is the total (random) number of BPs in $[0,T]$. Note that the start of $B_{m,n}$, $\tau_{m,n}(\theta)$, is an e_3 event of node m. For notational economy, we will omit θ in $\tau_{m,n}(\theta)$ and $\sigma_{m,n}(\theta)$ in what follows, but will keep in mind that $\tau_{m,n}$ and $\sigma_{m,n}$ are generally functions of θ. Next, observe that NBPs and BPs appear alternately throughout $[0,T]$ and let

$$\overline{B}_{m,n} = (\sigma_{m,n-1}, \tau_{m,n})$$

denote the NBP that precedes $B_{m,n}$ (thus, $\overline{B}_{m,n} \cup B_{m,n}$ is what we called a "resetting cycle" in studying the single-node case). For convenience, we shall set $\sigma_{m,0} = 0$ and $\sigma_{m,N_m} = T$.

Depending on the value of $x_m(t;\theta)$ at the starting and ending points of a NBP $\overline{B}_{m,n} = (\sigma_{m,n-1}, \tau_{m,n})$, we define four types of NBPs ("E" stands for "Empty" and "F" stands for "Full"):

(i) (E,E): $x_m(\sigma_{m,n-1};\theta) = 0$ and $x_m(\tau_{m,n};\theta) = 0$,

(ii) (E,F): $x_m(\sigma_{m,n-1};\theta) = 0$ and $x_m(\tau_{m,n};\theta) = b_m$,

(iii) (F,E): $x_m(\sigma_{m,n-1};\theta) = b_m$ and $x_m(\tau_{m,n};\theta) = 0$, and

(iv) (F,F): $x_m(\sigma_{m,n-1};\theta) = b_m$ and $x_m(\tau_{m,n};\theta) = b_m$.

Switchover points. The switchover points of $\alpha_m(t;\theta)$ for $m > 1$, as seen in (6.19), occur as follows:

(i) Just before an EP of node $m-1$ starts, we have $\alpha_m(t;\theta) = \beta_{m-1}(t)$. When the EP starts, the output of $m-1$ switches from $\beta_{m-1}(t)$ to $\alpha_{m-1}(t;\theta)$.

(ii) When the EP of node $m-1$ ends, the output of $m-1$ switches once again from $\alpha_{m-1}(t;\theta)$ to $\beta_{m-1}(t)$.

(iii) The third instance is less obvious. During an EP at node $m-1$, it is possible that an EP at node $m-2$ starts, in which case $\alpha_{m-1}(\theta;t)$ switches from $\beta_{m-2}(t)$ to $\alpha_{m-2}(\theta;t)$. When this happens, the output of $m-1$ switches from $\alpha_{m-1}(t)$ to $\alpha_{m-2}(t;\theta)$, therefore, $\alpha_m(t;\theta)=\alpha_{m-1}(t)=\alpha_{m-2}(t)$. Clearly, it is possible that a sequence of j such events occurs so that $\alpha_m(t;\theta)=\alpha_{m-1}(t)=\ldots=\alpha_{m-j}(t)$, where $j=1,\ldots,m-1$. In this case, all nodes $m-j,\ldots,m-1$ are empty and m inherits all switchovers experienced by these upstream nodes as each one starts an EP.

The following lemma asserts that switchover points of $\alpha_m(t;\theta)$ under case (ii) above are locally independent of θ (the proof may be found in [22]):

LEMMA 6.3 *Let σ_{m-1}, $m>1$, be a switchover point of $\alpha_m(t;\theta)$ with*

$$\alpha_m(\sigma_{m-1}^-;\theta)=\alpha_{m-1}(\sigma_{m-1}^-;\theta) \quad \text{and} \quad \alpha_m(\sigma_{m-1}^+;\theta)=\beta_{m-1}(\sigma_{m-1}^+). \tag{6.22}$$

Then, σ_{m-1} is locally independent of θ.

Thus, as in the single-node case, the end of an EP is independent of θ. Moreover, for $m>2$, during an EP of node $m-1$ we can see in (6.19) that $\alpha_m(t;\theta)=\alpha_{m-1}(t;\theta)$, which implies that if a switchover occurs at $\alpha_{m-1}(t;\theta)$, this switchover will be inherited by $\alpha_m(t;\theta)$, as well as the θ-dependence of it. This discussion motivates our definition of an *active* switchover point, which is generally a function of θ and is denoted by $s_{m,i}(\theta)$, $m>2$, $i=1,2,\ldots$:

DEFINITION 6.1 *A switchover point $s_{m,i}(\theta)$ of $\alpha_m(t;\theta)$ is termed* active *if:*

(i) $s_{m,i}(\theta)$ is the time when an EP at node $m-1$ starts; or

(ii) $s_{m,i}(\theta)$ is the time when $\alpha_{m-1}(t;\theta)$ experiences an active switchover within an EP of node $m-1$.

An active switchover point $s_{m,i}(\theta)$ at node m may belong to a BP $B_{m,n}$ or to a NBP $\overline{B}_{m,n}$. We define the following index sets that will help differentiating between different types of active switchover points depending on the type of interval they belong to:

$$\Psi_{m,n} \equiv \{i : s_{m,i} \in B_{m,n}\}, \tag{6.23}$$

$$\Psi^o_{m,n} \equiv \{i : s_{m,i} \in (\tau_{m,n}, \sigma_{m,n})\}, \tag{6.24}$$

$$\overline{\Psi}_{m,n} \equiv \{i : s_{m,i} \in \overline{B}_{m,n}\}. \tag{6.25}$$

Note that $B_{m,n}=[\tau_{m,n},\sigma_{m,n}]$, so we differentiate between open and closed intervals that define BPs in defining the sets $\Psi_{m,n}$ and $\Psi^o_{m,n}$. As we will see, of particular interest are active switchover points that coincide with the end of a FP, so we define the set of all BP indices that include such a point, Φ_m, as well as $\Gamma_m \subseteq \Phi_m$, a subset that includes those FPs that are followed by a NBP of type (F,E):

$$\Phi_m \equiv n : \sigma_{m,n} \text{ is an active switchover point, } n = 1,\ldots,N_m,$$
$$\Gamma_m \equiv \{n : n \in \Phi_m \text{ and } \overline{B}_{m,n+1} \text{ is of type } (F,E)\}.$$

Before proceeding, let us recall Assumption 6.1 and make some minor modifications to it to accommodate a multi-node model.

ASSUMPTION 6.2

a. W.p.1, the functions $\alpha_1(t)$, and $\beta_m(t)$, $m = 1,\ldots,M$ are piecewise analytic in the interval $[0,T]$.

b. For every $\theta \in \Theta$, w.p.1 no two events of a certain node m occur at the same time.

c. W.p.1, no two processes $\{\alpha_1(t)\}$, $\{\beta_m(t),\ m = 1,\ldots,M\}$ have identical values during any open subinterval of $[0,T]$.

Regarding part c, note that $\alpha_m(t;\theta)$, through (6.19), ultimately depends on one or more of the processes $\{\alpha_1(t)\}$, $\{\beta_i(t)\}$, $i = 1,\ldots,m$, therefore the requirement $A_m(t;\theta) \neq 0$ is reflected by the general statement under c.

Recall that a switchover point of $\alpha_m(t;\theta)$ is the time it switches among $\{\alpha_1(t)\}$ and $\{\beta_i(t)\}$, $i = 1,\ldots,m$. It is possible that a switchover may not cause a jump (discontinuity) in $\alpha_m(t;\theta)$. The following lemma (the proof may be found in [22]) is a consequence of Assumption 6.2 and shows that for an *active* switchover point, $\alpha_m(t;\theta)$ must experience a jump.

LEMMA 6.4 *If an active switchover point of $\alpha_m(t;\theta)$ occurs at $t = s_{m,i}$, then w.p. 1 it is an e_1 event of node m.*

We now proceed by determining the derivative $x_m^{'}(t;\theta)$ with respect to the controllable parameter θ and will show that it depends exclusively on the way that θ affects the active switchover points of $\alpha_m(t;\theta)$. We define the following two quantities for $m > 1$ that turn out to be crucial in our analysis:

$$\psi_{m,i} \equiv [\alpha_m(s_{m,i}^{+};\theta) - \alpha_m(s_{m,i}^{-};\theta)]s_{m,i}^{'}, \tag{6.26}$$

and, for $n \in \Phi_m$:

$$\phi_{m,n} \equiv [\alpha_m(\sigma_{m,n}^{+};\theta) - \beta_m(\sigma_{m,n})]\sigma_{m,n}^{'}. \tag{6.27}$$

Here, $s_{m,i}^{'}$ denotes the derivative of the active switchover time $s_{m,i}(\theta)$ with respect to θ and $\sigma_{m,n}^{'}$ denotes the derivative of a BP ending time $\sigma_{m,n}\ (\theta)$ with respect to θ. As the following lemma shows, $\psi_{m,i}$ and $\phi_{m,n}$ are crucial in evaluating the derivative $x_m^{'}(t;\theta)$.

LEMMA 6.5 *If $m = 1$, for $n = 1, ..., N_1$*

$$x_1'(t;\theta) = \begin{cases} 1 & \text{if } t \in B_{1,n} \cup \overline{B}_{1,n+1} \text{ and } x_1(\sigma_{1,n};\theta) = \theta \\ 0 & \text{otherwise.} \end{cases} \tag{6.28}$$

If $m > 1$, then for $n = 1, ..., N_m$

$$x_m'(t;\theta) = \begin{cases} 0 & \text{if } t \in B_{m,n} \\ -\sum_{k=1}^{K_{m,n}(t)} \psi_{m,k} - \mathbf{1}[n \in \Phi_m] \cdot \phi_{m,n} & \text{if } t \in \overline{B}_{m,n+1} \end{cases} \tag{6.29}$$

where $K_{m,n}(t)$ is the number of active switchover points in the interval $(\sigma_{m,n}, t) \subset \overline{B}_{m,n+1}$.

The proof of this result is given in [22]. It is easy to check that (6.28) is equivalent to our analysis of the single-node case, i.e., the effect of changing $\theta = b_1$ is to generate a state perturbation at node 1 when some FP ends at $t = \sigma_{m,n}$ which remains present for the ensuing NBP and is elminated when the next EP takes place. On the other hand, (6.29) shows the role of $\psi_{m,k}$ and $\phi_{m,n}$. The next two lemmas (whose proofs are also in [22]) provide the means to connect $x_m'(t;\theta)$ to $x_{m-1}'(t;\theta)$ and hence shed light into the way in which state perturbations propagate across nodes.

LEMMA 6.6 *For $m > 1$, let $s_{m,i}$ be an active switchover point of $\alpha_m(t;\theta)$. If it is the start of an EP at node $m-1$, then*

$$\psi_{m,i} = -x_{m-1}'(s_{m,i}^-;\theta). \tag{6.30}$$

Otherwise, if $s_{m,i}$ occurs during an EP of node $m-1$, then

$$\psi_{m,i} = \psi_{m-1,j} \tag{6.31}$$

for some j such that $s_{m,i} = s_{m-1,j}$.

Next, for $m > 1$, we define:

$$R_{m,n}(\theta) \equiv \frac{\alpha_m(\sigma_{m,n}^+;\theta) - \beta_m(\sigma_{m,n})}{\alpha_m(\sigma_{m,n}^+;\theta) - \alpha_m(\theta;\sigma_{m,n}^-)}. \tag{6.32}$$

By definition, $\sigma_{m,n}$ is the end of a BP at node m. We will make use of $R_{m,n}(\theta)$ when $n \in \Phi_m$, i.e., when $\sigma_{m,n}$ happens to be an active switchover point. If this is the case, then it follows from Lemma 6.4 and Assumption 6.2(b) that $\beta_m(t)$ is continuous at $t = \sigma_{m,n}$. Note that this quantity involves the processing rate information $\beta_m(\sigma_{m,n})$ (typically known, otherwise measurable) at $t = \sigma_{m,n}$, and the values of the inflow rates before and after a BP ends at $t = \sigma_{m,n}$. Using this definition, the next lemma allows us to obtain a simple relationship between the two crucial quantities $\psi_{m,i}$ and $\phi_{m,n}$.

LEMMA 6.7 *Let $n \in \Phi_m$ and $\sigma_{m,n} = s_{m,i}$ for some active switchover point of $\alpha_m(\theta;t)$. Then,*

$$\phi_{m,n} = R_{m,n}(\theta) \cdot \psi_{m,i} \tag{6.33}$$

where

$$0 < R_{m,n}(\theta) \leq 1. \tag{6.34}$$

Combining Lemmas 6.5-6.7 we obtain the following (a detailed proof is given in [22]):

THEOREM 6.3 *For $m > 1$ and $n = 1, \ldots, N_m$:*

$$x'_m(t;\theta) = \begin{cases} 0 & \text{if } t \in B_{m,n} \\ \sum_{k=1}^{K_{m,n}(t)} x'_{m-i^*}(s^-_{m,k};\theta) + & \\ \mathbf{1}[n \in \Phi_m] \cdot R_{m,n}(\theta) x'_{m-i^*}(\sigma^-_{m,n};\theta) & \text{if } t \in \overline{B}_{m,n+1} \end{cases} \tag{6.35}$$

where

$$i^* = \min_{j=1,\ldots,m-1} \{ j : x_{m-j}(s_{m,k};\theta) > 0 \} \tag{6.36}$$

and $K_{m,n}(t)$ is the number of active switchover points in the interval $(\sigma_{m,n}, t) \subset \overline{B}_{m,n+1}$.

Let us take a closer look at (6.35) in order to better understand the process through which changes in the state of one node affect the state of downstream nodes. For any $m > 1$, let us view $x'_m(t;\theta)$ as a perturbation in $x_m(\theta;t)$. For simplicity, let us initially ignore the case where $n \in \Phi_m$ and assume $i^* = 1$. Thus, we have

$$x'_m(t;\theta) = \sum_{k=1}^{K_{m,n}(t)} x'_{m-1}(s^-_{m,k};\theta)$$

if $t \in \overline{B}_{m,n+1}$. We can see that node $m-1$ only affects node m at time $s_{m,k}$ when an EP at node $m-1$ starts (recalling our definition of an active switchover point). In simple terms: *whenever node $m-1$ becomes empty, it propagates downstream to m its current perturbation*. These perturbations accumulate at m over all $K_{m,n}(t)$ active switchover points contained in a NBP $\overline{B}_{m,n+1}$. For example, in Figure 6.7, $s_{m,i+1}$ is a point where an EP starts at node $m-1$ while node m is in a NBP; at that time we get $x'_m(t;\theta) = x'_{m-1}(s^-_{m,i+1};\theta)$.

Moreover, when the NBP ends at $\tau_{m,n+1}$, the value of $x'_m(\tau^-_{m,n+1};\theta)$ will in turn be propagated downstream to $m+1$, before setting $x'_m(\tau^+_{m,n+1};\theta) = 0$ at the start of the ensuing EP at m.

Any cumulative perturbation at m is *eliminated by the presence of any BP*, i.e., when $t \in B_{m,n}$ as indicated by (6.35). For example, in Figure 6.7, $s_{m,i-1}$ is a point where an EP starts at node $m-1$ while node m is in a FP; therefore, it has no effect on $x_m(t;\theta)$, i.e., $x'_m(t;\theta) = 0$. The conclusion is that in order for a node to have a

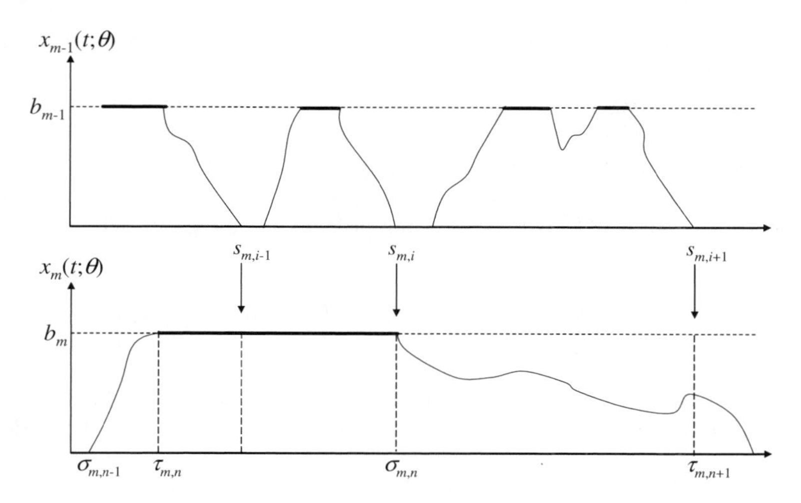

FIGURE 6.7: A sample path of two nodes in series. State perturbations propagate from $m-1$ to m at the start of an EP at $m-1$, provided that m is not in a BP at the time (as in the case of $s_{m,i-1}$).

chance to propagate a perturbation downstream, *it must become empty before it becomes full*. In view of this fact, we can argue that control at the edge of a tandem network is generally expected to have a limited impact on nodes that are several hops away, since propagating perturbations requires the combination of several events: a perturbation to be present and to be propagated at the start of an EP before it is eliminated by a FP; moreover this has to be true for a sequence of nodes. The probability of such a joint event is likely to be small as the number of hops increases. This provides an analytical substantiation to the conjecture that *congestion in a network cannot be easily regulated through control exercised several hops away*, unless the intermediate nodes experience frequent EPs providing the opportunity for perturbation propagation events.

Let us now look at the two aspects that were ignored in the discussion above. First, suppose that $i^* > 1$. This means that an EP occurs not just at node $m-1$, but also nodes $m-2,\ldots,m-i^*$, all at the same time. Thus, instead of propagating a perturbation from $m-1$ to m, the propagation now takes place from $m-i^*$ to m. Second, let us consider the case where $n \in \Phi_m$ in (6.35). This allows an EP that starts at $m-1$ to cause the end of a FP at node m. When this occurs, only a fraction, given by $R_{m,n}(\theta)$, of the perturbation at $m-1$ is propagated to node m. For example, in Figure 6.7, the point $s_{m,i}$ coincides with $\sigma_{m,n}$ and it therefore contributes another term scaled by $R_{m,n}$ as seen in (6.35).

Finally, note that the discussion above is independent of the way in which the controllable parameter affects the buffer content at $m=2$ and subsequently all downstream nodes through (6.35). In the particular case we are considering, however, we can see from (6.28) that the derivatives at node 1 are always given by 1. Thus, *the entire perturbation analysis process here reduces to counting EP events at all nodes that cause propagations* through (6.35). The only exception is for those events that start an EP at some $m-1$ and at the same time end a FP at m; in this case, the derivative at node m is affected by some amount dependent on $R_{m,n}(\theta) \in (0,1]$.

To summarize, IPA allows us to visualize the process of generating state perturbations at node 1 when $\theta = b_1$ is perturbed and propagating them through the system as follows:

- Perturbations are generated at $m = 1$ when a FP occurs. They are subsequently eliminated after an EP starts.
- Perturbations are fully propagated from $m-1$ to m when an EP starts at $m-1$ (more generally, at $m-i^*$ as defined in (6.36)) provided that m is in a NBP.
- Perturbations are partially propagated (by a fraction $R_{m,n}(\theta)$) from $m-1$ to m when an EP starts at $m-1$ and causes a FP at m to end.
- Perturbations at $m > 1$ are eliminated wnen a BP occurs.

Let us now see how the recursive evaluation of $x_m^{'}(t;\theta)$ in (6.35) can be used to obtain IPA estimators for the loss volume and workload performance metrics at each node defined for $m = 1,\dots,M$ as

$$L_m(\theta;T) = \int_0^T \gamma_m(t;\theta)dt, \tag{6.37}$$

$$Q_m(\theta;T) = \int_0^T x_m(t;\theta)dt. \tag{6.38}$$

The case of $L_1(\theta;T)$ and $Q_1(\theta;T)$ was considered in the last section, so we will focus on $m > 1$.

Let us define $\mathcal{F}_m$ to be the set of all indices of BPs that happen to be FPs at node m over $[0,T]$, i.e.,

$$\mathcal{F}_m \equiv \{n : x_m(t;\theta) = b_m \text{ for all } t \in B_{m,n},\ n = 1,\dots,N_m\}.$$

Observing that only FPs at node m will experience loss, we have

$$L_m(\theta;T) = \sum_{n\in\mathcal{F}_m} \int_{\tau_{m,n}}^{\sigma_{m,n}} \gamma_m(t;\theta)dt.$$

In addition, let us define the set

$$\Omega_{m,n} \equiv \Psi_{m,n} \cup \overline{\Psi}_{m,n} \tag{6.39}$$

which, recalling (6.23) and (6.25), includes the indices i of all active switchover points in the BP $B_{m,n} = [\tau_{m,n}(\theta),\sigma_{m,n}(\theta)]$ and the NBP that precedes it $\overline{B}_{m,n} =$

$(\sigma_{m,n-1}, \tau_{m,n})$. The following gives an explicit expression for the IPA estimator, $L_m'(\theta;T)$, of the expected loss volume over an interval $[0,T]$.

THEOREM 6.4 *The loss volume IPA derivative, $L_m'(\theta;T)$, $m = 2,\ldots,M$, is:*

$$L_m'(\theta;T) = -\sum_{n\in F_m}\sum_{i\in\Omega_{m,n}} \psi_{m,i} + \sum_{n\in\Gamma_m} \phi_{m,n} \tag{6.40}$$

where $\psi_{m,i}$ and $\phi_{m,n}$ are given by (6.30)–(6.31) and (6.33).

The proof of this result is given in [22]. In simple terms, to obtain $L_m'(\theta;T)$ we accumulate terms $-\psi_{m,i}$ over all active switchover points $s_{m,i}$ for each interval $(\sigma_{m,n-1}, \sigma_{m,n}]$, $n = 1,2,\ldots$ However, the result contributes to $L_m'(\theta;T)$ only if $\sigma_{m,n}$ ends a FP. The second term of (6.40) modifies the accumulation process as follows: Occasionally, $\sigma_{m,n}$ is followed by a NBP $(\sigma_{m,n}, \tau_{m,n+1})$ of type (F,E), i.e., the buffer at node m becomes empty. When this event takes place, the contribution $-\psi_{m,i}$ for $s_{m,i} = \sigma_{m,n}$ is modified by adding $\phi_{m,n}$ to it. In the example shown in Figure 6.7, there are two active switchover points in the interval $(\sigma_{m,n-1}, \sigma_{m,n}]$ at $s_{m,i-1}$ and at $s_{m,i}$. These contribute terms $-\psi_{m,i-1}$ and $-\psi_{m,i}$ to $L_m'(\theta;T)$ since the BP that ends at $\sigma_{m,n}$ is a FP. The second one happens to coincide with the end of the FP, i.e., $s_{m,i} = \sigma_{m,n}$. Since the next NBP is of type (F,E), we have $n \in \Gamma_m$ and a term $\phi_{m,n}$ is contributed to $L_m'(\theta;T)$. In addition, the active switchover point at $s_{m,i+1}$ does not contribute to $L_m'(\theta;T)$.

The terms $\psi_{m,i}$ and $\phi_{m,n}$ are given in Lemmas 6.6 and 6.7, where we can see that they depend on the derivatives $x_{m-1}'(s_{m,i}^-;\theta)$ propagated from the upstream node $m-1$ through every EP event that occurs at $m-1$. These derivatives are in turn provided by (6.35) in Theorem 6.3. We emphasize the fact that, as in the case of a single node, the IPA estimator *does not involve any knowledge of the stochastic processes characterizing arriving traffic or node processing* and allows for the possibility of correlations. The only information involved is the one required to calculate $R_{m,n}$ in (6.35), which, incidentally, occurs only when the end of a FP happens to be an active switchover point. If this contribution is negligible, (6.40) becomes a simple counter, since the values of $\psi_{m,i}$ are originally given by -1 at node 1, as seen in (6.28).

Turning our attention to the workload, we can partition $[0,T]$ into NBPs and BPs and get

$$Q_m(\theta;T) = \sum_{n=1}^{N_m}\left[\int_{\sigma_{m,n-1}}^{\tau_{m,n}} x_m(t;\theta)dt + \int_{\tau_{m,n}}^{\sigma_{m,n}} x_m(t;\theta)dt\right]$$

where N_m is the total number of BPs in $[0,T]$. We can then obtain (for a proof see [22]) the following IPA estimator.

THEOREM 6.5 *The workload IPA derivative, $Q_m'(\theta;T)$, $m = 2,\ldots,M$, is:*

$$Q_m'(\theta;T) = -\sum_{n=1}^{N_m} \sum_{i\in\overline{\Psi}_{m,n}} [\tau_{m,n} - s_{m,i}]\psi_{m,i} - \sum_{n\in\Phi_m} [\tau_{m,n+1} - \sigma_{m,n}]\phi_{m,n} \tag{6.41}$$

where $\psi_{m,i}$ and $\phi_{m,n}$ are given by (6.30)–(6.31) and (6.33).

This IPA estimator, similar to the IPA estimator in (6.40), involves accumulating terms $-\psi_{m,i}$ over active switchover points $s_{m,i}$. In this case, however, we are only interested in $s_{m,i}$ contained in NBPs $(\sigma_{m,n-1}, \tau_{m,n})$, $n = 1,\ldots,N_m$. The accumulation is done at $\tau_{m,n}$ with each such term scaled by $[\tau_{m,n} - s_{m,i}]$ measuring the time elapsed since the switchover point took place. The second term in (6.41) adds similar contributions made at the end of a NBP of type (F,E) due to active switchover points that coincide with the end of a FP at some time $\sigma_{m,n}$.

Both estimators (6.40) and (6.41) can be shown to be unbiased by establishing the Lipschitz continuity of $L_m(\theta;T)$ and $Q_m(\theta;T)$ with Lipschitz constant having a finite first moment. The proof is a direct extension of the single-node case (details are given in [22].)

6.6 Conclusions

This chapter has concentrated on stochastic flow systems, a class of stochastic hybrid systems with components interacting with each other through flows whose rates generally vary randomly. For highly complicated discrete event systems processing large numbers of discrete entities (packets in communication networks, parts in manufacturing systems, vehicles in transportation networks, etc.), Stochastic Flow Models (SFMs) of this type are extremely useful as abstractions of the underlying processes. One of the attractive features of these models is that they enable the use of very efficient sensitivity analysis methods. We have discussed Infinitesimal Perturbation Analysis (IPA) as one such method. IPA estimators are based on observed sample path data and yield unbiased gradient estimators of interesting performance metrics of the system with respect to various model parameters. In this chapter, we showed how to derive such estimators for some stochastic flow systems without feedback control. The next chapter will address this issue. We have also discussed, in Section 6.4, how these gradient estimators can be used for on-line optimization without any distributional knowledge of the stochastic processes involved and, in some cases, without even knowing any model parameter values. Extensive examples of such optimization processes may be found in [9],[22],[25],[13],[20].

Acknowledgments. The IPA results in this chapter were obtained through collaborative research with Benjamin Melamed, Christos Panayiotou, Gang Sun, Yorai Wardi, and Haining Yu.

References

[1] D. Anick, D. Mitra, and M. M. Sondhi, "Stochastic theory of a data-handling system with multiple sources," *The Bell System Technical Journal*, vol. 61, pp. 1871–1894, 1982.

[2] H. Kobayashi and Q. Ren, "A mathematical theory for transient analysis of communications networks," *IEICE Transactions on Communications*, vol. E75-B, pp. 1266–1276, 1992.

[3] R. L. Cruz, "A calculus for network delay, Part I: Network elements in isolation," *IEEE Transactions on Information Theory*, 1991.

[4] G. Kesidis, A. Singh, D. Cheung, and W. Kwok, "Feasibility of fluid-driven simulation for ATM network," in *Proceedings of IEEE Globecom*, vol. 3, 1996, pp. 2013–2017.

[5] K. Kumaran and D. Mitra, "Performance and fluid simulations of a novel shared buffer management system," in *Proceedings of IEEE INFOCOM*, March 1998.

[6] B. Liu, Y. Guo, J. Kurose, D. Towsley, and W. B. Gong, "Fluid simulation of large scale networks: Issues and tradeoffs," in *Proceedings of the International Conference on Parallel and Distributed Processing Techniques and Applications*, June 1999, Las Vegas, Nevada.

[7] A. Yan and W. Gong, "Fluid simulation for high-speed networks with flow-based routing," *IEEE Transactions on Information Theory*, vol. 45, pp. 1588–1599, 1999.

[8] S. Bohacek, J. Hespanha, J. Lee, and K. Obraczka, "A hybrid systems modeling framework for fast and accurate simulation of data communication networks," in *Proc. of the ACM Intl. Conf. on Measurements and Modeling of Computer Systems SIGMETRICS*, June 2003.

[9] C. G. Cassandras, Y. Wardi, B. Melamed, G. Sun, and C. G. Panayiotou, "Perturbation analysis for on-line control and optimization of stochastic fluid models," *IEEE Transactions on Automatic Control*, vol. AC-47, no. 8, pp. 1234–1248, 2002.

[10] R. Suri and B. Fu, "On using continuous flow lines for performance estimation of discrete production lines," in *Proceedings of 1991 Winter Simulation Conference*, Piscataway, NJ, 1991, pp. 968–977.

[11] R. Akella and P. R. Kumar, "Optimal control of production rate in a failure prone manufacturing system," *IEEE Transactions on Automatic Control*, vol. AC-31, pp. 116–126, Feb. 1986.

[12] J. R. Perkins and R. Srikant, "Failure-prone production systems with uncertain demand," *IEEE Transactions on Automatic Control*, vol. AC-46, pp. 441–449, March 2001.

[13] H. Yu and C. G. Cassandras, "Perturbation analysis for production control and optimization of manufacturing systems," *Automatica*, vol. 40, pp. 945–956, 2004.

[14] B. Mohanty and C. G. Cassandras, "The effect of model uncertainty on some optimal routing problems," *Journal of Optimization Theory and Applications*, vol. 77, pp. 257–290, 1993.

[15] S. Meyn, "Sequencing and routing in multiclass networks. Part I: Feedback regulation," *SIAM J. Control and Optimization*, vol. 43, no. 3, pp. 741–776, 2001.

[16] C. G. Cassandras and S. Lafortune, *Introduction to Discrete Event Systems*. Boston: Kluwer Academic Publishers, 1999.

[17] P. Glasserman, *Gradient Estimation via Perturbation Analysis*. Dordrecht, Holland: Kluwer Academic Publishers, 1991.

[18] Y. C. Ho and X. Cao, *Perturbation Analysis of Discrete Event Dynamic Systems*. Dordrecht, Holland: Kluwer Academic Publishers, 1991.

[19] H. J. Kushner and D. Clark, *Stochastic Approximation for Constrained and Unconstrained Systems*. Berlin, Germany: Springer-Verlag, 1978.

[20] H. Yu and C. G. Cassandras, "Perturbation analysis and multiplicative feedback control in communication networks," in *Performance Evaluation and Planning Methods for the Next Generation Internet*, A. Girard, B. Sansò, and F. Vázquez-Abad, Eds. Berlin, Germany: Springer-Verlag, 2005, pp. 297–332.

[21] Y. Wardi and B. Melamed, "Variational bounds and sensitivity analysis of traffic processes in continuous flow models," *J. Discrete Event Dynamic Systems*, vol. 11, pp. 249–282.

[22] G. Sun, C. G. Cassandras, and C. G. Panayiotou, "Perturbation analysis and optimization of stochastic flow networks," *IEEE Transactions on Automatic Control*, vol. 49, no. 12, pp. 2113–2128, 2004.

[23] R. Y. Rubinstein and A. Shapiro, *Discrete Event Systems: Sensitivity Analysis and Stochastic Optimization by the Score Function Method*. New York, New York: John Wiley and Sons, 1993.

[24] Y. Wardi, B. Melamed, C. G. Cassandras, and C. G. Panayiotou, "IPA gradient estimators in single-node stochastic fluid models," *Journal of Optimization Theory and Applications*, vol. 115, no. 2, pp. 369–406, 2002.

[25] G. Sun, C. G. Cassandras, and C. G. Panayiotou, "Perturbation analysis of multiclass stochastic fluid models," *Journal of Discrete Event Dynamic Systems: Theory and Applications*, vol. 14, pp. 267–307, 2004.

Chapter 7

Perturbation Analysis for Stochastic Flow Systems with Feedback

Yorai Wardi, George Riley, and Richelle Adams
Georgia Institute of Technology

7.1 Introduction 169
7.2 SFM with Flow Control 171
7.3 Retransmission-based Model 178
7.4 Simulation Experiments 186
7.5 Conclusions 188
References 189

7.1 Introduction

Fluid models long have been investigated as models for performance evaluation of queueing networks in various application domains, like telecommunications, manufacturing, and transportation. Unlike the established, essentially-discrete queueing models that capture the movement and storage of each individual entity (packet, job, vehicle), fluid models are inherently continuous, and by foregoing the identity of each entity, they focus instead on the aggregate flow. This could result in less detailed models, and hence possibly in faster simulation and new analysis techniques for performance evaluation. From a modelling standpoint, the aggregation of discrete entities into continuous flow has been justified in high-speed telecommunications networks, where traffic analyses have been carried out in [1, 9, 6]. Subsequently, the issue of simulation has been examined in [2, 8, 10, 17], and tradeoffs between simulations of the discrete models vs. their continuous-flow counterparts have been identified in [11]. A hybrid, discrete/fluid network simulator has been developed for networking applications in [12].

As already mentioned in the previous chapter, another reason to study fluid queues is their suitability to sensitivity analysis by Infinitesimal Perturbation Analysis (IPA). Developed during the nineteen-eighties as a gradient estimation technique for Discrete Event Systems (DES), IPA computes the derivatives (gradients) of sample performance functions with respect to continuous control parameters [7, 3]. It has been formulated mainly in the context of queueing networks, but also in a broader setting

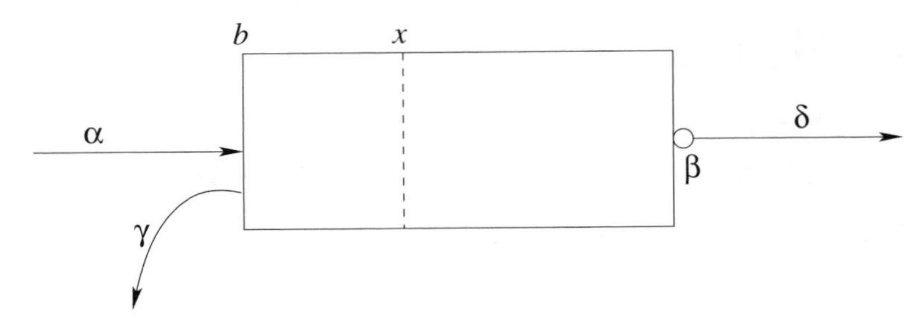

FIGURE 7.1: Basic stochastic flow model.

of DES. However, its scope has been limited to fairly simple models and traffic rules, excluding virtual-path routing, multiclass networks, and loss due to buffer overflow. The reason is that in such systems the IPA derivative is statistically biased, hence giving "wrong" (unreliable) results. In contrast, fluid queues appear to circumvent these limitations, and thus to extend the application domain of IPA by providing statistically unbiased derivatives. In particular, [15, 4] have developed unbiased IPA estimators for the derivatives of loss-rate and delay-related performance functions with respect to buffer size and inflow- and service-rate parameters, and [5, 14] extended the results to multiclass systems and to networks of fluid queues. Moreover, the above IPA estimators were shown to be quite simple to compute, and to either admit or be approximated by nonparametric and model-free formulas, namely formulas that are computable directly from a sample path, while being independent of its underlying probability law (nonparametric) and functional dependence on the control parameter (model-free). Consequently, the resulting IPA estimators can be computed on-line from the sample path of an actual system and potentially be used in network management and control (see [15, 4]).

Most of the above results have been derived for the single-server fluid queue shown in Figure 7.1, called the *basic Stochastic Flow Model (SFM)*. As defined in the last chapter, the basic SFM consists of a fluid tank followed by a server, and it is characterized by the buffer size (queue capacity) b and the following five random processes, defined over a given time-interval $[0, T_f]$ on a common probability space $(\Omega, \mathcal{F}, P)$: $\{\alpha(t)\}$ is the arrival (inflow) rate process; $\{\beta(t)\}$ is the service rate process; $\{x(t)\}$ is the workload (queue contents, buffer occupancy) process; $\{\gamma(t)\}$ is the spillover (overflow) rate process, and $\{\delta(t)\}$ is the outflow (output) rate process. The inflow rate process and the service rate process, together with the buffer size are jointly referred to as the *defining processes* since they determine the workload, overflow, and outflow processes which therefore are referred to as the *derived processes*; see [4, 15] for equations. The defining processes are assumed to be unaffected by the derived processes, and hence the basic SFM is said to be an *open-loop system*.

The previous chapter summarized the various IPA derivatives of the loss-rate and workload-rate functions, denoted respectively by L and Q, with respect to the buffer size and other parameters. We mention here the first, and most elegant result derived for SFM: The IPA derivative of the loss rate with respect to the buffer size [15, 4]. In common with the literature on IPA, we denote the variational (control) parameter by θ, and hence $b = \theta$ (the buffer size) in the present discussion. The loss rate function has the following form,

$$L(\theta) = \frac{1}{T_f}\int_0^{T_f} \gamma(t;\theta)dt, \tag{7.1}$$

where the dependence of the overflow process $\{\gamma(t)\}$ on θ is made explicit in the notation used. Then the IPA derivative, denoted by $L^{'}(\theta)$ ("prime" denoting derivative with respect to θ) is given by

$$L^{'}(\theta) = -\frac{1}{T_f}\mathcal{M}, \tag{7.2}$$

where $\mathcal{M}$ is the number of lossy non-empty periods in the time-interval $[0, T_f]$ (a *non-empty period* is a supremal interval during which the buffer is not empty, and a non-empty period is *lossy* if it incurs some loss). Note that this formula is nonparametric and model free, and it does not require any knowledge of the probability law underlying the defining processes. Therefore it can, and has been applied to a packet-based queueing model with good results. Although the packet-based model does not admit unbiased IPA estimators, we could apply to it (with success) the nonparametric and model-free formula (7.2) derived in the setting of SFM [15, 4].

The above results and their extensions to networks have established SFM as a natural modelling framework for IPA. However, within this framework the scope of IPA has been limited to open-loop systems. This chapter reports on some results pertaining to closed-loop feedback systems, namely systems whose defining processes are affected by the values of the processes derived from them. Two kinds of such systems will be considered: A flow-control SFM, and a loss-induced retransmission model. The first system has the form of the SFM shown in Figure 6.1 of the last chapter, where the inflow rate is tuned according to the value of the workload. The second system involves delayed retransmission of overflow fluid upon expiration of a congestion-control timer. Both systems constitute basic models whose purpose is to capture some salient features of telecommunications systems, and extensions of the results discussed here to realistic networks and protocols remains the subject of on-going research.

The rest of the chapter is organized as follows. Section 7.2 discusses the flow control model, and Section 7.3 concerns the timer-based model. Section 7.4 presents simulation results for the timer-based model, and Section 7.5 concludes the chapter.

7.2 SFM with Flow Control

The feedback system considered in this section has been analyzed by Yu and Cassandras in [19]. We present here only one of the fundamental results and provide a proof (under weaker assumptions) in order to illustrate the main arguments.

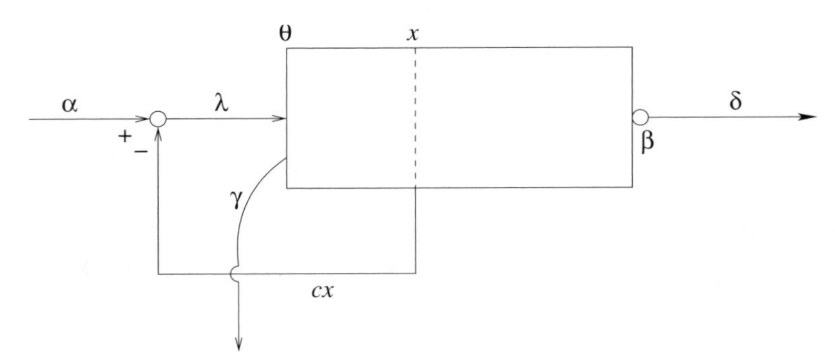

FIGURE 7.2: Flow-control model.

Consider the SFM shown in Figure 7.2, where the variable parameter θ is the buffer size. We are concerned with the loss-rate performance measure, $L(\theta)$, for which we will develop the IPA derivative $L^{'}(\theta)$ (throughout the chapter, "prime" will denote derivative with respect to θ). Fix $\theta > 0$. Let $\alpha(t)$ denote the external fluid arrival rate, and following Figure 7.1 we denote the service rate, workload, and spillover rate by $\beta(t)$, $x(t;\theta)$, and $\gamma(t;\theta)$. We note from Figure 7.2 that the inflow rate to the buffer, denoted by $\lambda(t;\theta)$, is given by

$$\lambda(t;\theta) = \alpha(t) - cx(t;\theta), \tag{7.3}$$

for a given constant $c > 0$. The defining processes are $\{\alpha(t)\}$ and $\{\beta(t)\}$, and the processes $\{x(t;\theta)\}$, $\{\gamma(t;\theta)\}$, and $\{\lambda(t;\theta)\}$ are derived from them and they also depend on θ. Let us define $\sigma(t)$ by

$$\sigma(t) = \alpha(t) - \beta(t), \tag{7.4}$$

and further define $A(t;\theta)$ by

$$A(t;\theta) = \sigma(t) - cx(t;\theta). \tag{7.5}$$

For a given $\theta > 0$ the workload $x(t;\theta)$ is then given by the following equation,

$$\frac{dx(t;\theta)}{dt^{+}} = \begin{cases} 0, & \text{if } x(t;\theta) = 0 \text{ and } A(t;\theta) \leq 0 \\ 0, & \text{if } x(t;\theta) = \theta \text{ and } A(t;\theta) \geq 0 \\ A(t;\theta), & \text{otherwise,} \end{cases} \tag{7.6}$$

whose initial condition is assumed to be $x(0;\theta) = 0$ for simplicity, and the loss rate $\gamma(t;\theta)$ is given by

$$\gamma(t;\theta) = \begin{cases} A(t;\theta), & \text{if } x(t;\theta) = \theta \\ 0, & \text{otherwise.} \end{cases} \tag{7.7}$$

The loss-rate performance function is given by

$$L(\theta) = \frac{1}{T_f}\int_0^{T_f} \gamma(t;\theta)dt. \tag{7.8}$$

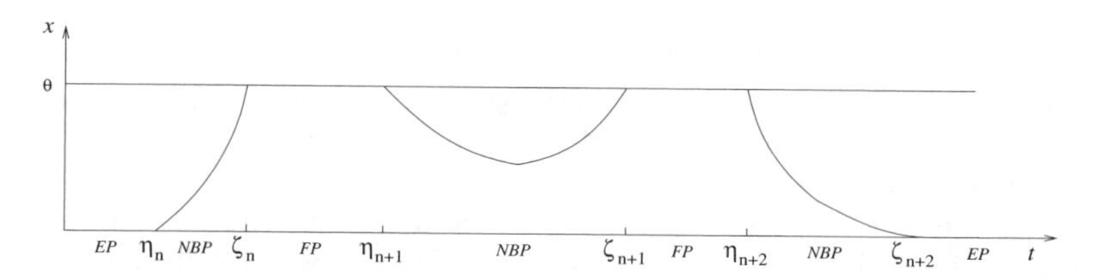

FIGURE 7.3: A typical state trajectory of the flow-control model.

The SFM, at a given fixed $\theta > 0$, can be viewed as a hybrid dynamical system whose input is comprised of the processes $\{\alpha(t)\}$ and $\{\beta(t)\}$, its state is $\{x(t;\theta)\}$, and its output is the spillover process $\{\gamma(t;\theta)\}$. The following assumption will be made throughout this section.

ASSUMPTION 7.1 (i) With probability 1 (w.p.1), the realization of the function $\sigma(t)$ is piecewise continuously differentiable on the interval $[0,T_f]$. (ii) There exists a constant $C > 0$ such that, for every realization of $\sigma(t)$, $|\sigma(t)| \leq C$ and $|\frac{d\sigma}{dt}(t)| \leq C$ (whenever this derivative exists) for every $t \in [0,T_f]$. (iii) With K_1 denoting the number of discontinuities of $\sigma(t)$ in the interval $[0,T_f]$, $E[K_1] < \infty$ ("E" denoting expectation). (iv) With K_2 denoting the number of times $\frac{d\sigma}{dt}(t)$ switches sign in the interval $[0,T_f]$, $E[K_2] < \infty$.[1]

A part of a typical trajectory (realization) of the workload process $x(t;\theta)$ is shown in Figure 7.3, and following the notation and taxonomy established in [19], we can partition it into the following three types of segments: (i) Full periods (FP), empty periods (EP), and periods that are neither full nor empty. Full periods and empty periods we commonly called boundary periods (BP), and the *nth* BP is denoted by BP_n. In contrast, the periods that are neither full nor empty are called non-boundary periods (NBP), and we denote the *nth* NBP by NBP_n. Furthermore, we denote the lower boundary point and the upper boundary point of BP_n by ζ_n and η_{n+1}, respectively, so that $NBP_n = (\eta_n, \zeta_n)$ and $BP_n = [\zeta_n, \eta_{n+1}]$; see Figure 7.3 for an illustration. We classify as *events* the occurrence of either a jump (discontinuity) in $\sigma(t)$ or the start of a BP. The former kind of events are called *exogenous* since their timing is independent of θ, while the latter kind of events are called *endogenous* by dint of the dependence of their timing on θ via the state variable $x(t;\theta)$. We make the following assumption.

ASSUMPTION 7.2 For a given fixed $\theta > 0$, w.p.1 no two or more events occur at the same time.

We next address the development of the IPA derivative $L'(\theta)$. Figure 7.3 and

[1] By "sign" we mean positive, negative, or zero.

Equations (7.6) - (7.8) indicate that the derivative term $L^{'}(\theta)$ exists as long as the following two conditions are satisfied: (i) No BP constitutes a single point, and (ii) the equality $A(t;\theta) = 0$ does not hold during any open subset of a BP. This will become evident from the following analysis that derives the above derivative term. Moreover, if one or both of these conditions fail to be satisfied, then the one-sided derivatives $\frac{d}{d\theta^{+}}(L(\theta))$ and $\frac{d}{d\theta^{-}}(L(\theta))$ still exist. Thus, to simplify the discussion we assume that the above two conditions are in force w.p.1 at a given $\theta > 0$.

Let N denote the number of BPs in the interval $[0, T_f]$, and define Ψ_F by

$$\Psi_F := \{n = 1, \ldots, N \ : \ BP_n \text{ is an FP}\}. \tag{7.9}$$

Examining Equations (7.7) and (7.8) and using Figure 7.3 as a visual aid, we see that

$$L(\theta) = \frac{1}{T_f} \sum_{n \in \Psi_F} \int_{\zeta_n}^{\eta_{n+1}} A(t;\theta)dt, \tag{7.10}$$

and hence,

$$L^{'}(\theta) \ = \ \frac{1}{T_f} \sum_{n \in \Psi_F} \frac{d}{d\theta}\Big(\int_{\zeta_n}^{\eta_{n+1}} A(t;\theta)dt\Big). \tag{7.11}$$

Let us examine each one of the derivative terms in the Right-Hand Side (RHS) of (7.11). Note that η_n and ζ_n generally are functions of θ as well, and recall that "prime" denotes derivative with respect to θ. Thus, for every $n \in \Psi_F$ we have that

$$\frac{d}{d\theta}\Big(\int_{\zeta_n}^{\eta_{n+1}} A(t;\theta)dt\Big) \ = \ \int_{\zeta_n}^{\eta_{n+1}} A^{'}(t;\theta)dt \ + A(\eta_{n+1}^{-};\theta)\eta_{n+1}^{'}(\theta) - A(\zeta_n^{+};\theta)\zeta_n^{'}(\theta). \tag{7.12}$$

LEMMA 7.1 *For every $n = 1, \ldots, N$, we have that $A(\eta_n^{-};\theta)\eta_n^{'}(\theta) = 0$ and $A(\eta_n^{+};\theta)\eta_n^{'}(\theta) = 0$.*

PROOF η_n is the time at which BP_n ends. This happens as a result of a change in the sign of $A(t;\theta)$, either from positive to negative when $x(\eta_n;\theta) = \theta$ (end of an FP), or from negative to positive when $x(\eta_n;\theta) = 0$ (end of an EP). Now such a change in sign can occur either abruptly, due to a jump in $\sigma(t)$, or continuously, when $A(t;\theta)$ is continuous at $t = \eta_n$. In the first case the jump is the result of an exogenous event, and hence $\eta_n^{'}(\theta) = 0$. In the second case, $A(\eta_n;\theta) = 0$ since $A(t;\theta)$ is continuous at $t = \eta_n$ and it change sign there. In either case, $A(\eta_n^{-};\theta)\eta_n^{'}(\theta) = A(\eta_n^{+};\theta)\eta_n^{'}(\theta) = 0$. ∎

The point $t = \zeta_n$ is the time of an endogenous event and by Assumption 7.2 there is no exogenous event at that time, and hence $A(\zeta_n^{+};\theta) = A(\zeta_n^{-};\theta) = A(\zeta_n;\theta)$. Consequently, and by Lemma 7.1, we have that

$$\frac{d}{d\theta}\Big(\int_{\zeta_n}^{\eta_{n+1}} A(t;\theta)dt\Big) \ = \ \int_{\zeta_n}^{\eta_{n+1}} A^{'}(t;\theta)dt \ - A(\zeta_n;\theta)\zeta_n^{'}(\theta). \tag{7.13}$$

Next, consider the term $A(\zeta_n;\theta)\zeta_n^{'}(\theta)$ in the RHS of (7.13). Examining Figure 7.3 and Equation (7.6), and recalling that $n \in \Psi_F$ (meaning that $BP_n = [\zeta_n, \eta_{n+1}]$ is an FP), it is apparent that

$$\int_{\eta_n}^{\zeta_n} A(t;\theta)dt \;=\; \begin{cases} \theta, \text{ if } x(\eta_n;\theta) = 0 \\ 0, \text{ if } x(\eta_n;\theta) = \theta. \end{cases} \tag{7.14}$$

Taking derivatives with respect to θ,

$$\int_{\eta_n}^{\zeta_n} A^{'}(t;\theta)dt \;+\; A(\zeta_n^{-};\theta)\zeta_n^{'}(\theta) \;-\; A(\eta_n^{+};\theta)\eta_n^{'}(\theta) = \begin{cases} 1, \text{ if } x(\eta_n;\theta) = 0 \\ 0, \text{ if } x(\eta_n;\theta) = \theta. \end{cases} \tag{7.15}$$

By Lemma 7.1 and Assumption 7.2 (implying that $A(t;\theta)$ is continuous at $t = \zeta_n$, and hence, as we have seen, $A(\zeta_n^{-};\theta) = A(\zeta_n;\theta)$), we get that

$$A(\zeta_n;\theta)\zeta_n^{'}(\theta) \;=\; -\int_{\eta_n}^{\zeta_n} A^{'}(t;\theta)dt \;+\; \begin{cases} 1, \text{ if } x(\eta_n;\theta) = 0 \\ 0, \text{ if } x(\eta_n;\theta) = \theta. \end{cases} \tag{7.16}$$

Plugging this in (7.13) we obtain,

$$\frac{d}{d\theta}\Big(\int_{\zeta_n}^{\eta_{n+1}} A(t;\theta)dt\Big) \;=\; \int_{\eta_n}^{\zeta_n} A^{'}(t;\theta)dt + \int_{\zeta_n}^{\eta_{n+1}} A^{'}(t;\theta)dt - \begin{cases} 1, \text{ if } x(;\theta\eta_n) = 0 \\ 0, \text{ if } x(\eta_n;\theta) = \theta. \end{cases} \tag{7.17}$$

Next, we consider the two integrals in the RHS of (7.17), and to this end, we first evaluate the terms $A^{'}(t;\theta)$. Recall (7.5) that $A(t;\theta) = \sigma(t) - cx(t;\theta)$, and hence

$$A^{'}(t;\theta) = -cx^{'}(t;\theta). \tag{7.18}$$

For the term $x^{'}(t;\theta)$, we have the following results.

LEMMA 7.2 *For every $n \in \Psi_F$, and for every $t \in (\zeta_n, \eta_{n+1})$, $x^{'}(t;\theta) = 1$.*

PROOF The proof is immediate from the fact that $x(t;\theta) = \theta$ for all $t \in (\zeta_n, \eta_{n+1})$. ∎

LEMMA 7.3 *For every $n \in \Psi_F$, and for every $t \in (\eta_n, \zeta_n)$, (i) if $x(\eta_n;\theta) = 0$ then $x^{'}(t;\theta) = 0$, and (ii) if $x(\eta_n;\theta) = \theta$ then*

$$x^{'}(t;\theta) \;=\; e^{-c(t-\eta_n)}. \tag{7.19}$$

PROOF Part (i) is obvious (see Figure 7.3) since following an EP the state variable $x(t;\theta)$ will be independent of θ until the buffer becomes full. Regarding Part (ii), let $x(\eta_n;\theta) = \theta$. By (7.6), for all $t \in (\eta_n, \zeta_n)$,

$$\frac{d}{dt}x(t;\theta) \;=\; A(t;\theta) \;=\; \sigma(t) - cx(t;\theta), \tag{7.20}$$

and hence,

$$x(t;\theta) \; = \; \theta e^{-c(t-\eta_n)} + \int_{\eta_n}^{t} e^{-c(t-\tau)}\sigma(\tau)d\tau. \tag{7.21}$$

Taking derivatives with respect to θ,

$$\begin{aligned} x^{'}(t;\theta) \; &= \; \theta c e^{-c(t-\eta_n)}\eta_n^{'}(\theta) + e^{-c(t-\eta_n)} - e^{-c(t-\eta_n)}\sigma(\eta_n^{+})\eta_n^{'}(\theta) \\ &= \; e^{-c(t-\eta_n)} - \big(\sigma(\eta_n^{+}) - c\theta\big)\eta_n^{'}(\theta)e^{-c(t-\eta_n)}. \end{aligned} \tag{7.22}$$

But $\sigma(\eta_n^{+}) - c\theta = A(\eta_n^{+};\theta)$, and hence, and by Lemma 7.1, the last term in (7.22) is 0. This establishes (7.19). ∎

Putting it all together, we now have the following result.

PROPOSITION 7.1 *For every $n \in \Psi_F$,*

$$\frac{d}{d\theta}\Big(\int_{\zeta_n}^{\eta_{n+1}} A(t;\theta)dt\Big) \; = \; \begin{cases} -c(\eta_{n+1} - \zeta_n) - 1, & \text{if } x(\eta_n;\theta) = 0 \\ -c(\eta_{n+1} - \zeta_n) - (1 - e^{-c(\zeta_n - \eta_n)}), & \text{if } x(\eta_n;\theta) = \theta. \end{cases} \tag{7.23}$$

PROOF The proof is immediate from (7.17), (7.18), Lemma 7.2, and Lemma 7.3. ∎

Equation (7.11) together with Proposition 7.1 provide a nonparametric and model-free formula for the IPA derivative $L^{'}(\theta)$. Recalling that a non-empty period is a supremal time-interval during which the buffer is not empty, we see that an algorithm for computing $L^{'}(\theta)$ need not have any information about the underlying probability law; it only has to track the beginning and end of FPs and EPs and to know whether an FP is the first one in the non-empty period containing it.

References [18, 19] have derived analogous IPA estimators for the loss rate and delay-related performance measures with respect to various feedback laws and variational parameters, and [20] extended the analysis to networks of SFMs. These references also contain simulation experiments of optimization problems defined on packet-based, high-speed network models, where the IPA derivatives were used in stochastic approximation algorithms. The experiments were successful in the sense that the algorithms typically appeared to converge towards optimal parameter points.

Finally, a word must be said about the unbiasedness of the IPA estimator. Consider the abstract setting where $G(\theta)$ is a random function of a one-dimensional variable θ, defined on a common probability space $(\Omega, \mathcal{F}, P)$, where $\theta \in \Theta$, a given closed and bounded interval. Suppose that, for a given $\theta \in \Theta$, the derivative $G^{'}(\theta)$ exists w.p.1 (the appropriate one-sided derivative, if θ is an end-point of Θ). Let $g(\theta)$ denote the expected value of $G(\theta)$, namely, $g(\theta) := E[G(\theta)]$, with $E[\cdot]$ denoting expectation. The sample derivative $G^{'}(\theta)$ is said to be *unbiased* (see [7, 3]) if the operators of expectation and differentiation are interchangeable, namely,

$$E[G^{'}(\theta)] \; = \; g^{'}(\theta). \tag{7.24}$$

Obviously unbiasedness is a useful property when the purpose of the sample derivatives $G'(\theta)$ is to estimate, by Monte Carlo simulation, the derivative term $g'(\theta)$ whenever it is unavailable by analytical means.

The problem with IPA in the setting of queueing networks is that often it gives biased derivatives, namely Equation (7.24) is not satisfied. This would be the case for the loss function $L(\theta)$ defined in the context of packet-based models. In contrast, we mentioned earlier that the analogous IPA derivatives in the SFM setting often are unbiased, and we next prove this point for the loss-rate function $L(\theta)$ defined in (7.8). Recall from [13] that the derivative $g'(\theta)$ exists and the sample derivative $G'(\theta)$ is unbiased if the following two conditions are in force: (i) For every $\theta \in \Theta$, the derivative $G'(\theta)$ exists w.p.1, and (ii) w.p.1, the sample-based function $G(\cdot)$ is Lipschitz continuous on Θ, and the Lipschitz constant, K, has a finite first moment, namely $E[K] < \infty$. The first condition is not crucial: If it does not hold true, but if the one-sided derivatives $\frac{dG(\theta)}{d\theta^+}$ and $\frac{dG(\theta)}{d\theta^-}$ exist w.p.1 for every given $\theta \in \Theta$, and if condition (ii) holds, then the one-sided derivatives $\frac{dg(\theta)}{d\theta^+}$ and $\frac{dg(\theta)}{d\theta^-}$ exist and the following equations hold, $E\left[\frac{dG(\theta)}{d\theta^+}\right] = \frac{dg(\theta)}{d\theta^+}$ and $E\left[\frac{dG(\theta)}{d\theta^-}\right] = \frac{dg(\theta)}{d\theta^-}$. On the other hand the second condition is crucial, and its absence is what renders problematic the application of IPA to queueing networks.

Now considering the sample performance function $L(\theta)$ defined in (7.8), suppose that $\theta \in \Theta$ where Θ is a closed bounded interval whole left boundary point is positive, and suppose that Assumption 7.1 and Assumption 7.2 are in force. We have mentioned that $L'(\theta)$ exists as long as the equality $A(t;\theta) = 0$ does not hold true on an open subset of a BP, and no BP is a singleton; but the one-sided derivatives $\frac{dL(\theta)}{d\theta^+}$ and $\frac{dL(\theta)}{d\theta^-}$ always exist. Thus, unbiasedness of the IPA derivatives (or one-sided derivatives) hinges on the following result.

PROPOSITION 7.2 *W.p.1, the function $L(\theta)$ has a Lipschitz constant K throughout Θ, and $E[K] < \infty$.*

PROOF Let $M(\theta)$ denote the number of FPs in the interval $[0, T_f]$. If $t = \eta$ is the end point of an FP then (by Equation (7.6))

$$A(\eta^+;\theta) < 0 \leq A(\eta^-;\theta). \tag{7.25}$$

Now $x(\eta;\theta) = \theta$, and hence, and by Equation (7.5),

$$\sigma(\eta^+) < c\theta \leq \sigma(\eta^-). \tag{7.26}$$

This implies that, at time $t = \eta$, either there is a jump (discontinuity) of $\sigma(\cdot)$, or the sign of $\frac{d\sigma}{dt}(\cdot)$ is changed from non-negative to negative. In any case, by Assumption 7.1 (iii) and (iv), there exists $M > 0$ such that $M(\theta) \leq M$ for every $\theta \in \Theta$, and $E[M] < \infty$.

Next, by Proposition 7.1, for every FP $[\zeta_n, \eta_{n+1}]$,

$$\left|\frac{d}{d\theta}\Big(\int_{\zeta_n}^{\eta_{n+1}} A(t;\theta)dt\Big)\right| \leq c(\eta_{n+1}-\zeta_n)+1. \tag{7.27}$$

Summing up in the last equation over all $n \in \Psi_F$ and using (7.11), we have the inequality $|L'(\theta)| \leq T_f^{-1}(cT_f + M(\theta))$ and hence

$$|L'(\theta)| \leq c + T_f^{-1}M \tag{7.28}$$

whenever the sample derivative $L'(\theta)$ exists; otherwise similar inequalities hold for the one-sided derivatives. Now the sample performance function is continuous and has bounded one-sided derivatives and hence it is Lipschitz continuous, with the Lipschitz constant $K := c + T_f^{-1}M$. Finally, $E[K] < \infty$ since $E[M] < \infty$. ■

7.3 Retransmission-based Model

This section presents an example of an SFM with feedback which captures an essential behavior of congestion control mechanisms,[2] while the next chapter will present a more detailed hybrid-system model for studying congestion control. The underlying system consists of a basic fluid model that captures the essence of delayed retransmission due to buffer overflow. The SFM, shown in Figure 7.4, has a finite buffer, and fluid "molecules" that are lost to overflow attempt to re-enter the buffer T seconds after their loss. T represents a timeout parameter that is in the heart of flow control and congestion control schemes. Unlike many studies of congestion control, we do not assume unlimited data at the input, but rather a random external inflow-rate process, and the purpose of the congestion control is to defer the admission of the input flow from a time of high demand to a later time of lesser demand. Thus, the congestion control mechanism is defined by two parameters, namely the timeout parameter T and the buffer size. The purpose of this section is to investigate the effect of the buffer size on the loss volume, which represents a measure of the system's inefficiency due to retransmissions. We recognize that in practice (e.g., TCP) the retransmission timers are state dependent whereas we use a constant timer, the reason is that our objective is but to analyze a simple model in order to study a fundamental behavior. The algorithm that we develop for this model could be used to gauge sensitivity analysis in more realistic networks, and on-going research carries out extensive simulation experiments.

[2]The results presented in this section have appeared in [16].

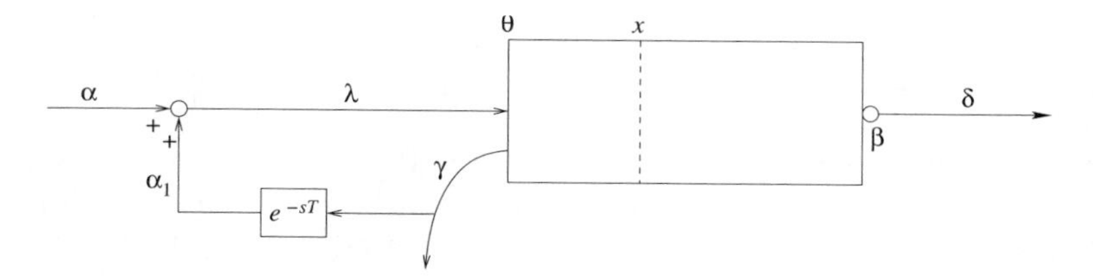

FIGURE 7.4: Retransmission model.

The SFM that we consider is shown in Figure 7.4, where the variable parameter under consideration is the buffer size θ. Its defining processes are the external arrival rate $\{\alpha(t)\}$ and the service rate $\{\beta(t)\}$, which are independent of θ. Based on them we have the following two derived processes: (i) The workload process $\{x(t;\theta)\}$, and (ii) the overflow process $\{\gamma(t;\theta)\}$. The feedback arrival process due to retransmission is denoted by $\{\alpha_1(t;\theta)\}$. All of these processes are assumed defined on a common probability space $(\Omega, \mathcal{F}, P)$, and on a given time interval $[0, T_f]$. For a fixed $\theta > 0$, the various derived processes are defined by the following equations. First, the total input flow rate is defined by

$$\lambda(t;\theta) := \alpha(t) + \alpha_1(t;\theta), \tag{7.29}$$

and in analogy with the last section, we define

$$\sigma(t) := \alpha(t) - \beta(t), \tag{7.30}$$

and

$$A(t;\theta) := \sigma(t) + \alpha_1(t;\theta). \tag{7.31}$$

Next, the workload $x(t;\theta)$ and the spillover rate $\gamma(t;\theta)$ are defined by Equations (7.6) and (7.7), respectively, reproduced here:

$$\frac{dx(t;\theta)}{dt^+} = \begin{cases} 0, & \text{if } x(t;\theta) = 0 \text{ and } A(t;\theta) \leq 0, \\ 0, & \text{if } x(t;\theta) = \theta \text{ and } A(t;\theta) \geq 0 \\ A(t;\theta) & \text{otherwise.} \end{cases} \tag{7.32}$$

with the initial condition $x(0;\theta) = 0$ for simplicity, and

$$\gamma(t;\theta) = \begin{cases} A(t;\theta), & \text{if } x(t;\theta) = \theta, \\ 0, & \text{if } x(t;\theta) < 0. \end{cases} \tag{7.33}$$

Finally, the feedback input rate is given by

$$\alpha_1(t;\theta) = \gamma(t-T;\theta). \tag{7.34}$$

The performance measure of interest is the loss volume throughout the interval $[0, T_f]$, denoted by $L(\theta)$, and defined by

$$L(\theta) = \int_0^{T_f} \gamma(t;\theta)dt. \tag{7.35}$$

As in the last section we distinguish between various segments in the state trajectory: Empty Periods (EP) are maximal intervals during which $x(t;\theta) = 0$; Full Periods (FP) are maximal intervals during which $x(t;\theta) = \theta$; Boundary Periods (BP) are either EPs or FPs; and Nonboundary Periods (NBP) are supremal intervals during which $0 < x(t;\theta) < \theta$. We further denote the *nth* NBP by NBP_n, and its end points, by η_n and ζ_n (both generally functions of θ) so that $NBP_n = (\eta_n, \zeta_n)$. The boundary period immediately following NBP_n is $[\zeta_n, \eta_{n+1}]$. For an illustration, see Figure 7.3. As in the last section, we denote by Ψ_F the integer-set n such that the BP following NBP_n is an FP, namely, $\Psi_F := \{n = 1, 2, \ldots, \; : \; [\zeta_n, \eta_{n+1}] \text{ is an FP}\}$. With this definition, and by (7.35) and (7.33), the sample performance function assumes the following form,

$$L(\theta) \;=\; \sum_{n \in \Psi_F} \int_{\zeta_n}^{\eta_{n+1}} A(t;\theta) dt. \tag{7.36}$$

Embedded in the SFM processes is a sequence of events, and these are classified into three types: (i) An *exogenous event* is a jump (discontinuity) in $\sigma(t)$; (ii) an *induced event* is defined as a jump in $\alpha_1(t;\theta) = \gamma(t - T;\theta)$; and (iii) an *endogenous event* amounts to the end of an NBP, when the buffer becomes either full or empty. Similarly to the last section, we make the following assumptions.

ASSUMPTION 7.3 (i) W.p.1, the realization of the function $\sigma(t)$ is piecewise continuously differentiable on the interval $[0, T_f]$. (ii) There exists a constant $C > 0$ such that, for every realization of $\sigma(t)$, $|\sigma(t)| \leq C$ and $|\frac{d\sigma}{dt}(t)| \leq C$ for every $t \in [0, T_f]$. (iii) With K_1 denoting the number of discontinuities of $\sigma(t)$ in the interval $[0, T_f]$, $E[K_1] < \infty$. (iv) With K_2 denoting the number of times $\frac{d\sigma}{dt}(t)$ switches sign in the interval $[0, T_f]$, $E[K_2] < \infty$.

ASSUMPTION 7.4 For every $\theta > 0$, no two or more events occur at the same time $t \in [0, T_f]$.

Fix $\theta > 0$. We next analyze the sample path in order to derive a formula for the IPA derivative $L'(\theta)$. To this end, we assume that all BPs are of a positive length (i.e., none is a singleton), and that it does not happen that $A(t;\theta) = 0$ on an open time-interval contained in a BP. We point out that if one of these assumptions is not satisfied then the one-sided derivatives $\frac{dL(\theta)}{d\theta^-}$ and $\frac{dL(\theta)}{d\theta^+}$ still exist as long as Assumption 7.3 and Assumption 7.4 are in force.

According to (7.36), we have that

$$L'(\theta) \;=\; \sum_{n \in \Psi_F} \frac{d}{d\theta} \Big(\int_{\zeta_n}^{\eta_{n+1}} A(t;\theta) dt \Big). \tag{7.37}$$

Taking derivative for each term in the RHS of (7.37), and noting that $A(\zeta_n^+;\theta) = A(\zeta_n^-;\theta) = A(\zeta_n;\theta)$ by dint of Assumption 7.4, we obtain the following equation for every $n \in \Psi_F$,

$$\frac{d}{d\theta}\Big(\int_{\zeta_n}^{\eta_{n+1}} A(t;\theta)dt\Big) = \int_{\zeta_n}^{\eta_{n+1}} A^{'}(t;\theta)dt + A(\eta_{n+1}^{-};\theta)\eta_{n+1}^{'}(\theta) - A(\zeta_n;\theta)\zeta_n^{'}(\theta). \tag{7.38}$$

It remains to evaluate each one of the terms in the RHS of (7.38), and for the purpose of that we first investigate terms like $\big(A(\tau^{-};\theta) - A(\tau^{+};\theta)\big)\tau^{'}(\theta)$, where $\tau(\theta)$ is the time of an event.

LEMMA 7.4 *If $\tau(\theta)$ is the time of an exogenous event then $\tau^{'}(\theta) = 0$.*

PROOF Immediate from the fact that exogenous events and their timing are independent of θ. ∎

LEMMA 7.5 *Let $\tau(\theta)$ be the time of an induced event. If $A(\tau(\theta)^{-};\theta) > (A(\tau(\theta)^{+};\theta)$, then $\tau^{'}(\theta) = 0$.*

PROOF By definition of induced events, $\tau(\theta)$ is the time of a jump in $\alpha_1(t;\theta) = \gamma(t-T;\theta)$, which could be caused by either an exogenous event, an endogenous event, or an induced event at that time. In case of an exogenous event, $\tau^{'}(\theta) = 0$ by Lemma 7.4. An endogenous event means that the buffer became full, and hence $\gamma(t^{+} - T;\theta) \geq 0 = \gamma(t^{-} - T;\theta)$. This means that $A(\tau(\theta)^{-};\theta) \leq A(\tau(\theta)^{+};\theta)$, contradicting the assumption of this lemma. Finally, in case of an induced event, we carry the argument recursively backwards until, at some time $\tau(\theta) - kT$, $k = 2,3,\ldots$, there was an exogenous event. This recursive process is finite since there are no induced events in the interval $[0,T]$, and hence $\tau^{'}(\theta) = 0$. ∎

Recall that η_{n+1} is the end point of BP_n.

LEMMA 7.6 *For every $n \in \Psi_F$,*

$$A(\eta_{n+1}^{-};\theta)\eta_{n+1}^{'}(\theta) = A(\eta_{n+1}^{+};\theta)\eta_{n+1}^{'}(\theta) = 0. \tag{7.39}$$

PROOF $\eta_{n+1}(\theta)$ is the end point of an FP whose termination at that time can be brought about by either one of the following three causes: (i) A jump in $A(t;\theta)$ due to an exogenous event; (ii) a jump in $A(t;\theta)$ due to an induced event; and (iii) a continuous decline in the values of $A(t;\theta)$ from positive to negative as t passes through $\eta_{n+1}(\theta)$. In case (i) $\eta_{n+1}^{'}(\theta) = 0$ by Lemma 7.4; in case (ii) $\eta_{n+1}^{'}(\theta) = 0$ by Lemma 7.5; and in case (iii) $A(\eta_{n+1}^{-};\theta) = A(\eta_{n+1}^{+};\theta) = A(\eta_{n+1};\theta) = 0$. In either case, (7.39) is satisfied. ∎

By (7.38) and Lemma 7.6 it now follows that for every $n \in \Psi_F$,

$$\frac{d}{d\theta}\Big(\int_{\zeta_n}^{\eta_{n+1}} A(t;\theta)dt\Big) \;=\; \int_{\zeta_n}^{\eta_{n+1}} A^{'}(t;\theta)dt \;-\; A(\zeta_n;\theta)\zeta_n^{'}(\theta). \tag{7.40}$$

We next consider the term $A^{'}(t;\theta)$ in the RHS of (7.40).

LEMMA 7.7 *If t is neither the time of an event nor the end time of an FP, then $A^{'}(t;\theta) = 0$.*

PROOF Let t satisfy the above condition. Since $A(t;\theta) = \sigma(t) + \gamma(t-T;\theta)$, we have that $A^{'}(t;\theta) = \gamma^{'}(t-T;\theta)$. If the point $t-T$ was not in an FP then clearly $A^{'}(t;\theta) = 0$. Suppose then that $t-T$ was in an FP. By (7.33), $\gamma(t-T;\theta) = A(t-T;\theta)$. Repeating the argument recursively backwards for $t-kT$, $k = 2,3,\ldots$, we conclude that $A^{'}(t;\theta) = 0$. ∎

Lemma 7.7 implies that the integral term in the RHS of (7.40) depends on the timing of events occurring during the FP $[\zeta_n(\theta), \eta_{n+1}(\theta)]$. To formalize, we denote by $z_k = z_k(\theta)$ the time of the kth induced event in the interval $[0, T_f]$, and for every $n \in \Psi_F$, we define Φ_n by

$$\Phi_n := \{k = 1,2,\ldots, | z_k(\theta) \in (\zeta_n(\theta), \eta_{n+1}(\theta))\}. \tag{7.41}$$

We now have the following result.

PROPOSITION 7.3 *For every $n \in \Psi_F$, the following equation is in force.*

$$\frac{d}{d\theta}\Big(\int_{\zeta_n}^{\eta_{n+1}} A(t;\theta)dt\Big) \;=\; \sum_{k\in\Phi_n} \big(A(z_k(\theta)^-;\theta) - A(z_k(\theta)^+;\theta)\big) z_k^{'}(\theta) - A(\zeta_n;\theta)\zeta_n^{'}(\theta). \tag{7.42}$$

PROOF By Lemma 7.7, the only nonzero terms contributing to the integral term in the RHS of (7.40) are due to jumps in $A(t;\theta)$ during the FP $[\zeta_n, \eta_{n+1}]$. These jumps correspond to exogenous events or induced events. Equation (7.42) now follows from Lemma 7.4. ∎

We next derive an expression for the last term in the RHS of (7.42), $A(\zeta_n;\theta)\zeta_n^{'}(\theta)$. Define Ξ_n by

$$\Xi_n := \{k = 1,2,\ldots, | z_k(\theta) \in (\eta_n(\theta), \zeta_n(\theta))\}, \tag{7.43}$$

namely the index of the times of induced events occurring in $NBP_n = (\eta_n(\theta), \zeta_n(\theta))$.

PROPOSITION 7.4 *For every $n \in \Psi_F$, the term $A(\zeta_n;\theta)\zeta_n^{'}(\theta)$ has the* *following form,*

$$A(\zeta_n;\theta)\zeta_n'(\theta) = \begin{cases} 1 + A(\eta_n^+;\theta)\eta_n'(\theta) - \sum_{k\in\Xi_n}\Big(A(z_k(\theta)^-;\theta) - A(z_k(\theta)^+;\theta)\Big)z_k'(\theta), \\ \qquad \text{if } x(\eta_n(\theta);\theta) = 0 \\ -\sum_{k\in\Xi_n}\Big(A(z_k(\theta)^-;\theta) - A(z_k(\theta)^+;\theta)\Big)z_k'(\theta), \\ \qquad \text{if } x(\eta_n(\theta);\theta) = \theta. \end{cases} \tag{7.44}$$

PROOF Consider first the case where $x(\eta_n(\theta);\theta) = 0$. Then, by (7.32),

$$\int_{\eta_n(\theta)}^{\zeta_n(\theta)} A(t;\theta)dt = \theta. \tag{7.45}$$

Taking derivatives with respect to θ and using Lemma 7.7, we obtain that

$$\sum_{k\in\Xi_n} \big(A(z_k^-;\theta) - A(z_k^+;\theta)\big)z_k'(\theta) + A(\zeta_n;\theta)\zeta_n'(\theta) - A(\eta_n^+;\theta)\eta_n'(\theta) = 1. \tag{7.46}$$

On the other hand, if $x(\eta_n(\theta);\theta) = \theta$ then (by (7.32)),

$$\int_{\eta_n(\theta)}^{\zeta_n(\theta)} A(t;\theta)dt = 0, \tag{7.47}$$

and taking derivatives with respect to θ,

$$\sum_{k\in\Xi_n} \big(A(z_k^-;\theta) - A(z_k^+;\theta)\big)z_k'(\theta) + A(\zeta_n;\theta)\zeta_n'(\theta) - A(\eta_n^+;\theta)\eta_n'(\theta) = 0; \tag{7.48}$$

since $\eta_n(\theta)$ is the end of an FP, Lemma 7.6 implies that $A(\eta_n^+;\theta)\eta_n'(\theta) = 0$, and hence, and by (7.48),

$$\sum_{k\in\Xi_n} \big(A(z_k^-;\theta) - A(z_k^+;\theta)\big)z_k'(\theta) + A(\zeta_n;\theta)\zeta_n'(\theta) = 0. \tag{7.49}$$

Finally, (7.44) follows from (7.46) and (7.49). ■

To put it all together, recall (from Section 7.2) that a non-empty period is defined as a supremal subinterval of $[0,T_f]$ where the buffer is not empty, and a non-empty period is lossy if it incurs some loss. Let $\mathcal{N}_m$, $m = 1,\ldots,M$ denote the lossy non-empty periods in increasing order; both M and $\mathcal{N}_m$ are θ-dependent and random. Furthermore, define $L_m(\theta) := \int_{\mathcal{N}_m}\gamma(t;\theta)dt$, and note that

$$L'(\theta) = \sum_{m=1}^{M} L_m'(\theta). \tag{7.50}$$

Let b_m denote the beginning time of $\mathcal{N}_m$, and let e_m denote the last time in $\mathcal{N}_m$ when the buffer is full. Define $\Lambda_m := \{k = 1,2,\ldots,|z_k(\theta) \in (b_m,e_m]\}$. We now have the following result.

PROPOSITION 7.5 *$L_m^{'}(\theta)$ has the following form,*

$$L_m^{'}(\theta) = \sum_{k\in\Lambda_m} \big(A(z_k^-;\theta) - A(z_k^+;\theta)\big)z_k^{'}(\theta) - 1 - A(b_m^+;\theta)b_m^{'}(\theta). \tag{7.51}$$

PROOF Follows immediately from Proposition 7.3 and Proposition 7.4. ∎

To gauge the effect of Proposition 7.5 on the computation of $L^{'}(\theta)$, consider first a jump in $\gamma(\cdot;\theta)$ at time t. This jump will induce an event T seconds later, at time $t+T$. At this time there are three possibilities: (i) The buffer is full; (ii) the buffer is empty; and (iii) the buffer is neither full nor empty. In the first case, another event will be induced T seconds later. In the second case, the induced event at time $t+T$ will not induce any future events. In the third case $t+T$ is in an NBP, and there are two possibilities regarding it: (1) The NBP ends in an EP; and (2) the NBP ends in an FP. In the first case the event at time $t+T$ will not induce any future events, and in the second case, it will induce an event T seconds after the end of the NBP.

Next, consider the terms $\big(A(z_k^-;\theta) - A(z_k^+;\theta)z_k^{'}(\theta)$ in the RHS of Equation (7.51). $z_k(\theta)$ is the time of an event induced by a jump in $\gamma(\cdot;\theta)$ T seconds earlier. The inducing event at that time was either (i) induced; (ii) exogenous; or (iii) endogenous. In the first case $t-T = z_\ell(\theta)$ for some $\ell < k$, and hence (and by (7.33)), $\big(A(z_k^-;\theta) - A(z_k^+;\theta)\big)z_k^{'}(\theta) = \big(A(z_\ell^-;\theta) - A(z_\ell^+;\theta)\big)z_\ell^{'}(\theta)$. In case (ii), $\big(A(z_k^-;\theta) - A(z_k^+;\theta)\big)z_k^{'}(\theta) = 0$. In case (iii), $t-T = \zeta_n(\theta)$ for some $n \in \Psi_F$, and $\big(A(z_k^-;\theta) - A(z_k^+;\theta)\big)z_k^{'}(\theta) = A(\zeta_n;\theta)\zeta_n^{'}(\theta)$, where the latter term is given by Proposition 7.4.

We discern a recursive structure whereby each lossy non-empty period $\mathcal{N}_m = (b_m, e_m)$ contributes to $L^{'}(\theta)$ the term $-1 - A(b_m;\theta)b_m^{'}(\theta)$, and all other terms in (7.51) correspond to induced events. The IPA derivative $L^{'}(\theta)$ can be computed by Algorithm 1, below. This algorithm uses two accumulators, respectively denoted by $\mathcal{A}$ and $\mathcal{A}_x$; $\mathcal{A}$ is used to compute $L^{'}(\theta)$, and $\mathcal{A}_x$ is used to accumulate the effects of the sum-terms in the RHS of (7.51) during a NBP to see if it would end in an EP or an FP.

The algorithm has the structure of a DES whose events are embedded in the trajectory of the workload process, and thus it jumps forward in time from one event to the next. We classify three types of events: (i) End of an EP; (ii) buffer becoming full; and (iii) induced event of the system. Note a slight departure from the taxonomy of events as earlier defined for the system, since we include here the end of an EP as an event. The event-type occurring at time t is denoted by $\varepsilon(t)$. At every time t the algorithm has a list of future induced events that have been previously scheduled, and their future occurrence times. In addition, the sample path will yield the next end-of-EP event or buffer-becoming-full event. To borrow terminology from the literature on discrete event simulation, we denote by $\mathcal{E}$ the set of enabled events, namely the events which may occur, and we note that $\mathcal{E}$ depends on t as well as on the state of the system. Moreover, for every $e \in \mathcal{E}$, we denote the next occurrence time of e by t_e. We assume that the algorithm, in conjunction with the sample path, updates the set $\mathcal{E}$ and the clock variables t_e (for all $e \in \mathcal{E}$) and we will not mention this action explicitly in its formal description. Associated with each pending induced event at a future time t there is a quantity $c(t)$ that will be added to either one of the accumulators. The accumulator $\mathcal{A}$ is updated whenever the buffer becomes full or an induced event occurs during an FP. However, induced events during a NBP may or may not eventually be added to $\mathcal{A}$ depending on whether the NBP will be followed by an FP or an EP, and therefore, the temporary accumulator $\mathcal{A}_x$ stores the effects of the induced events occurring during the NBP until its end.

ALGORITHM 7.1 *Initialization:* Set $t = 0$, $\mathcal{A} = 0$, and $\mathcal{A}_x = 0$.
MainLoop: While ($t < T_f$) {
  Compute $t := \min\{t_e : e \in \mathcal{E}\}$ and advance time to t
  If ($\varepsilon(t)$ = end-of-EP; i.e., $t = b_m(\theta)$ for some $m \geq 1$) {
    Set $A_x = -1 - A(b_m^+;\theta)b_m^{'}(\theta)$
  }
  Else if ($\varepsilon(t)$ = buffer-becoming-full) {
    Set $A = A + A_x$
    Set $c(t+T) = \mathcal{A}_x$
    Set $A_x = 0$
  }
  Else (i.e., $\varepsilon(t)$ = induced-event) {
    If ($t \in$ FP) {
      Set $A = A + c(t)$
      Set $c(t+T) = c(t)$
    }
    else if ($t \in$ NBP) {
      Set $A_x = A_x + c(t)$
    }
  }
}
Set $L^{'}(\theta) = \mathcal{A}$.

The algorithm would be based on nonparametric and model-free formulas (amounting to a counting process), and hence computable in real time, if the term

$$A(b_m(\theta)^+;\theta)b_m^{'}(\theta)$$

were always equal to 0; see Equation (7.51). The latter term, if non-zero, disturbs the model-free structure. It certainly can be computed from simulation, but its evaluation in real time may be problematic. However, we next argue that it would not arise frequently under certain traffic conditions. Recall that $b_m(\theta)$ is the time an EP ends and hence an NBP begins, and this means a transition of $A(\cdot;\theta)$ from negative to positive at time $t = b_m(\theta)$. If this transition is continuous then $A(b_m(\theta);\theta) = 0$, and hence $A(b_m(\theta)^+;\theta)b_m^{'}(\theta) = 0$. If this is an abrupt change (jump) due to an exogenous event, then $b_m^{'}(\theta) = 0$, and again $A(b_m(\theta)^+;\theta)b_m^{'}(\theta) = 0$. Only if this transition is due to an induced event, is it possible that $A(b_m(\theta)^+;\theta)b_m^{'}(\theta) \neq 0$. This means that a jump in $\gamma(\cdot;\theta)$ at time $b_m(\theta)-T$, when the buffer was full, causes the termination of an EP at time $b_m(\theta)$. This, we believe, would occur infrequently except under widely fluctuating network traffic conditions, and hence we neglected this term in the algorithm's implementation whose results are presented in the next section.

Finally, the unbiasedness of the IPA estimator $L^{'}(\theta)$ follows in the same way as in the last section, and hence the next proposition is stated without a proof. Let θ be constrained to a closed and bounded interval Θ whose left end-point is positive.

PROPOSITION 7.6 *Suppose that Assumption 7.3 and Assumption 7.4 are satisfied. Then, w.p.1, the function $L(\theta)$ has a Lipschitz constant K throughout Θ, and $E[K]<\infty$.*

PROOF The proof is analogous to that of Proposition 7.2. ∎

7.4 Simulation Experiments

In order to verify the theoretical results developed in the last section, we ran some simulations of a discrete, packet-based queue. The input process consists of the superposition of 100 on-off sources. The durations of both the *on* and *off* periods are independent, exponentially distributed with mean of 50 msec each. During an *on* period, data flows from a source at the rate of 190 Kbps, and then assembled into 512-byte packets. Only full packets are transmitted; those that do not fill up during an *on* period have to wait for the next *on* period to be completed. All of the packets are multiplexed for transmission on a 10-Mbps line according to a FIFO queueing discipline, so that the nominal traffic intensity is $\rho=0.95$. The buffer holds complete packets, and hence its size, θ, has integer values in terms of packets. The retransmission timer has the value $T=0.14$ seconds, about 312 times the packets' transmission time.

We considered the optimization problem of balancing network inefficiency with maximum packet delay. Inefficiency is characterized by the loss volume, $L(\theta)$, since any loss requires retransmission, while the maximum packet transmission time is proportional to the buffer size, θ. Therefore, we defined the optimization problem as minimizing the function $f(\theta)$ defined by

$$f(\theta) \;=\; E\big(L(\theta)\big)+\theta, \tag{7.52}$$

where we chose $T_f=1.0$ second. We used IPA in conjunction with a stochastic approximation algorithm to solve this problem dynamically. The algorithm updates the parameter θ by computing the sample derivative $L^{'}(\theta)+1$ and then taking a

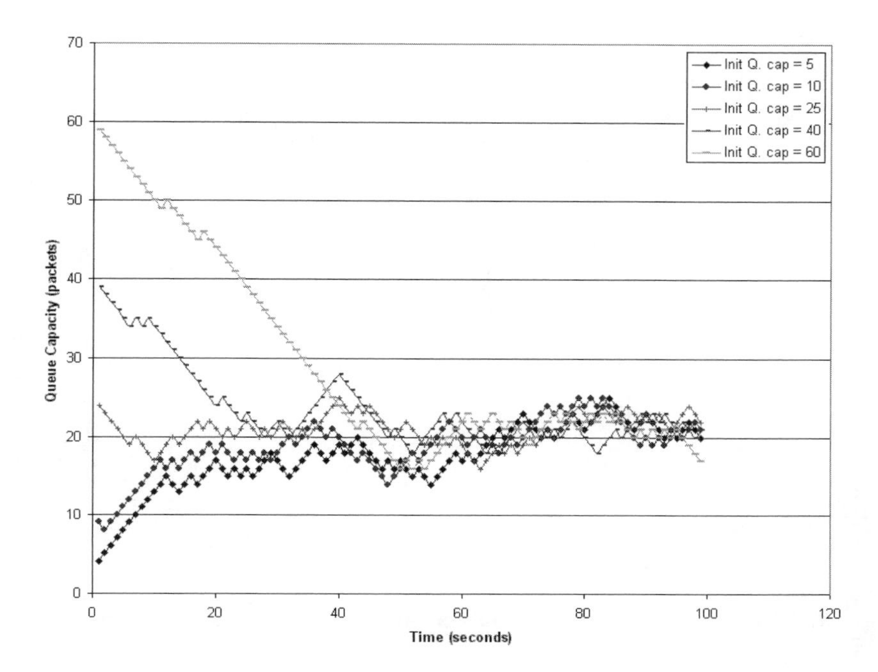

FIGURE 7.5: Simulation results of the optimization algorithm with various starting parameter points.

step in the opposite direction. The step has the smallest-possible value, namely one packet. Thus, denoting by θ_k the value of θ in the kth iteration, the algorithm has the following form,

$$\theta_{k+1} = \theta_k - sgn\big(L^{'}(\theta) + 1\big), \tag{7.53}$$

where $sgn(\cdot)$ is the sign function. We point out that the simulation/optimization program did not reset the queue to empty whenever a new iteration was computed. Instead, it kept the final state (including all variables used to compute L and $L^{'}$) of the simulation at θ_k to be used as the initial state of the simulation at θ_{k+1}. This corresponds more closely to what would take place in real-time control than resetting the initial state to that of an empty queue at each new iteration.

Results of the optimization algorithm are shown in Figure 7.5. The various graphs correspond to various initial iteration points, θ_1, and they all used different random seeds. They show the progression of the iterates θ_k as functions of the iteration index $k = 1, \ldots, 100$. Thus, since each iteration was simulated for one second (i.e., $T_f = 1.0$), each run of the optimization algorithm corresponded to 100 seconds of simulated time. The results indicate convergence to an optimal value of θ at slightly above 20. These results are corroborated by Figure 7.6, which contains a plot of the function $f(\theta) = L(\theta) + 1$ as a function of θ, computed by extensive and independent

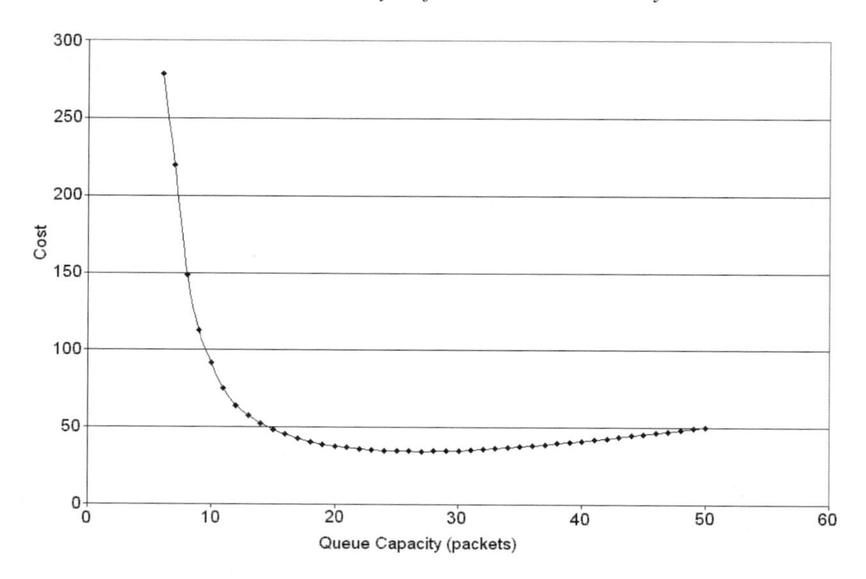

FIGURE 7.6: Graph of the sample performance function as a function of θ.

simulation runs.

7.5 Conclusions

This chapter concerns the application of infinitesimal perturbation analysis to fluid queues with feedback. It derives basic results for two kinds of single-server systems: One implements flow control whereby the inflow rate is tuned by the queue's content, and the other models timer-based retransmission of fluid that is lost due to buffer overflow. In either case, we derived the IPA derivative of the loss volume as a function of the buffer size, and showed it to admit nonparametric and model-free approximations. Some extensions of these results have appeared in [19, 18, 20], and they suggest further extensions to a large class of networks and performance measures that could be of interest in telecommunications applications. The nonparametric and model-free nature of the IPA estimators, or adequate approximations thereof, hold promise of the development of a novel technique for network management and control. While most management techniques are based on observation and monitoring of network performance, the availability of nonparametric and model-free IPA can add derivative information to the decision-making process, thereby resulting in potentially more effective management and control techniques. Future research will focus on the development of these ideas.

Acknowledgments. The section on flow control summarizes research performed by Haining Yu and Christos G. Cassandras. The section on the retransmission-based model presents results developed by the authors of this chapter.

References

[1] D. Anick, D. Mitra, and M. Sondhi, "Stochastic Theory of a Data-Handling System with Multiple Sources," *The Bell System Technical Journal*, Vol. 61, pp. 1871-1894, 1982.

[2] S. Bohacek, J. Hespanha, J. Lee, and K. Obraczka, "A Hybrid System Modeling Framework for Fast and Accurate Simulation of Data Communication Networks," in *Proc. ACM International Conference on Measurements and Modeling of Computer Systems (SIGMETRICS)*, June 2003.

[3] C.G. Cassandras and S. Lafortune, *Introduction to Discrete Event Systems*, Kluwer Academic Publishers, Boston, Massachusetts, 1999.

[4] C.G. Cassandras, Y. Wardi, B. Melamed, G. Sun, and C.G. Panayiotou, "Perturbation Analysis for On-Line Control and Optimization of Stochastic Fluid Models," *IEEE Transactions on Automatic Control*, Vol. AC-47, No. 8, pp. 1234-1248, 2002.

[5] C.G. Cassandras, G. Sun, C.G. Panayiotou, and Y. Wardi, "Perturbation Analysis and Control of Two-Class Stochastic Fluid Models for Communication Networks," *IEEE Transactions on Automatic Control*, Vol. 48, pp. 770-782, 2003.

[6] R. Cruz, "A Calculus for Network Delay, Part I: Network Elements in Isolation," *IEEE Transactions on Information Theory*, 1991.

[7] Y.C. Ho and X.R. Cao, *Perturbation Analysis of Discrete Event Dynamic Systems*, Kluwer Academic Publishers, Boston, Massachusetts, 1991.

[8] G. Kesidis, A. Singh, D. Cheung, and W. Kwok, "Feasibility of Fluid-Driven Simulation for ATM Networks," in *Proc. IEEE Globecom*, Vol. 3, pp. 2013-2017, 1996.

[9] H. Kobayashi and Q. Ren, "A Mathematical Theory for Transient Analysis of Communications Networks," *IEICE Transactions on Communications*, Vol. E75-B, pp. 1266-1276, 1992.

[10] K. Kumaran and D. Mitra, "Performance and Fluid Simulations of a Novel Shared Buffer Management Systems", in *Proc. IEEE INFOCOM*, March 1998.

[11] B. Liu, Y. Guo, J. Kurose, D. Towsley, and W.-B. Gong, "Fluid Simulation of Large-Scale Networks: Issues and Tradeoffs," in *Proc. the International Conerence on Parallel and Distributed Processing Techniques and Applications*, Las Vegas, Nevada, June 1999.

[12] B. Melamed, S. Pan, and Y. Wardi, "HNS: A Streamlined Hybrid Network Simulator," *ACM Transactions on Modeling and Simulation*, Vol. 14, no. 3, pp. 1-27, 2004.

[13] R.Y. Rubinstein and A. Shapiro, *Discrete Event Systems: Sensitivity Analysis and Stochastic Optimization by the Score Function Method*, John Wiley and Sons, New York, New York, 1993.

[14] G. Sun, C.G. Cassandras, Y. Wardi, C.G. Panayiotou, and G. Riley, "Perturbation Analysis and Optimization of Stochastic Flow Networks," *IEEE Transactions on Automatic Control*, Vol. 49, No. 12, pp. 2143-2159, 2004.

[15] Y. Wardi, B. Melamed, C.G. Cassandras, and C.P. Panayiotou, "On-Line IPA Gradient Estimators in Stochastic Continuous Fluid Models," *Journal of Optimization Theory and Applications*, Vol. 115, No. 2, pp. 369-406, 2002.

[16] Y. Wardi and G. Riley, "IPA for Spillover Volume in a Fluid Queue with Retransmissions," in *Proc. 43rd IEEE Conference on Decision and Control*, pp. 3756-3761, Nassau, Bahamas, December 2004.

[17] A. Yan and W.-B. Gong, "Fluid Simulation for High-Speed Networks with Flow-Based Routing," *IEEE Transactions on Information Theory*, Vol. 45, pp. 1588-1599, 1999.

[18] H. Yu and C.G. Cassandras, "Perturbation Analysis of Feedback-Controlled Stochastic Flow Systems," *IEEE Transactions on Automatic Control*, Vol. 49, No. 8, pp. 1317-1332, 2004.

[19] H. Yu and C.G. Cassandras, "A New Paradigm for an On-Line Management of Communication Networks with Multiplicative Feedback Control," in *Performance Evaluation and Planning Methods for the Next Generation Internet*, Eds. A. Girard, B. Sanso', and F. Vasquez-Abad, pp. 297-332, Springer-Verlag, New York, 2005.

[20] H. Yu and C.G. Cassandras, "Perturbation Analysis and Feedback Control of Communication Networks Using Stochastic Hybrid Models," to appear in *Nonlinear Analysis*, 2006.

Chapter 8

Stochastic Hybrid Modeling of On-Off TCP Flows

João Hespanha
University of California, Santa Barbara

8.1 Related Work 193
8.2 A Stochastic Model for TCP 199
8.3 Analysis of the TCP SHS Models 203
8.4 Reduced-order Models 204
8.5 Conclusions 211
Appendix 212
References 216

Most of today's Internet traffic uses the Transmission Control Protocol (TCP) and it is therefore not surprising that modeling TCP's behavior has been attracting the attention of academia and industry ever since this protocol was proposed.

TCP is responsible for regulating the rate at which a source of data transmits packets, a task known as *congestion control*. TCP follows a greedy algorithm that constantly attempts to increase the rate at which data is transmitted. Eventually, the network is unable to carry the data and packets are dropped. When this is detected, the TCP source decreases its sending rate by roughly dividing it in half. However, shortly after that, the TCP source returns to its continuous attempt to increase the sending rate and the cycle repeats itself. During the first cycle (i.e., until the first drop) the rate is increased in an exponential fashion to rapidly reach an adequate value. This is called TCP's *slow-start* mode. However, in subsequent cycles, the increase is more gentle and roughly follows a linear law. This more gentle increase is crucial for stability (see, e.g., [6]). This second phase is called *congestion-avoidance*. The reader may want to consult a textbook, such as [33], for a more detailed description of TCP.

As noted by Bohacek et al. [7], hybrid systems provide a natural framework to model TCP because this protocol is characterized by different modes of operation with distinct dynamics. The need for stochastic hybrid systems (SHS) arises because many of the events that drive traffic models — such as packet drops and the start/termination of transmissions — are well modeled by stochastic processes.

In this chapter, we construct a SHS model for TCP. Each realization of this model is meant to represent the traffic generated by a single user that initiates a TCP session, waits until the transfer terminates (the "transmission-on" period), spends some time "processing" the file received (the "transmission-off" period), and then initiates a new TCP session. This behavior is continuously repeated, but the transfer-sizes and the durations of the off-periods are selected from pre-specified distributions. The transfer-sizes implicitly determine the duration of the on-periods, taking into account TCP's instantaneous transfer rate. This type of on-off model was also considered by [2].

Since we do not restrict our attention to infinitely long TCP sessions (usually called long-lived flows), we need to explicitly model TCP's slow-start mode that dominates short transfers [23]. Moreover, we take into account the delay between the time instant at which a drop is detected and the corresponding reaction by TCP, typically one round-trip time later.

Using moment closure techniques inspired by Bohacek [5] and further discussed in [13], we build systems of ordinary differential equations (ODEs) that provide accurate approximations to the dynamics of the average sending rate as well as its higher order moments, including the standard deviation.

When our results are specialized to long-lived flows (always-on), the average sending rates are consistent with previously established models (at least for slowly varying drop-probabilities), which validates our modeling methodology.

The most surprising results are obtained for on-off flows. For a few transfer-size distributions reported in the literature, we show that the standard deviation is much larger than the average sending rate of individual TCP flows. Moreover, the packet drop probability has a surprisingly small effect on the average sending rate, but provides a strong control on its standard deviation. The explanation seems to be that, even with a heavy tail, the bulk of the data "slips-through" with very few drops. In practice, either all packets are sent during the slow-start mode or shortly after TCP enters the congestion-avoidance mode. The precise time at which the first drop occurs has a tremendous influence on the average sending rate of the flow, thus the very high standard deviation.

The fact that the "heaviness" of the tail seems to have little impact on Internet's performance was also pointed out by Liu et al. [17]. However, the fact that the packet drop probability exerts a much larger impact on the standard deviation of the sending rate than on its average has significant implications for congestion control. In particular, it seems to indicate that previously used long-lived flow models may not be suitable for the analysis and design of congestion control algorithms for on-off TCP flows. It also questions the validity of the "TCP-friendly" formula for aggregate on-off TCP flows.

The remainder of this chapter is organized as follows. In Section 8.1 we discuss our overall modeling approach to TCP in the context of alternative modeling techniques. In Section 8.2 we present the formal Stochastic Hybrid Systems (SHS) model for a single-user on-off TCP flow. A short introduction to the class of SHS considered in this chapter is provided in the Appendix for the interested reader. In Section 8.3 we derive an infinite-dimensional system of ordinary differential equations that describe the dynamics of the statistical moments of TCP's sending rate. In Section 8.4 we show how this model can be truncated to obtain approximate models that are amenable to the investigation of TCP's behavior. We examine the properties of the model obtained for long-lived flows (always-on) as well as for on-off flows with two realistic transfer-size distributions.

8.1 Related Work

There is an extensive literature on models that describe the properties of TCP congestion control. We start by presenting several widely used models for long-lived TCP flows (corresponding to infinitely long transfers). We then discuss a few more recent models for finite TCP flows.

8.1.1 Models for Long-lived Flows

A great deal of effort has been placed in characterizing the steady-state behavior of long-lived TCP flows [26, 20, 22, 27, 28, 31]. In particular, in studying the relation between the average transmission rate μ, the average round-trip time RTT, and the per-packet drop probability p_{drp}. The so called "TCP-friendly" formula

$$\mu = \frac{c}{RTT\sqrt{p_{\text{drp}}}} \tag{8.1}$$

has been derived by several authors, with small variations on the value of the constant c: Ott et al. [26] obtained $c = 1.310$; Mahdavi and Floyd [20] and Mathis et al. [22] obtained $c = 1.225$; and Bohacek et al. [6] obtained $c = 1.270$ for small values of p_{drp}. This formula reflects the fact that drops are the sole mechanism that keeps TCP's transmission rate bounded, when there is an infinite amount of data to transmit. Moreover, it specifies that as the drop probability goes to zero, the transmission rate should grow to infinity inversely proportional to $\sqrt{p_{\text{drp}}}$.

In general, the receiver acknowledges the arrival of each data packet by sending a short ACK packet back to the source. However, the protocol allows for the transmission of acknowledgments to be delayed, so that a single ACK packet can acknowledge the arrival of multiple data packets. Padhye et al. [27, 28] considered the general case of delayed acknowledgments and obtained $c = 1.225/\sqrt{n_{\text{ack}}}$ for (8.1), where n_{ack} denotes the number of acknowledgments per ACK packet. Typically $n_{\text{ack}} = 2$ when there are delayed acknowledgments and $n_{\text{ack}} = 1$ in their absence.

The primary mechanism for detection of drops is through the *triple duplicate ACKs mechanism*. In essence, a source declares a data packet dropped if it did not yet receive an acknowledgment for that packet, but it already received ACK packets for three data packets that were subsequently transmitted. However, since an ACK packet is only generated when a data packet successfully reaches the destination, this mechanism for drop detection fails when not enough data packets reach the destination (after a packet is dropped). A timeout mechanism is used to recover from this situation. Padhye et al. [27, 28] further improved (8.1) by considering the effect of timeouts due to insufficient ACKs to detect a drop through the triple duplicate ACKs mechanism. They also took into account receiver-imposed limitations on the congestion window size. This led to

$$\mu = \min\left\{\frac{W_{\max}}{RTT}, \left(RTT\sqrt{\frac{2n_{\text{ack}}\,p_{\text{drp}}}{3}} + T_0\,p_{\text{drp}}(1+32p_{\text{drp}}^2)\min\{1, \sqrt{6n_{\text{ack}}\,p_{\text{drp}}}\}\right)^{-1}\right\}, \quad (8.2)$$

where $W_{\max}$ denotes the maximum congestion window size and T_0 the period during which no packets are sent following a timeout (typically determined experimentally). This formula proved reasonably accurate even when a significant portion of drops are detected through timeouts, for which (8.1) is not. Sikdar et al. [31] derived an alternative formula to (8.2), which proved equally accurate. However, both derivations assume that losses within a window are strongly correlated. In particular that when one drop occurs all subsequent packets in the same window are also dropped. This assumption is reasonable for drop-tail queuing, but not for active queuing policies such as Random Early Detection (RED) [10].

The derivations of (8.1) and (8.2) in the papers mentioned above analyze the evolution of the congestion window size between consecutive drop events for a single flow. Therefore, μ should be understood as a *time-average for a single TCP flow*. This type of approach was pursued in [27, 28, 30, 19, 15, 16, 18] to derive dynamic models for the congestion-avoidance stage of long-lived TCP flows. Kunniyur and Srikant [15] and Lakshmikantha et al. [16] proposed the continuous-time model

$$\dot{\mu}(t) = \frac{1}{RTT^2} - \frac{2}{3}p_{\text{drp}}(t-RTT)\mu(t)\mu(t-RTT) \quad (8.3)$$

where $\mu(t)$ denotes an "instantaneous" average sending rate, RTT the window size, and $p_{\text{drp}}(t)$ the per-packet drop probability; whereas Low et al. [19] proposed the continuous-time model

$$\dot{\omega}(t) = \frac{(1-p_{\text{drp}}(t-\tau))\,\omega(t-\tau)}{RTT(t-\tau)\omega(t)} - \frac{1}{2}\frac{p_{\text{drp}}(t-\tau)\omega(t)\omega(t-\tau)}{RTT(t-\tau)}, \qquad \mu(t) = \frac{\omega(t)}{RTT(t)},$$

where ω denotes the "instantaneous" average congestion window size, and τ a constant "equilibrium" round-trip time. Shakkottai and Srikant [30] proposed the discrete-time model

$$\mu(t+1) = \mu(t) + \frac{1}{RTT^2} - \frac{2}{3}p_{\text{drp}}(t-RTT)\mu(t-RTT)\big(\mu(t-RTT)+a\big) \quad (8.4)$$

where μ denotes the average congestion window size, RTT the average window size (in discrete-time units), and a the average of the "uncontrolled" competing flow; whereas Low [18] proposed the discrete-time model

$$\mu(t+1) = \mu(t) + \frac{1-p_{\text{drp}}(t)}{RTT^2} - \frac{2}{3}p_{\text{drp}}(t)\mu(t)^2 \quad (8.5)$$

In these discrete-time models, the time units must be sufficiently large for the time average to be meaningful.

In the models (8.3)–(8.5) the state should be understood as a time averaged quantity over an interval of time for which the fluid approximation is meaningful. In particular, for a "drop-rate" to be meaningful the averaging period T_{ave} must be sufficiently large to include several drops. Using (8.1), we can compute the (steady-state) average number of drops that occur in T_{ave}, which should be much larger than one. Since the average number of drops in an interval of length T_{ave} is given by the number of packets $\mu\, T_{\text{ave}}$ sent in this period times the drop probability p_{drp}, we conclude that we must have

$$p_{\text{drp}}\, \mu\, T_{\text{ave}} = \frac{c\sqrt{d}T_{\text{ave}}}{RTT} \gg 1 \Leftrightarrow T_{\text{ave}} \gg \frac{RTT}{c\sqrt{p_{\text{drp}}}} \; (\gg RTT).$$

One should therefore not expect these models to be valid over time scales of the order of $\frac{RTT}{c\sqrt{p_{\text{drp}}}}$ or smaller. It turns out that if one linearizes[1] (8.3) around its equilibrium point, one obtains a stable system with a time constant of $\frac{1.633RTT}{\sqrt{p_{\text{drp}}}}$, which is never significantly larger than $\frac{RTT}{c\sqrt{p_{\text{drp}}}}$. Similar conclusions can be obtained for the other models discussed so far. At least from a theoretical perspective, this seems to compromise the validity of these single-flow models. However, we will see shortly that these models can be re-interpreted as multi-flow ensemble models, for which these concerns do not necessarily arise.

When one wants to examine the dynamics of TCP flows for time scales on the order of only a few round-trip times, the time averaging period must be shortened. Any fluid approximation must use an averaging period no smaller than one round-trip time because otherwise this would violate approximating the sending rate by the congestion window size divided by the round-trip time. Bohacek et al. [7] proposed a modeling framework where quantities are only averaged over time periods of roughly one round-trip time. Drops were kept as discrete-events and were modeled explicitly, leading to a *hybrid control system.* The models developed were shown to be very accurate, even looking at time-traces of individual flows. The models developed for TCP flows correspond to the transfer of finitely many packets, capturing slow-start, congestion-avoidance, fast-recovery, and timeouts.

Misra et al. [24, 25] took a different approach to avoid averaging over long time intervals. They still consider averaging over time periods of roughly one round-trip time, but utilize *ensemble averages* for drop events. Using Itô Calculus they derive the following approximate model for the congestion-avoidance stage of long-lived TCP flows:

$$\dot{\omega}(t) = \frac{1}{RTT(t)} - \frac{\omega(t)p_{\text{drp}}\big(t-RTT(t)\big)\,\omega\big(t-RTT(t)\big)}{2RTT\big(t-RTT(t)\big)}, \quad \mu(t) = \frac{\omega(t)}{RTT(t)}, \tag{8.6}$$

[1] For simplicity, we ignore the delays which are much smaller than T_{ave} as well as the dependence of RTT on μ.

where ω should be understood as the *ensemble average* of the congestion window size and p_{drp} the per-packet drop probability (also in an ensemble sense). Interestingly, aside from the different interpretation of the quantities involved, the ensemble-average model (8.6) resembles very much the time-average models (8.3)–(8.5). This model was refined to include the effect of slow-start and timeouts, leading to

$$\dot{\omega}(t) = \frac{p_{\text{ss}}\omega(t) + (1 - p_{\text{ss}})}{RTT(t)} - \left(\frac{\omega(t)}{2}(1 - p_{\text{to}}) + (\omega(t) - 1)p_{\text{to}}\right) \frac{p_{\text{drp}}\big(t - RTT(t)\big)\,\omega\big(t - RTT(t)\big)}{RTT\big(t - RTT(t)\big)}, \quad (8.7)$$

where p_{ss} denotes the probability that a flow is in slow-start mode and p_{to} the probability that a drop leads to timeout. However, (8.7) still ignores the zeroing of the sending rate during the timeout period and does not provide values for the probability p_{ss}. It is suggested that p_{to} can be approximated by $\min\{1, 3/\omega\}$ based on an argument by Padhye et al. [28] that assumes that losses within a window are strongly correlated. As mentioned above, this assumption seems reasonable for drop-tail queuing but probably not for RED .

Shakkottai and Srikant [30] also used stochastic aggregation to reduce the time-scales over which a model is valid. In particular, they showed that the aggregation of n discrete-time models like (8.3) can described by

$$\dot{\mu}_n(t) = \frac{1}{RTT^2} - \frac{2}{3}p_{\text{drp}}(t - RTT)\mu_n(t - RTT)\big(\mu_n(t - RTT) + a\big), \quad (8.8)$$

where μ_n denotes the average congestion window size with respect to the ensemble of n flows and the parameter a models the effect of competing uncontrolled background traffic. They showed that for large n this model is valid over time scales n times smaller than those of the original discrete-time model (8.4), which was only valid for time scales much larger than one round-trip time.

A key feature of TCP's behavior is the existence of a delay between the occurrence of a drop and its detection and eventual reaction by TCP. This delay has been identified as one of the causes for queue-length instabilities (cf., e.g., the survey [19]). The models (8.3)–(8.8) attempt to capture this by introducing delays in all the right-hand-side terms related to the detection of drops. However, since delayed differential equations are difficult to analyze, these models are usually simplified by solely considering a delay in the term p_{drp}, which turns these equation back into (time-varying) ordinary differential equations.

We pursue here a stochastic version of the hybrid models proposed in [7]. As in [24, 25] time averaging is done over intervals of roughly one round-trip time to obtain continuously varying sending rates, and we then investigate the dynamics of ensemble averages. However, we do not just consider the evolution of averages as in [24, 25, 30]. Instead, we also study the dynamics of high-order statistical moments. Our models explicitly considers the delay between drop occurrence and detection by TCP and take this into account in the ensemble averaging.

8.1.2 Models for On-Off Flows

Analytical studies of finite TCP flows have been pursued in relatively few papers. Mellia et al. [23] proposed a stochastic model for short-lived flows that predicts the flow's completion time as a function of the transfer-size, drop probability, and round-trip time. This model ignores congestion-avoidance altogether and is therefore only valid for very short flows, for which most data is sent during the slow-start mode. Zheng et al. [34] proposed an improved model to predict a flow's completion time that considers congestion-avoidance and is therefore applicable both to short and long-lived flows. However, neither [23] nor [34] provide explicit dynamic models for TCP traffic, such as the ones described before for long-lived flows.

In [2], Baccelli and Hong extended their models for long-lived TCP flows proposed in [3] to on-off flows. For each source, they extract a sequence of random independent and identically distributed (i.i.d) transfer-sizes and a sequence of random i.i.d. "think times." Each source then alternates between on-periods (during which data is transmitted) and off-periods (the think times). The duration of the on-periods is determined by the transfer-sizes and the corresponding throughput. We will borrow this type of on-off behavior for our SHS model of TCP. Baccelli and Hong [2] stop short of analyzing the resulting stochastic model and simply use it to produce Monte Carlo simulations that run much faster than packet-level simulators. The computational gains are achieved by considering fluid models and by foregoing the identity of individual packets, very much as discussed in the previous two chapters.

Marsan et al. [21] proposed a fairly sophisticated model for aggregate TCP flows of several classes, each with different routes through the network. The state of their model consists of functions $P^i_{ss}(w,t,\ell)$ that keep track of the number of flows of class i in slow-start mode at time t, with congestion window size no larger than w and remaining transfer-size no larger than ℓ. Similarly, functions $P^i_{ca}(w,t,\ell)$ keep track of the number of flows in the congestion-avoidance mode. The evolution of all the $P_{ss}(\cdot)$ and $P_{ca}(\cdot)$ are governed by a system of Partial Differential Equations (PDEs).[2] Their model can also capture fast-recovery, the maximum window size of the TCP sources, and both RED and drop-tail queues. This is achieved by introducing additional states, which keep track of how many flows are in fast recovery, how many flows have reached the maximum window size, etc. This modeling framework is very powerful but since there is no explicit characterization of individual flows, it is not possible to build the type of on-off models proposed by Baccelli and Hong [2]. Instead, the model takes as inputs the rates at which new TCP flows start and finish. To emulate the type of on-off behavior in [2], Marsan et al. [21] propose to use heuristic rules to adjust these rates as a function of the loss probability and other model parameters. The main shortcoming of the models proposed by Marsan et al. [21] is an infinite-dimensional state-space that makes it challenging to use these models to gain insight into TCP's behavior or to analyze the stability of congestion control algorithms.

[2]For exponentially distributed transfer-sizes, the ℓ dependence can be dropped and the resulting PDEs are one-dimensional.

The models proposed in this chapter consider ensembles of single-user on-off TCP flows, similar to the ones proposed by Baccelli and Hong [2]. The off-periods are assumed exponentially distributed whereas the on-periods are determined by the amount of data being transfered. We take as given the probability distribution of the transfer-sizes and this implicitly determines the distribution of the on-periods, by taking into account the (time-varying) sending rate, which is implicitly determined by the drop probability. This allows us to obtain directly from the model parameters such as the probability p_{ss} in (8.7) of a flow being in slow-start mode.

A key difference between our work and [21] is that we use our SHS model to construct a *system of ODEs* that describes the evolution of the mean and higher order moments for TCP's sending rate, whereas Marsan et al. [21] construct a *system of PDEs* to describe what essentially amounts to the whole distribution of the congestion window size. Although this distribution could also be computed for our SHS model [12], we opted to work with a more parsimonious state-space that only describes first and second order moments. Even though the resulting models are more complex than the ones discussed in Section 8.1.1 for long-lived flows, the closed-form computation of steady-state throughput and a stability analysis is still tractable.

As mentioned above, we combine the hybrid modeling framework for network modeling introduced by [7] with the stochastic drop models used by [24, 25] for RED. Both these models have been validated and do not seem too controversial. The third crucial element to an on-off TCP model is the distribution of the transfer-sizes, which is a much more controversial issue. Some studies seem to indicate that these distributions are heavy-tailed (cf., the survey [29]) but others conclude that there is little evidence to support such claim [17, 8]. Settling this issue is beyond the scope of this chapter so we opted to present results for two distributions: One with and another without a significant tail. Both distributions approximate data found in the literature. One consists of a mixture of two exponentials that approximately models the file distribution observed in the UNIX file system [14]. The parameters chosen for the mixture of exponentials model fairly well the "waist" of the distribution but somewhat underestimate its tail. The second distribution consists of a mixture of three exponentials and approximates the data reported by Arlitt et al. [1] obtained from monitoring transfers from a world-wide web proxy within an Internet Service Provider. This approximation captures fairly well the tail of the distribution (at least up to 100 MB, for which data is available). The idea of approximating heavy tail distributions for transfer-sizes using a mixture of exponentials was proposed by Feldmann and Whitt [9]. Like us, Marsan et al. [21] also used a mixture of exponentials to model TCP transfer-sizes.

8.2 A Stochastic Model for TCP

In this section we present a SHS model for single-user on-off TCP flows. Many-user models can be obtained by aggregating several of these single-user models. Our model is based on the hybrid modeling framework proposed by Bohacek et al. [7] and the stochastic models by Misra et al. [24, 25].

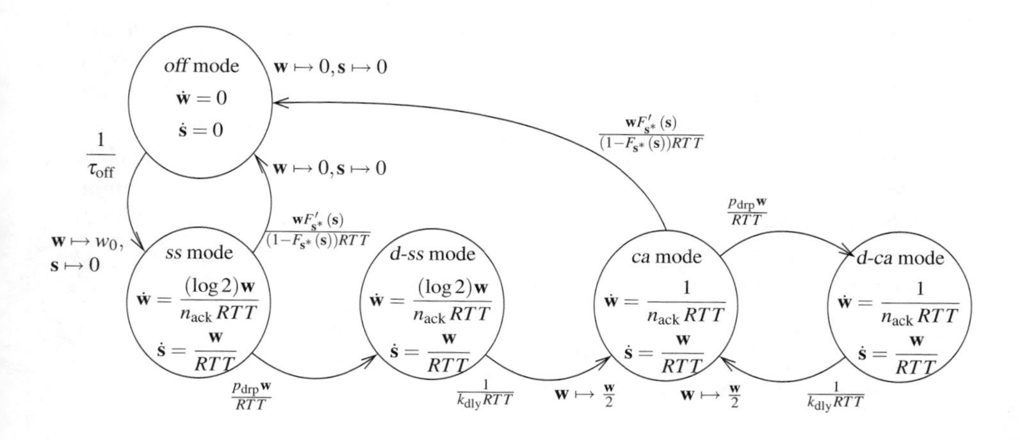

FIGURE 8.1: Stochastic hybrid model for a single-user TCP flow.

The SHS that we use to model a single-user on-off TCP flow is represented graphically in Figure 8.1. This model has five modes: *off* which corresponds to flow inactivity; *ss* which corresponds to slow-start; *ca* which corresponds to congestion-avoidance; and two other modes (*d-ss* and *d-ca*) that will be explained shortly. Each mode corresponds to a node in the graph in Figure 8.1.

As time progresses, the model transitions from one mode to another according to specific rules that will be discussed below. While on each mode, two continuous variables evolve according to the differential equations displayed inside the corresponding node. These variables are TCP's congestion window size $\mathbf{w}$ and the cumulative number of packets $\mathbf{s}$ sent so far in a particular connection. The evolution of these variables obeys the following rules:

(i) During the *off* mode the flow is inactive and we simply have $\mathbf{w} = \mathbf{s} = 0$.

(ii) The *ss* mode corresponds to TCP's slow-start. In this mode, $\mathbf{w}$ packets are sent each round-trip time RTT and the congestion window size $\mathbf{w}$ increases by one for each ACK packet received. Until a drop occurs this means that the rate at which packets are sent is equal to $\mathbf{r} = \frac{\mathbf{w}}{RTT}$ and the number of ACK packets received is equal to $\frac{\mathbf{r}}{n_{\text{ack}}}$, where n_{ack} denotes the number of data packets acknowledged per each ACK packet received. This can be modeled by

$$\dot{\mathbf{w}} = \frac{(\log 2)\mathbf{r}}{n_{\text{ack}}} = \frac{(\log 2)\mathbf{w}}{n_{\text{ack}}RTT}, \qquad \dot{\mathbf{s}} = \mathbf{r} = \frac{\mathbf{w}}{RTT}. \tag{8.9}$$

The $(\log 2)$ factor compensates for the error introduced by approximating the discrete increments by a continuous increase.

Without delayed ACKs ($n_{\text{ack}} = 1$), it is straightforward to check that if the round-trip time RTT were approximately constant this model would lead to the usual doubling of $\mathbf{w}$ every RTT. Typical implementations of TCP that use delayed ACKs set $n_{\text{ack}} = 2$. In this case, (8.9) leads to a multiplication of $\mathbf{w}$ by $\sqrt{2}$ every RTT. This is consistent with the analysis by Sikdar et al. [32], which shows that for $n_{\text{ack}} = 2$ the number of packets sent in the nth round-trip time of slow-start is approximately equal to $(1 + \frac{\sqrt{2}}{2})\sqrt{2}^n$. This formula is matched exactly by the fluid model in (8.9) when one sets $\mathbf{w} = 1.428$ in the beginning of *ss*. On the other hand, for $n_{\text{ack}} = 1$, the number of packets sent in the nth round-trip time of slow-start should be equal to 2^{n-1}. This matches the fluid model by making $\mathbf{w} = .693$ at the beginning of *ss*.

(iii) The *ca* mode corresponds to TCP's congestion-avoidance. In this mode $\mathbf{w}$ increases by $1/\mathbf{w}$ for each ACK packet received and, as in slow-start, $\mathbf{w}$ packets are sent each round-trip time RTT. This can be modeled by

$$\dot{\mathbf{w}} = \frac{1}{\mathbf{w}} \frac{\mathbf{r}}{n_{\text{ack}}} = \frac{1}{n_{\text{ack}}RTT}, \qquad \dot{\mathbf{s}} = \mathbf{r} = \frac{\mathbf{w}}{RTT}. \tag{8.10}$$

In practice, both in (8.9) and in (8.10), the round-trip time RTT is generally a time-varying quantity.

When a drop occurs, this event will not be immediately detected by TCP. To account for the delay, we consider two additional "delay" modes: *d-ss* and *d-ca*. When a drop occurs while the system is in *ss* or in *ca* it immediately transitions to *d-ss* or to *d-ca*, respectively. However, the dynamics of $\mathbf{w}$ and $\mathbf{r}$ remain unchanged. Only after a time period of roughly one round-trip time, TCP will react to the drop and adjust its congestion window size $\mathbf{w}$ and the sending rate $\mathbf{r}$.

The transitions between modes, which are represented by arrows in the graph in Figure 8.1, are stochastic events that occur at specific "rates." Informally, the rate at which a transition occurs corresponds to the expected number of times that transition will take place in a unit of time (cf. the Appendix and [11]). In Figure 8.1, these rates are shown as labels next to the start of the arrows. Some transitions trigger instantaneous changes (resets) of $\mathbf{w}$ and/or $\mathbf{s}$. These are represented near the end of the arrow. The following events are associated with transitions:

(i) Drop occurrences correspond to transitions from the *ss* and *ca* modes to the *d-ss* and *d-ca* modes. These events occur at a rate given by $p_{\text{drp}}\mathbf{r}$, where p_{drp} denotes the per-packet drop probability and $\mathbf{r} := \frac{\mathbf{w}}{RTT}$ the packet sending rate. The drop rate p_{drp} will generally be time-varying.

(ii) Drop detections correspond to transitions from the *d-ss* and *d-ca* modes to the *ca* mode. These events occur at a rate given by $\frac{1}{k_{\text{dly}}RTT}$, which is consistent with an exponentially-distributed average delay of $k_{\text{dly}}RTT$. Typically $k_{\text{dly}} = 1$. The detection of a drop leads to a division of the congestion window size $\mathbf{w}$ by two.

(iii) The start of new flows corresponds to transitions from the *off* to the *ss* mode. These events occur at a rate given by $\frac{1}{\tau_{\text{off}}}$, which is consistent with an exponentially distributed duration of the off-periods with average τ_{off}. At the start of each new flow the congestion window size $\mathbf{w}$ is set equal to

$$w_0 := \begin{cases} .693 & n_{\text{ack}} = 1 \\ 1.428 & n_{\text{ack}} = 2. \end{cases}$$

and the number of packets sent $\mathbf{s}$ is reset to zero.

(iv) The termination of flows correspond to transitions from the *ss* and *ca* modes to the *off* mode. These events occur at a rate given by

$$\frac{\mathbf{r}F'_{\mathbf{s}^*}(\mathbf{s})}{1 - F_{\mathbf{s}^*}(\mathbf{s})}, \qquad \mathbf{r} := \frac{\mathbf{w}}{RTT}, \tag{8.11}$$

which is consistent with a distribution $F_{\mathbf{s}^*} : [0, \infty) \to [0, 1]$ for the number $\mathbf{s}^*$ of packets sent in each TCP session (cf. Appendix). When a flow terminates both $\mathbf{w}$ and $\mathbf{s}$ are reset to zero. The rate in (8.11) is usually called the *hazard rate* of the transfer-size distribution $F_{\mathbf{s}^*}$.

Two main simplifications were considered: We ignored fast-recovery after a drop is detected by three duplicate ACKs and we ignored timeouts. Fast-recovery takes relatively little time and has little impact on the overall throughput unless the number of drops is very high [6]. As mentioned before, timeouts typically occur when several consecutive packets are dropped, preventing the detection of drops by the triple duplicate ACKs mechanism. Therefore, timeouts have an especially severe impact on the throughput when drops are highly correlated. However, here we are mostly interested in RED for which high correlations are unlikely and therefore timeouts only occur under very high drop rates.
A few specific instances of the general model in Figure 8.1 are of interest.

Long-lived flows This model is obtained by assuming that the number of packets to transmit is infinitely large, i.e., that $F_{\mathbf{s}^*}(s) = 0, \forall s < \infty$. In this case, we can ignore the *off*, *ss*, and *d-ss* modes since they will only be active for a brief initial period and after this the SHS will continuously switch between the *ca* and *d-ca* modes.

Exponential transfer-sizes This model is obtained by assuming that the number of packets to transmit is exponentially distributed with average k, i.e., $F_{\mathbf{s}^*}(s) = 1 - e^{-\frac{s}{k}}$, $\forall s \geq 0$. In this case, the termination of flows occurs at a rate given by

$$\frac{\mathbf{w}F'_{\mathbf{s}^*}(\mathbf{s})}{(1 - F_{\mathbf{s}^*}(\mathbf{s}))RTT} = \frac{k^{-1}\mathbf{w}}{RTT},$$

which is independent of the continuous state $\mathbf{s}$. For exponential transfer-sizes we can therefore ignore this state variable.

Pareto transfer-sizes It has been observed that modeling the distribution of transfer-sizes as an exponential is an over-simplification. For example, it has been argued that heavy-tail models are more fitting to experimental data (cf., e.g., [1, 4, 29]). Inspired by this work, we also consider a model for which the distribution of the number of packets to transmit is a shifted-Pareto[3] with shape parameter a and scale parameter b, i.e., $F_{\mathbf{s}^*}(s) = 1 - \left(\frac{b}{s+b}\right)^a$, $\forall s \geq 0$. Assuming that $a > 1$, this corresponds to an average number of packets equal to $\frac{ab}{a-1}$. The existence of high-order statistical moments depends on the value of a and the kth moment exists only for $k < a$. For this distribution, the termination of flows occurs at a rate given by

$$\frac{\mathbf{w}F'_{\mathbf{s}^*}(\mathbf{s})}{(1 - F_{\mathbf{s}^*}(\mathbf{s}))RTT} = \frac{a\,\mathbf{w}}{(\mathbf{s}+b)RTT}.$$

Mixed-exponential transfer-sizes An alternative to a Pareto distribution that turns out to be computationally more attractive and still fits well with experimental data is a mixture of exponentials [9]. According to this model, transfer-sizes are sampled from a family of M exponential random variables $\mathbf{s}_i$, $i \in \{1, 2, \ldots, M\}$ by selecting a sample from the ith random variable $\mathbf{s}_i$ with probability p_i. Each $\mathbf{s}_i$ corresponds to a distinct average transfer-size k_i.

To model this as a SHS, we consider M alternative

$$\{ss_i, d\text{-}ss_i, ca_i, d\text{-}ca_i : i = 1, 2, \ldots, M\}$$

modes, each corresponding to a specific exponential distribution for the transfer-sizes. The transitions from the inactive mode *off* to the slow-start mode ss_i corresponding to an average transfer-size of k_i occurs with probability p_i, which corresponds to a transition rate given by $\frac{p_i}{\tau_{\text{off}}}$. To obtain the desired distribution for the transfer-size, the transitions from ca_i and ss_i to the inactive mode *off* occur at a rate given by $\frac{k_i^{-1} w}{RTT}$. A similar technique could be used to obtain a mixture of exponentials for the distribution of the off-periods.

In this chapter, we will mostly focus our attention on long-lived flows and mixed-exponential transfer-sizes (with exponential as the special case $M = 1$). The resulting SHS have polynomial vector fields, reset maps, and transition intensities, which facilitates their analysis [13]. Pareto transfer-sizes appear to be more difficult to analyze, but can be well approximated by the more tractable mixed-exponential model [9].

[3] The usual Pareto distribution takes values on (b, ∞) whereas the shifted-Pareto distribution used here takes values from $(0, \infty)$.

8.3 Analysis of the TCP SHS Models

To investigate the dynamics of the moments of the sending rate $\mathbf{r}(t) = \frac{\mathbf{w}(t)}{RTT(t)}$, we denote by $\mu_{q_0,n}(t)$ the nth-order (uncentered) statistical moment of $\mathbf{r}(t)$ restricted to a particular mode $\mathbf{q} = q_0 \in \mathscr{Q} := \{\text{off}, ss_i, \text{d-}ss_i, ca_i, \text{d-}ca_i : i = 1,2,\ldots,M\}$. This is captured by the following definition

$$\mu_{q_0,n}(t) := \mathrm{E}\left[\psi_{q_0,n}(\mathbf{q}(t),\mathbf{w}(t),t)\right], \quad \psi_{q_0,n}(q,w,t) := \begin{cases} \frac{w^n}{RTT(t)^n} & q = q_0 \\ 0 & \text{otherwise,} \end{cases} \tag{8.12}$$

$\forall n \geq 0,\ q_0 \in \mathscr{Q}$. The probability that the flow is in mode $q_0 \in \mathscr{Q}$ at time t is then given by

$$\mathrm{P}(\mathbf{q}(t) = q_0) = \mu_{q_0,0}(t), \quad \forall t \geq 0;$$

the nth-order statistical moment of $\mathbf{r}(t)$, conditioned to the flow being in mode $q_0 \in \mathscr{Q}$ is given by

$$\mathrm{E}[\mathbf{r}^n(t) \mid \mathbf{q}(t) = q_0] = \frac{\mu_{q_0,n}(t)}{\mathrm{P}(\mathbf{q}(t) = q_0)} = \frac{\mu_{q_0,n}(t)}{\mu_{q_0,0}(t)}, \quad \forall t \geq 0; \tag{8.13}$$

and the nth-order statistical moment of the overall sending rate is given by

$$\mathrm{E}[\mathbf{r}^n(t)] = \sum_{q \in \mathscr{Q}} \mu_{q,n}(t), \quad \forall t \geq 0.$$

The following result shows that these moments are the solution of an infinite-dimensional system of ODEs that can be obtained by direct application of results in [11, 13]. Details are provided in the appendix.

***THEOREM 8.1* Infinite-dimensional models.** *The statistical moments of the long-lived flows model satisfy the following equations*[4]*:*

$$\dot{\mu}_{\text{ca}_i,n} = \frac{n\mu_{\text{ca}_i,n-1}}{n_{\text{ack}}RTT^2} - \frac{n\dot{RTT}\mu_{\text{ca}_i,n}}{RTT} - p_{\text{drp}}\mu_{\text{ca}_i,n+1} + \frac{\mu_{\text{d-ca}_i,n}}{2^n k_{\text{dly}} RTT}, \tag{8.14a}$$

$$\dot{\mu}_{\text{d-ca}_i,n} = \frac{n\mu_{\text{d-ca}_i,n-1}}{n_{\text{ack}}RTT^2} - \frac{n\dot{RTT}\mu_{\text{d-ca}_i,n}}{RTT} + p_{\text{drp}}\mu_{\text{ca}_i,n+1} - \frac{\mu_{\text{d-ca}_i,n}}{k_{\text{dly}} RTT}. \tag{8.14b}$$

The statistical moments of the mixed-exponentials transfer-sizes model:

$$\dot{\mu}_{\text{off},0} = -\frac{\mu_{\text{off},0}}{\tau_{\text{off}}} + \sum_{j=1}^{M} k_j^{-1}(\mu_{\text{ss}_j,1} + \mu_{\text{ca}_j,1}), \tag{8.15a}$$

[4] To simplify the notation, we omit the time-dependence of RTT and p_{drp}.

$$\dot{\mu}_{\text{ss}_i,n} = \frac{p_i w_0^n \mu_{\text{off},0}}{\tau_{\text{off}} RTT^n} + n\frac{(\log 2) - n_{\text{ack}}R\dot{T}T}{n_{\text{ack}}RTT}\mu_{\text{ss}_i,n} - (p_{\text{drp}} + k_i^{-1})\mu_{\text{ss}_i,n+1}, \tag{8.15b}$$

$$\dot{\mu}_{\text{d-ss}_i,n} = n\frac{(\log 2) - n_{\text{ack}}R\dot{T}T}{n_{\text{ack}}RTT}\mu_{\text{d-ss}_i,n} + p_{\text{drp}}\mu_{\text{ss}_i,n+1} - \frac{\mu_{\text{d-ss}_i,n}}{k_{\text{dly}}RTT}, \tag{8.15c}$$

$$\dot{\mu}_{\text{ca}_i,n} = \frac{n\mu_{\text{ca}_i,n-1}}{n_{\text{ack}}RTT^2} - \frac{nR\dot{T}T\mu_{\text{ca}_i,n}}{RTT} - (p_{\text{drp}} + k_i^{-1})\mu_{\text{ca}_i,n+1} + \frac{\mu_{\text{d-ss}_i,n} + \mu_{\text{d-ca}_i,n}}{2^n k_{\text{dly}}RTT}, \tag{8.15d}$$

$$\dot{\mu}_{\text{d-ca}_i,n} = \frac{n\mu_{\text{d-ca}_i,n-1}}{n_{\text{ack}}RTT^2} - \frac{nR\dot{T}T\mu_{\text{d-ca}_i,n}}{RTT} + p_{\text{drp}}\mu_{\text{ca}_i,n+1} - \frac{\mu_{\text{d-ca}_i,n}}{k_{\text{dly}}RTT}. \tag{8.15e}$$

The statistical moments of the Pareto transfer-sizes model:

$$\dot{v}_{\text{off},0,0} = -\frac{v_{\text{off},0,0}}{\tau_{\text{off}}} + a(v_{\text{ss},1,1} + v_{\text{ca},1,1}), \tag{8.16a}$$

$$\dot{v}_{\text{ss},n,m} = \frac{w_0^n v_{\text{off},0,0}}{b^m RTT^n \tau_{\text{off}}} + n\frac{(\log 2) - n_{\text{ack}}R\dot{T}T}{n_{\text{ack}}RTT}v_{\text{ss},n,m} - p_{\text{drp}}v_{\text{ss},n+1,m} - (m+a)v_{\text{ss},n+1,m+1}, \tag{8.16b}$$

$$\dot{v}_{\text{d-ss},n,m} = n\frac{(\log 2) - n_{\text{ack}}R\dot{T}T}{n_{\text{ack}}RTT}v_{\text{d-ss},n,m} + p_{\text{drp}}v_{\text{ss},n+1,m} - \frac{v_{\text{d-ss},n,m}}{k_{\text{dly}}RTT} - mv_{\text{d-ss},n+1,m+1}, \tag{8.16c}$$

$$\begin{aligned}\dot{v}_{\text{ca},n,m} = {} & \frac{nv_{\text{ca},n-1,m}}{n_{\text{ack}}RTT^2} - \frac{nR\dot{T}Tv_{\text{ca},n,m}}{RTT} - p_{\text{drp}}v_{\text{ca},n+1,m} \\ & + \frac{v_{\text{d-ss},n,m} + v_{\text{d-ca},n,m}}{2^n k_{\text{dly}}RTT} - (m+a)v_{\text{ca},n+1,m+1},\end{aligned} \tag{8.16d}$$

$$\dot{v}_{\text{d-ca},n,m} = \frac{nv_{\text{d-ca},n-1,m}}{n_{\text{ack}}RTT^2} - \frac{nR\dot{T}Tv_{\text{d-ca},n,m}}{RTT} + p_{\text{drp}}v_{\text{ca},n+1,m} - \frac{v_{\text{d-ca},n,m}}{k_{\text{dly}}RTT} - mv_{\text{d-ca},n+1,m+1}, \tag{8.16e}$$

where $v_{q_0,n,0}$, $q_0 \in \mathcal{Q}, n \geq 0$ *is used to denote* $\mu_{q_0,n}$ *and, for every* $m > 0$,

$$v_{q_0,n,m}(t) := \mathrm{E}\left[\varphi_{q_0,n,m}(\mathbf{q}(t), \mathbf{w}(t), \mathbf{s}(t))\right],$$

$$\varphi_{q_0,n,m}(q,w,s,t) := \begin{cases} \frac{w^n}{RTT(t)^n(s+b)^m} & q = q_0 \\ 0 & \text{otherwise.} \end{cases}$$

8.4 Reduced-order Models

The systems of infinitely many ODEs[5] that appear in Theorem 8.1 describe *exactly* the evolution of the moments of the sending rate **r**, but finding a solution to these equations does not appear to be simple. However, as noted by Bohacek [5] and

[5] Notice that the integer n ranges from 0 to ∞.

others, Monte Carlo simulations reveal that the steady-state distribution of the sending rate is well approximated by a log-normal distribution. Assuming that on each mode the sending rate $\mathbf{r}$ approximately obeys a log-normal distribution even during transients, we can truncate the systems of infinitely many differential equations that appear in Theorem 8.1. This procedure is known as *moment closure*.

We recall that, if the random variable $\mathbf{x}$ has a log-normal distribution then $\mathrm{E}[\mathbf{x}^3] = \frac{\mathrm{E}[\mathbf{x}^2]^3}{\mathrm{E}[\mathbf{x}]^3}$. This means that if $\mathbf{r}$ is approximately log-normal distributed in the mode $q \in \mathcal{Q}$, we have that

$$\mu_{q,3} = \mu_{q,0}\,\mathrm{E}[\mathbf{r}^3 \mid \mathbf{q} = q] \approx \mu_{q,0}\frac{\mathrm{E}[\mathbf{r}^2 \mid \mathbf{q} = q]^3}{\mathrm{E}[\mathbf{r} \mid \mathbf{q} = q]^3} = \frac{\mu_{q,0}\,\mu_{q,2}^3}{\mu_{q,1}^3}, \tag{8.17}$$

where we used (8.13). Using (8.17) in (8.14)–(8.15), we can eliminate any terms $\mu_{q_0,n}$, $n \geq 3$ in the equations for $\dot{\mu}_{q_0,n}$, $n \leq 2$, thus constructing a finite-dimensional model to approximately describe the dynamics of the first two moments of the sending rate. The reader is referred to [13] for a more detailed treatment on the use of this type of truncations to analyze SHS.

8.4.1 Long-lived Flows

The following model for long-lived TCP flows can be obtained from (8.14) using the approximation in (8.17).

$$\dot{\mu}_{\text{ca},0} = \frac{1-\mu_{\text{ca},0}}{k_{\text{dly}}\,RTT} - p_{\text{drp}}\,\mu_{\text{ca},1}, \tag{8.18a}$$

$$\dot{\mu}_{\text{ca},1} = \frac{\mu_{\text{ca},0}}{n_{\text{ack}}\,RTT^2} - \frac{\dot{RTT}\,\mu_{\text{ca},1}}{RTT} - p_{\text{drp}}\,\mu_{\text{ca},2} + \frac{\mu_{\text{d-ca},1}}{2k_{\text{dly}}\,RTT}, \tag{8.18b}$$

$$\dot{\mu}_{\text{d-ca},1} = \frac{1-\mu_{\text{ca},0}}{n_{\text{ack}}\,RTT^2} + p_{\text{drp}}\,\mu_{\text{ca},2} - \frac{\dot{RTT}\,\mu_{\text{d-ca},1}}{RTT} - \frac{\mu_{\text{d-ca},1}}{k_{\text{dly}}\,RTT}, \tag{8.18c}$$

$$\dot{\mu}_{\text{ca},2} = \frac{2\mu_{\text{ca},1}}{n_{\text{ack}}\,RTT^2} - \frac{2\dot{RTT}\,\mu_{\text{ca},2}}{RTT} - \frac{p_{\text{drp}}\,\mu_{\text{ca},0}\,\mu_{\text{ca},2}^3}{\mu_{\text{ca},1}^3} + \frac{\mu_{\text{d-ca},2}}{4k_{\text{dly}}\,RTT}, \tag{8.18d}$$

$$\dot{\mu}_{\text{d-ca},2} = \frac{p_{\text{drp}}\,\mu_{\text{ca},0}\,\mu_{\text{ca},2}^3}{\mu_{\text{ca},1}^3} + \frac{2\mu_{\text{d-ca},1}}{n_{\text{ack}}\,RTT^2} - \frac{2\dot{RTT}\,\mu_{\text{d-ca},2}}{RTT} - \frac{\mu_{\text{d-ca},2}}{k_{\text{dly}}\,RTT}. \tag{8.18e}$$

This model has a stable equilibrium point at[6]

$$\mu_{\text{ca},0} = 1 + \frac{2k_{\text{dly}}^2\, p_{\text{drp}}}{n_{\text{ack}}} - k_{\text{dly}}\, p_{\text{drp}}\, W, \tag{8.19a}$$

$$\mu_{\text{ca},1} = \frac{W}{RTT} - \frac{2k_{\text{dly}}}{n_{\text{ack}}\, RTT}, \tag{8.19b}$$

$$\mu_{\text{ca},2} = \frac{2n_{\text{ack}} - (n_{\text{ack}}\, W - 2k_{\text{dly}})k_{\text{dly}}\, p_{\text{drp}}}{n_{\text{ack}}^2\, p_{\text{drp}}\, RTT^2}, \tag{8.19c}$$

$$\mu_{d\text{-ca},1} = \frac{2k_{\text{dly}}}{n_{\text{ack}}\, RTT}, \qquad \mu_{d\text{-ca},2} = \frac{8k_{\text{dly}}\, W}{3n_{\text{ack}}\, RTT^2}, \tag{8.19d}$$

$$p_{\text{drp}} \approx \frac{12n_{\text{ack}}}{n_{\text{ack}}\, W - 2k_{\text{dly}}} \frac{1}{15k_{\text{dly}} + \sqrt{48n_{\text{ack}}^2\, W^2 - 168k_{\text{dly}}\, n_{\text{ack}}\, W + 45k_{\text{dly}}^2}}, \tag{8.19e}$$

where W denotes the *delay-throughput product* defined by

$$W := RTT\,\mathrm{E}[\mathbf{r}] = (\mu_{\text{ca},1} + \mu_{d\text{-ca},1})RTT,$$

which is also equal to the average window size $\mathrm{E}[\mathbf{w}]$. Figure 8.2 compares the equilibrium points obtained from this reduced model with the steady-state values obtained from Monte Carlo simulations of the original SHS. The Monte Carlo simulations were obtained using the algorithm described in [12, Section 2.1]. The match is essentially perfect, confirming the validity of the log-normal approximation.

In the same plot we also included what the TCP-friendly formula (8.1) would have predicted for the total sending rate. We can see that both the Monte Carlo simulations of the original SHS and the reduced model (8.18) basically agree with the TCP-friendly formula (8.1) for the average sending rate. However, the stochastic models also provide information about the variability in the sending rate from flow to flow (measured by the standard deviation).

8.4.2 Mixed-exponential Transfer-sizes

We consider now a TCP SHS model for mixed-exponentially distributed transfer-sizes. To keep the model small, we assume that the delays are negligible $k_{\text{dly}} \approx 0$. This model can be obtained from (8.15) by making $k_{\text{dly}} \downarrow 0$ and realizing that in this case the variables $\mu_{d\text{-ss},n}$ and $\mu_{d\text{-ca},n}$ exhibit fast dynamics with equilibrium at

$$\frac{\mu_{d\text{-ss}_i,n}}{k_{\text{dly}} RTT} = \frac{n_{\text{ack}}\, p_{\text{drp}} \mu_{\text{ss}_i,n+1}}{n_{\text{ack}} - k_{\text{dly}} n\big((\log 2) - n_{\text{ack}} \dot{RTT}\big)}.$$

$$\frac{\mu_{d\text{-ca}_i,n}}{k_{\text{dly}} RTT} = \frac{n\mu_{d\text{-ca}_i,n-1}}{n_{\text{ack}} RTT^2 (1 + nk_{\text{dly}} \dot{RTT})} + \frac{p_{\text{drp}} \mu_{\text{ca}_i,n+1}}{1 + nk_{\text{dly}} \dot{RTT}}.$$

[6]The approximate value for p_{drp} was obtained by considering a quadratic approximation around $p_{\text{drp}} = 0$ to the equation $\dot{\mu}_{d\text{-ca},2} = 0$ with the variables $\mu_{\text{ca},0}$, $\mu_{\text{ca},1}$, $\mu_{\text{ca},2}$, $\mu_{d\text{-ca},1}$, $\mu_{d\text{-ca},2}$ eliminated through the use of the remaining equations. It is possible to compute the exact value of p_{drp} at equilibrium, but (8.19e) is much simpler and provides a very good approximation.

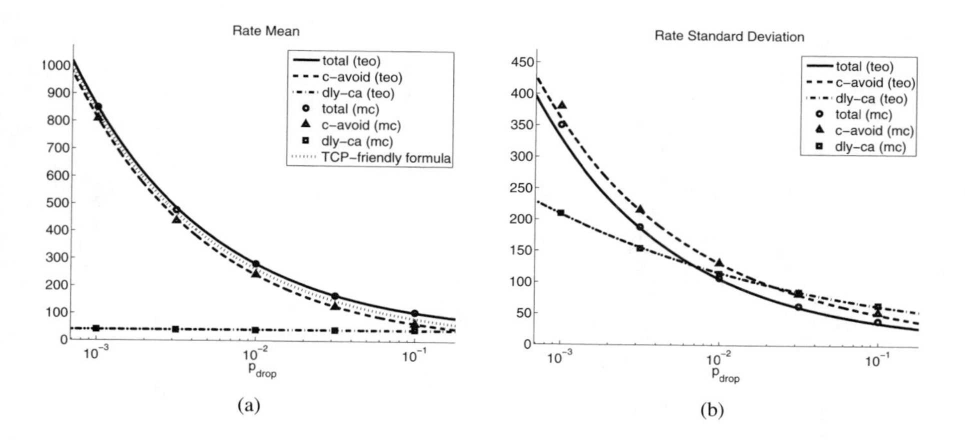

FIGURE 8.2: Steady-state values for the average (a) and standard deviation (b) of the sending rate as a function of the drop probability, for the long-lived SHS model. The dashed and dot-dashed lines provide the contributions to the sending rate from the *ca* and *d-ca* modes, respectively. All lines were obtained from (8.19) with $RTT =$ 50 ms, $k_{\rm dly} = 1$, $n_{\rm ack} = 1$. The circle, triangle, and square symbols were obtained from Monte Carlo simulations. The dotted line corresponds to the TCP-friendly formula (8.1).

As $k_{\rm dly} \downarrow 0$, we obtain

$$\frac{\mu_{d\text{-}ss_i,n}}{k_{\rm dly}RTT} \to p_{\rm drp}\mu_{ss_i,n+1}, \qquad \frac{\mu_{d\text{-}ca_i,n}}{k_{\rm dly}RTT} \to \frac{n\mu_{d\text{-}ca_i,n-1}}{n_{\rm ack}RTT^2} + p_{\rm drp}\mu_{ca_i,n+1} \to p_{\rm drp}\mu_{ca_i,n+1}.$$

Replacing this into (8.15), yields[7]

$$\dot{\mu}_{\rm off,0} = -\frac{\mu_{\rm off,0}}{\tau_{\rm off}} + \sum_{j=1}^{M} k_j^{-1}(\mu_{ss_j,1} + \mu_{ca_j,1}), \tag{8.20a}$$

$$\dot{\mu}_{ss_i,n} = \frac{p_i w_0^n \mu_{\rm off,0}}{\tau_{\rm off} RTT^n} + n\frac{(\log 2) - n_{\rm ack}R\dot{T}T}{n_{\rm ack}RTT}\mu_{ss_i,n} - (p_{\rm drp} + k_i^{-1})\mu_{ss_i,n+1}, \tag{8.20b}$$

$$\dot{\mu}_{ca_i,n} = \frac{n\mu_{ca_i,n-1}}{n_{\rm ack}RTT^2} - \frac{nR\dot{T}T\,\mu_{ca_i,n}}{RTT} - (p_{\rm drp} + k_i^{-1})\mu_{ca_i,n+1} + \frac{p_{\rm drp}}{2^n}(\mu_{ss_i,n+1} + \mu_{ca_i,n+1}). \tag{8.20c}$$

We can now use (8.17) to construct from (8.20) the following finite dimensional approximate model.

$$\dot{\mu}_{ss_i,0} = \frac{p_i\big(1 - \sum_{j=1}^{M}(\mu_{ss_i,0} + \mu_{ca_i,0})\big)}{\tau_{\rm off}} - (p_{\rm drp} + k_i^{-1})\,\mu_{ss_i,1}, \tag{8.21a}$$

[7]These equations could also be derived directly from a SHS model without the delay modes *d-ss* and *d-ca*.

$$\dot{\mu}_{\mathrm{ca}_i,0} = p_{\mathrm{drp}}\,\mu_{\mathrm{ss}_i,1} - k_i^{-1}\,\mu_{\mathrm{ca}_i,1}, \tag{8.21b}$$

$$\dot{\mu}_{\mathrm{ss}_i,1} = \frac{w_0\, p_i\big(1-\sum_{j=1}^{M}(\mu_{\mathrm{ss}_i,0}+\mu_{\mathrm{ca}_i,0})\big)}{\tau_{\mathrm{off}}\, RTT} + \frac{(\log 2) - n_{\mathrm{ack}}\, R\dot{T}T}{n_{\mathrm{ack}}\, RTT}\mu_{\mathrm{ss}_i,1} - (p_{\mathrm{drp}}+k_i^{-1})\,\mu_{\mathrm{ss}_i,2}, \tag{8.21c}$$

$$\dot{\mu}_{\mathrm{ca}_i,1} = \frac{\mu_{\mathrm{ca}_i,0}}{n_{\mathrm{ack}}\, RTT^2} - \frac{R\dot{T}T\,\mu_{\mathrm{ca}_i,1}}{RTT} + \frac{p_{\mathrm{drp}}\mu_{\mathrm{ss}_i,2}}{2} - \big(\frac{p_{\mathrm{drp}}}{2}+k_i^{-1}\big)\mu_{\mathrm{ca}_i,2}, \tag{8.21d}$$

$$\dot{\mu}_{\mathrm{ss}_i,2} = \frac{w_0^2\, p_i\big(1-\sum_{j=1}^{M}(\mu_{\mathrm{ss}_i,0}+\mu_{\mathrm{ca}_i,0})\big)}{\tau_{\mathrm{off}}\, RTT^2} + \frac{(\log 4)\,\mu_{\mathrm{ss}_i,2}}{n_{\mathrm{ack}}\, RTT} - \frac{2R\dot{T}T\,\mu_{\mathrm{ss}_i,2}}{RTT} - (p_{\mathrm{drp}}+k_i^{-1})\frac{\mu_{\mathrm{ss}_i,0}\,\mu^3_{\mathrm{ss}_i,2}}{\mu^3_{\mathrm{ss}_i,1}}, \tag{8.21e}$$

$$\dot{\mu}_{\mathrm{ca}_i,2} = \frac{2\,\mu_{\mathrm{ca}_i,1}}{n_{\mathrm{ack}}\, RTT^2} - \frac{2R\dot{T}T\,\mu_{\mathrm{ca}_i,2}}{RTT} + \frac{p_{\mathrm{drp}}}{4}\frac{\mu_{\mathrm{ss}_i,0}\,\mu^3_{\mathrm{ss}_i,2}}{\mu^3_{\mathrm{ss}_i,1}} - \Big(\frac{3\,p_{\mathrm{drp}}}{4}+k_i^{-1}\Big)\frac{\mu_{\mathrm{ca}_i,0}\,\mu^3_{\mathrm{ca}_i,2}}{\mu^3_{\mathrm{ca}_i,1}}. \tag{8.21f}$$

We present next simulations of these differential equations for a few representative parameter values.

Figure 8.3 corresponds to a transfer-size distribution that results from the mixture of two exponentials ($M = 2$) with parameters

$$k_1 = 3.5\text{ KB}, \qquad k_2 = 246\text{ KB},$$
$$p_1 = 88.87\%, \qquad p_2 = 11.13\%. \tag{8.22}$$

The first exponential corresponds to small "mice" transfers (3.5 KB average) and the second to "elephant" mid-size transfers (246 KB average.) The small transfers are assumed more common (88.87%.) These parameters result in a distribution with an average transfer-size of 30.58 KB and for which 11.13% of the transfers account for 89.7% of the total volume transfered. This is consistent with the file distribution observed in the UNIX file system [14]. However, it does not accurately capture the tail of the distribution (it lacks the "mammoth" files that will be considered later.) The results obtained with the reduced model (8.21) still match quite well those obtained from Monte Carlo simulations of the original SHS, especially taking into account the very large standard deviations. Also here, the Monte Carlo simulations were obtained using the algorithm described in [12, Section 2.1]. It is worth to point out that the simulation of (8.21) takes just a few seconds, whereas each Monte Carlo simulation takes orders of magnitude more time.

Two somewhat surprising conclusions can be drawn from Figure 8.3 for this distribution of transfer-sizes and off-periods:

(i) The *average total sending rate varies very little with the per-packet drop probability*, at least up to the drop probabilities of 33% shown in Figure 8.3(b). This is perhaps not surprising when most packets are transmitted in the slow-start mode, but this phenomenon persists even when a significant fraction of packets are sent in the congestion-avoidance mode [which occurs for drop probabilities above .8%, as seen in Figure 8.3(b)].

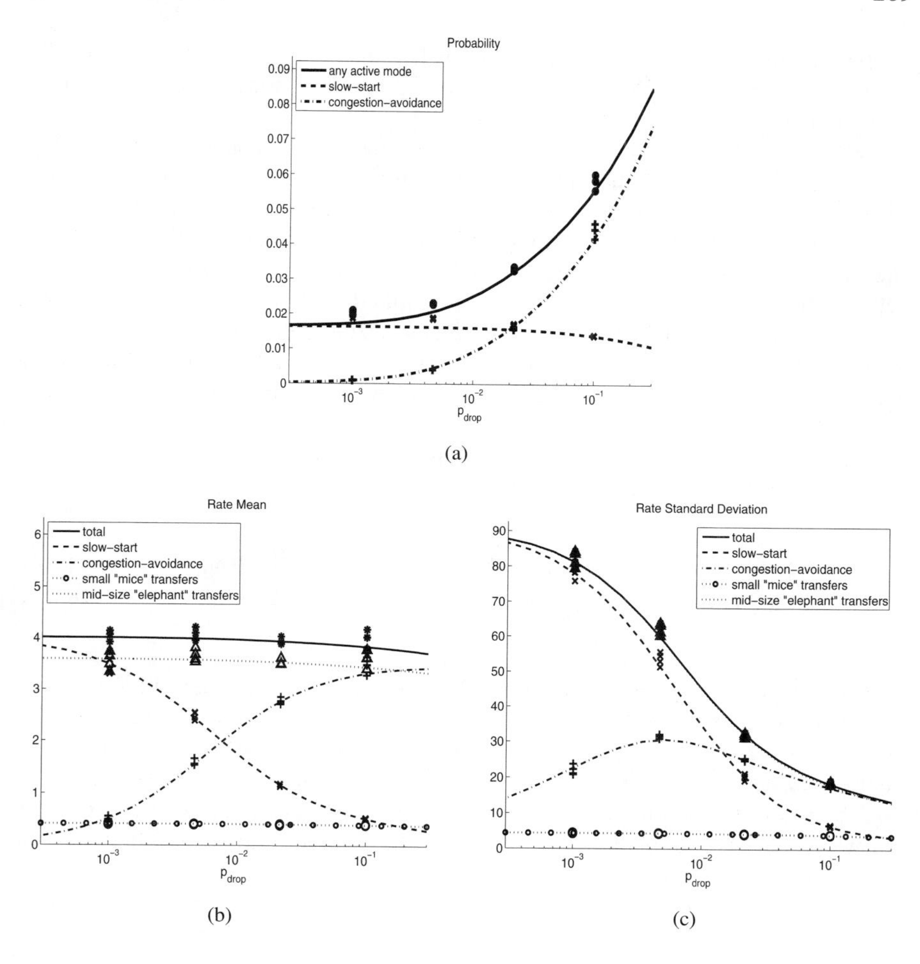

FIGURE 8.3: Steady-state values for the probability of a flow being on each mode (a) and the average (b) and standard deviation (c) of the sending rate as a function of the drop probability. The solid lines were obtained for the mixed-exponentials model (8.21) with $RTT = 50$ ms, $n_{\text{ack}} = 1$, and a transfer-size distribution that results from the mixture of two exponentials with the parameters in (8.22). The average off-period was set to $\tau_{\text{off}} = 5$ sec. The (larger) symbols were obtained from Monte Carlo simulations.

(ii) The *dynamics of individual TCP flows are dominated by second order moments.* In Figure 8.3, the standard deviation is 5 to 20 times larger than the average sending rate, which is very accurately predicted by the reduced model.

This behavior is quite different from the one observed for the long-lived flows considered in Section 8.4.1, where the average sending rate varies significantly with the drop probability and its standard deviation is less than half of its average value. The reader is encouraged to compare the plots in Figure 8.2 for long-lived flows with the corresponding plots in Figure 8.3 for on-off flows.

We recall that both figures correspond to the statistics of a single-user TCP flow. It is not surprising to find out that for a single-user and the same packet drop probability, a long-lived flow will utilize much more bandwidth than an on-off flow. This results in the observed difference in the vertical-axis scales between the plots in Figures 8.2 and 8.3. When one aggregates the flows of n (independent) users, the vertical scales in Figures 8.2 and 8.3 will appear multiplied by n (for the average rate) and by $\sqrt{n}$ (for its standard deviation.) However, the shape of the curves (as p_{drop} varies) will not change and we still conclude that for on-off flows the drop probability exercises a much tighter control on the standard deviation than on the average sending rate of n users.

We consider next a transfer-size distribution that results from a mixture of three exponentials ($M = 3$) with parameters

$$k_1 = 6 \text{ KB}, \qquad k_2 = 400 \text{ KB}, \qquad k_3 = 10 \text{ MB}, \tag{8.23a}$$

$$p_1 = 98\%, \qquad p_2 = 1.7\%, \qquad p_3 = .02\%. \tag{8.23b}$$

The first exponential corresponds to small "mice" transfers, the second to mid-size "elephant" transfers, and the third to large "mammoth" transfers. The resulting distribution, shown in Figure 8.4, approximates reasonably well the one reported by Arlitt et al. [1] obtained from monitoring transfers from a world-wide web proxy within an Internet Service Provider (ISP), at least for transfer-sizes up to 100 MB, for which data is available. This distribution has a much heavier tail than the one considered before.

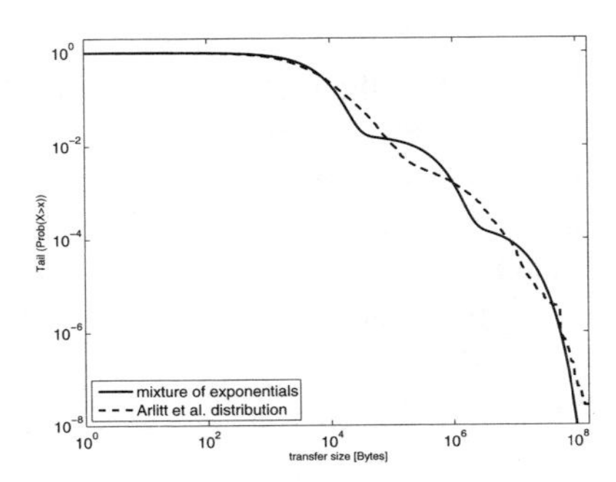

FIGURE 8.4: Complementary cumulative distribution function (ccdf) of transfer-sizes resulting from the mixture of three exponentials with the parameters in (8.23). This distribution was used in the simulations in Figure 8.5.

Figure 8.5 contains results obtained from the reduced model. We do not present Monte Carlo results because the simulation times needed to capture the tails of the transfer-size distribution are prohibitively large. It turns out that the main conclusions drawn before still hold: The average sending rate varies relatively little with the packet drop probability and the dynamics of TCP flows are dominated by second order moments. The mid-size "elephants" still dominate followed by the small "mice." The large "mammoth" transfers occur at a rate that is not sufficiently large to have a significant impact on the average sending rate.

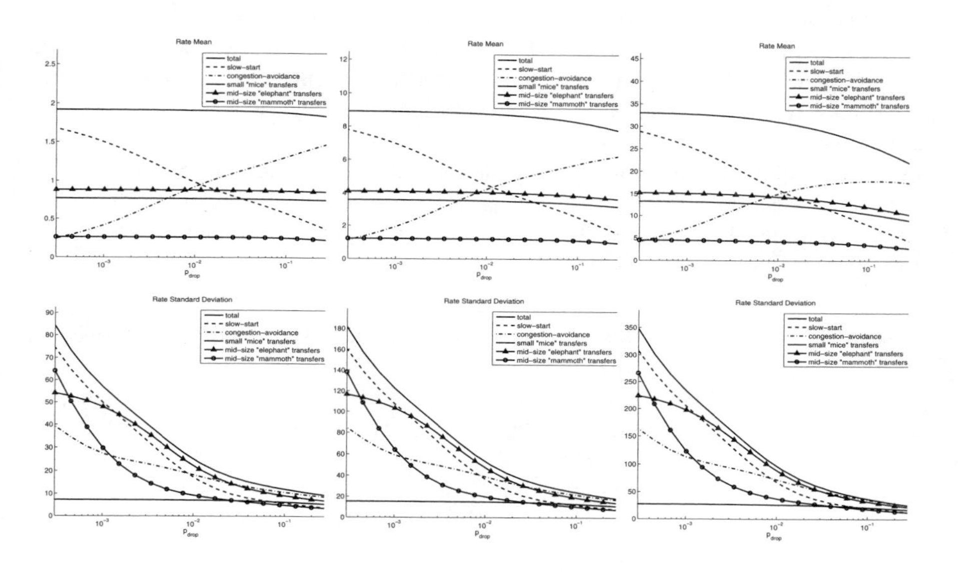

FIGURE 8.5: Steady-state values for the average (top) and standard deviation (bottom) of the sending rate as a function of the drop probability. These results were obtained from the mixed-exponentials model (8.21), with $RTT = 50$ ms, $n_{\text{ack}} = 1$, and a transfer-size distribution resulting from the mixture of three exponentials with the parameters in (8.23). The average off-period was set to $\tau_{\text{off}} = 5$ sec (left), 1 sec (middle), and 0.2 sec (right.)

8.5 Conclusions

We presented a stochastic model for on-off TCP flows that considers both slow-start and congestion-avoidance. This model takes directly into account the distribution of the transfer-sizes to determine the probability of flows being active.

One important observation that stems from this work is that for realistic transfer-size distributions, high-order statistical moments seem to dominate the dynamics of TCP flows, with a standard deviation of the sending rate much larger than its average value. Also, the drop probability appears to have a much more significant effect on the standard deviation of the sending rate than on its average value. This may have significant implications for the congestion control: It seems to indicate that previously used long-lived flow models are not suitable for the analysis and design of congestion control algorithms; and also questions the validity of the "TCP-friendly" formula for the aggregation of many single-user on-off TCP flows.

We are currently investigating how a maximum congestion window size imposed by the receiver affects the dynamics of TCP. It is straightforward to incorporate a maximum window size of $W_{\max}$ in the SHS model in Figure 8.1: In essence one would simply replace $\mathbf{r} = \mathbf{w}/RTT$ by $\mathbf{r} = \min\{\mathbf{w}, W_{\max}\}/RTT$. However, the repercussions of this modification will almost certainly be more complex than limiting the average sending rate to be below $W_{\max}/RTT$, especially in view of the large standard deviations. Future work also includes the analysis and design of active queue management algorithms, based on these models.

Acknowledgments. We thank Martin Arlitt for making available the processed data regarding the transfer-size distribution used in Section 8.4.2; and Stephan Bohacek and Katia Obraczka for insightful discussions. The material in this chapter is based upon work supported by the National Science Foundation under Grants No. CCR-0311084, ANI-0322476.

Appendix

Stochastic Hybrid Systems

For completeness we recall the definition of a *Stochastic Hybrid System* (SHS) introduced by Hespanha [11]. Formally, a SHS is defined by a differential equation

$$\dot{\mathbf{x}} = f(\mathbf{q}, \mathbf{x}, t), \tag{8.24}$$

a family of *m discrete transition/reset maps*

$$(\mathbf{q}, \mathbf{x}) = \phi_\ell(\mathbf{q}^-, \mathbf{x}^-, t), \qquad \ell \in \{1, \ldots, m\}, \tag{8.25}$$

and a family of *m transition intensities*

$$\lambda_\ell(\mathbf{q}, \mathbf{x}, t), \qquad \ell \in \{1, \ldots, m\}, \tag{8.26}$$

where $\mathcal{Q}$ denotes a (typically finite) set and $f : \mathcal{Q} \times \mathbb{R}^n \times [0,\infty) \to \mathbb{R}^n$, $\phi_\ell : \mathcal{Q} \times \mathbb{R}^n \times [0,\infty) \to \mathcal{Q} \times \mathbb{R}^n$, $\lambda_\ell : \mathcal{Q} \times \mathbb{R}^n \times [0,\infty) \to [0,\infty)$, $\forall \ell \in \{1,\ldots,m\}$. A SHS characterizes a jump process $\mathbf{q} : \Omega \times [0,\infty) \to \mathcal{Q}$ called the *discrete state*; a stochastic process $\mathbf{x} : \Omega \times [0,\infty) \to \mathbb{R}^n$ with piecewise continuous sample paths called the *continuous state* ; and m stochastic counters $\mathbf{N}_\ell : \Omega \times [0,\infty) \to \mathbb{N}_{>0}$ called the *transition counters*.

In essence, between transition counter increments the discrete state remains constant whereas the continuous state flows according to (8.24). At transition times the continuous and discrete states are reset according to (8.25). Each transition counter $\mathbf{N}_\ell$ counts the number of times that the corresponding discrete transition/reset map ϕ_ℓ is "activated." The frequency at which this occurs is determined by the transition intensities (8.26). In particular, the probability that the counter $\mathbf{N}_\ell$ will increment and therefore that the corresponding transition takes place in an "elementary interval" $(t, t+dt]$ is given by $\lambda_\ell(\mathbf{q}(t), \mathbf{x}(t), t)dt$. In practice, one can think of the intensity of a transition as the instantaneous rate at which this transition occurs. The reader is referred to [11] for a mathematically precise characterization of a SHS.

It is often convenient to represent a SHS by a directed graph as in Figure 8.6, where each node corresponds to a discrete mode and each edge to a transition between discrete modes. The nodes are labeled with the corresponding discrete mode and the vector fields that determines the evolution of the continuous state in that particular mode. The start of each edge is labeled with the corresponding transition intensity and the end is labeled with the reset map.

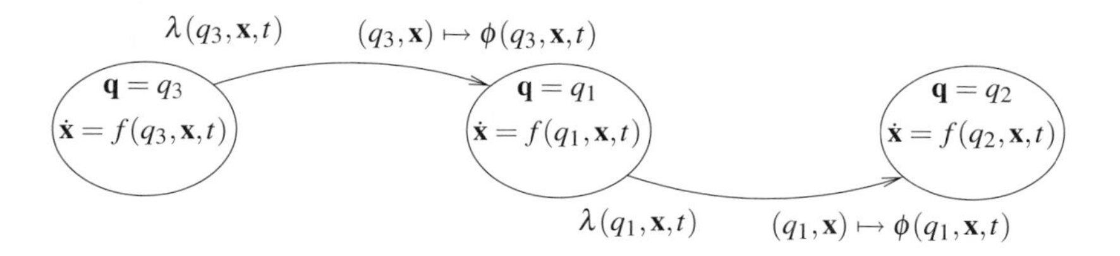

FIGURE 8.6: Graphical representation of a stochastic hybrid system.

The following result can be used to compute expectations on the state of a SHS. For simplicity of presentation we omit a few technical assumptions that are straightforward to verify for the SHS considered here:

THEOREM 8.2 *[11]* *For every function $\psi : \mathcal{Q} \times \mathbb{R}^n \times [0,\infty) \to \mathbb{R}$ that is continuously differentiable with respect to its second and third arguments, we have that*

$$\frac{\mathrm{d}}{\mathrm{d}t} \mathrm{E}[\psi(\mathbf{q}(t), \mathbf{x}(t), t)] = \mathrm{E}[(L\psi)(\mathbf{q}(t), \mathbf{x}(t), t)], \tag{8.27}$$

where, for every $(q,x,t) \in \mathcal{Q} \times \mathbb{R}^n \times [0,\infty)$,

$$(L\psi)(q,x,t) := \frac{\partial \psi(q,x,t)}{\partial x} f(q,x,t) + \frac{\partial \psi(q,x,t)}{\partial t} + \\ + \sum_{\ell=1}^{m} \Big(\psi\big(\phi_\ell(q,x,t),t\big) - \psi(q,x,t) \Big) \lambda_\ell(q,x,t), \quad (8.28)$$

and $\frac{\partial \psi(q,x,t)}{\partial x}$ *and* $\frac{\partial \psi(q,x,t)}{\partial t}$ *denote the gradient of* $\psi(q,x,t)$ *with respect to* x *and the partial derivative of* $\psi(q,x,t)$ *with respect to* t*, respectively. The operator* $\psi \mapsto L\psi$ *defined by* (8.28) *is called the* extended generator *of the SHS.* □

The transition intensities and reset maps for the TCP SHS model in Figure 8.1 are defined as follows:

$$\lambda_{\text{drp}}(q,w,s,t) := \begin{cases} \frac{p_{\text{drp}}(t)w}{RTT} & q \in \{ss, ca\} \\ 0 & \text{otherwise} \end{cases} \qquad \phi_{\text{drp}}(q,w,s,t) := \begin{cases} (d\text{-}ss, w, s) & q = ss \\ (d\text{-}ca, w, s) & q = ca \\ (q,w,s) & \text{otherwise} \end{cases}$$

$$\lambda_{\text{d2ca}}(q,w,s,t) := \begin{cases} \frac{1}{k_{\text{dly}} RTT} & q \in \{d\text{-}ss, d\text{-}ca\} \\ 0 & \text{otherwise} \end{cases} \qquad \phi_{\text{d2ca}}(q,w,s,t) := \begin{cases} (ca, \frac{w}{2}, s) & q \in \{d\text{-}ss, d\text{-}ca\} \\ (q,w,s) & \text{otherwise} \end{cases}$$

$$\lambda_{\text{start}}(q,w,s,t) := \begin{cases} \frac{1}{\tau_{\text{off}}} & q = off \\ 0 & \text{otherwise} \end{cases} \qquad \phi_{\text{start}}(q,w,s,t) := \begin{cases} (ss, w_0, 0) & q = off \\ (q,w,s) & \text{otherwise} \end{cases}$$

$$\lambda_{\text{end}}(q,w,s,t) := \begin{cases} \frac{wF'_{\mathbf{s}^*}(s)}{(1-F_{\mathbf{s}^*}(s))RTT} & q \in \{ss, ca\} \\ 0 & \text{otherwise} \end{cases} \qquad \phi_{\text{end}}(q,w,s,t) := \begin{cases} (\text{off}, 0, 0) & q \in \{ss, ca\} \\ (q,w,s) & \text{otherwise,} \end{cases}$$

where

$$w_0 := \begin{cases} .693 & n_{\text{ack}} = 1 \\ 1.428 & n_{\text{ack}} = 2. \end{cases}$$

For the mixed-exponentials model, the intensities and reset maps are given by

$$\lambda_{\text{drp}}(q,w,s,t) := \begin{cases} \frac{p_{\text{drp}}(t)w}{RTT} & q \in \{ss_i, ca_i : \forall i\} \\ 0 & \text{otherwise} \end{cases} \qquad \phi_{\text{drp}}(q,w,s,t) := \begin{cases} (d\text{-}ss_1, w, s) & q = ss_1 \\ \vdots & \vdots \\ (d\text{-}ss_M, w, s) & q = ss_M \\ (d\text{-}ca_1, w, s) & q = ca_1 \\ \vdots & \vdots \\ (d\text{-}ca_M, w, s) & q = ca_M \\ (q,w,s) & \text{otherwise} \end{cases}$$

$$\lambda_{\text{d2ca}}(q,w,s,t) := \begin{cases} \frac{k_{\text{dly}}^{-1}}{RTT} & q \in \{d\text{-}ss_i, d\text{-}ca_i : \forall i\} \\ 0 & \text{otherwise} \end{cases} \qquad \phi_{\text{d2ca}}(q,w,s,t) := \begin{cases} (ca_1, \frac{w}{2}, s) & q \in \{d\text{-}ss_1, d\text{-}ca_1\} \\ \vdots & \vdots \\ (ca_M, \frac{w}{2}, s) & q \in \{d\text{-}ss_M, d\text{-}ca_M\} \\ (q,w,s) & \text{otherwise} \end{cases}$$

$$\lambda_i(q,w,s,t) := \begin{cases} \frac{p_i}{\tau_{\text{off}}} & q = off \\ 0 & \text{otherwise} \end{cases} \qquad \phi_i(q,w,s,t) := \begin{cases} (ss_i, w_0, 0) & q = off \\ (q,w,s) & \text{otherwise} \end{cases}$$

$$\lambda_{\text{end}}(q,w,s,t) := \begin{cases} \frac{k_1^{-1}w}{RTT} & q \in \{ss_1, ca_1\} \\ \vdots & \vdots \\ \frac{k_M^{-1}w}{RTT} & q \in \{ss_M, ca_M\} \\ 0 & \text{otherwise} \end{cases} \qquad \phi_{\text{end}}(q,w,s,t) := \begin{cases} (\text{off}, 0, 0) & q \in \{ss_i, ca_i : \forall i\} \\ (q,w,s) & \text{otherwise,} \end{cases}$$

where the λ_i and ϕ_i, $i \in \{1,2,\ldots,M\}$ replace λ_{start} and ϕ_{start}, respectively, in the previous model.

Proofs

PROOF (of Theorem 8.1.) Applying to the functions $\psi_{q_0,n}$ defined in (8.12) the extended generator L for the mixed-exponentials SHS model yields

$$\begin{aligned} L\psi_{q_0,n}(q,w,s,t) &= \frac{\partial \psi_{q_0,n}(q,w,t)}{\partial w} f(q,w,s,t) + \frac{\partial \psi_{q_0,n}(q,w,t)}{\partial t} \\ &+ (\psi_{q_0,n}(\phi_{\text{drp}}(q,w,s,t),t) - \psi_{q_0,n}(q,w,t))\lambda_{\text{drp}}(q,w,s,t) \\ &+ (\psi_{q_0,n}(\phi_{\text{d2ca}}(q,w,s,t),t) - \psi_{q_0,n}(q,w,t))\lambda_{\text{d2ca}}(q,w,s,t) \\ &+ (\psi_{q_0,n}(\phi_{\text{end}}(q,w,s,t),t) - \psi_{q_0,n}(q,w,t))\lambda_{\text{end}}(q,w,s,t) \\ &+ \sum_{j=1}^{M} (\psi_{q_0,n}(\phi_j(q,w,s,t),t) - \psi_{q_0,n}(q,w,t))\lambda_j(q,w,s,t), \end{aligned}$$

from which we obtain by direct computation that

$$(L\psi_{\text{off},0})(q,w,t) = \begin{cases} \frac{k_1^{-1}w}{RTT} & q \in \{ss_1, ca_1\} \\ \vdots & \vdots \\ \frac{k_M^{-1}w}{RTT} & q \in \{ss_M, ca_M\} \\ -\frac{1}{\tau_{\text{off}}} & q = \text{off} \\ 0 & \text{otherwise} \end{cases}$$

$$= -\frac{\psi_{\text{off},0}(q,w,t)}{\tau_{\text{off}}} + \sum_{j=1}^{M} k_j^{-1}\left(\psi_{ss_j,1}(q,w,t) + \psi_{ca_j,1}(q,w,t)\right)$$

$$(L\psi_{ss_i,n})(q,w,t) =$$

$$= \begin{cases} \frac{n\left((\log 2)n_{\text{ack}}^{-1} - R\dot{T}T\right)w^n - (p_{\text{drp}} + k_i^{-1})w^{n+1}}{RTT^{n+1}} & q = ss_i \\ \frac{p_i w_0^n}{\tau_{\text{off}} RTT^n} & q = \text{off} \\ 0 & \text{otherwise} \end{cases}$$

$$= \frac{p_i w_0^n \psi_{\text{off},0}(q,w,t)}{\tau_{\text{off}} RTT^n} + n\frac{(\log 2) - n_{\text{ack}} R\dot{T}T}{n_{\text{ack}} RTT}\psi_{ss_i,n}(q,w,t) - (p_{\text{drp}} + k_i^{-1})\psi_{ss_i,n+1}(q,w,t)$$

$$(L\psi_{\text{d-}ss_i,n}(q,w,t)) = \begin{cases} \frac{p_{\text{drp}} w^{n+1}}{RTT^{n+1}} & q = ss_i \\ \frac{n\left((\log 2)n_{\text{ack}}^{-1} - R\dot{T}T\right)w^n - k_{\text{dly}}^{-1}w^n}{RTT^{n+1}} & q = \text{d-}ss_i \\ 0 & \text{otherwise} \end{cases}$$

$$= n\frac{(\log 2) - n_{\text{ack}} R\dot{T}T}{n_{\text{ack}} RTT}\psi_{\text{d-}ss_i,n}(q,w,t) + p_{\text{drp}}\psi_{ss_i,n+1}(q,w,t) - \frac{\psi_{\text{d-}ss_i,n}(q,w,t)}{k_{\text{dly}} RTT}$$

$$(L\psi_{\text{ca}_i,n}(q,w,t)) =$$
$$= \begin{cases} \frac{w^n}{2^n k_{\text{dly}} RTT^{n+1}} & q \in \{\text{d-ss}_i, \text{d-ca}_i\} \\ \frac{n n_{\text{ack}}^{-1} w^{n-1} - nR\dot{T}T w^n - (p_{\text{drp}} + k_i^{-1}) w^{n+1}}{RTT^{n+1}} & q = \text{ca}_i \\ 0 & \text{otherwise} \end{cases}$$
$$= \frac{n\psi_{\text{ca}_i,n-1}(q,w,t)}{n_{\text{ack}} RTT^2} - \frac{nR\dot{T}T\,\psi_{\text{ca}_i,n}(q,w,t)}{RTT} - (p_{\text{drp}} + k_i^{-1})\psi_{\text{ca}_i,n+1}(q,w,t)$$
$$+ \frac{\psi_{\text{d-ss}_i,n}(q,w,t) + \psi_{\text{d-ca}_i,n}(q,w,t)}{2^n k_{\text{dly}} RTT}$$
$$(L\psi_{\text{d-ca}_i,n}(q,w,t)) = \begin{cases} \frac{p_{\text{drp}} w^{n+1}}{RTT^{n+1}} & q = \text{ca}_i \\ \frac{n n_{\text{ack}}^{-1} w^{n-1} - nR\dot{T}T w^n - k_{\text{dly}}^{-1} w^n}{RTT^{n+1}} & q = \text{d-ca}_i \\ 0 & \text{otherwise} \end{cases}$$
$$= \frac{n\psi_{\text{d-ca}_i,n-1}(q,w,t)}{n_{\text{ack}} RTT^2} - \frac{nR\dot{T}T\,\psi_{\text{d-ca}_i,n}(q,w,t)}{RTT} + p_{\text{drp}}\psi_{\text{ca}_i,n+1}(q,w,t) - \frac{\psi_{\text{d-ca}_i,n}(q,w,t)}{k_{\text{dly}} RTT}.$$

To obtain (8.15), we use (8.27) to conclude that

$$\dot{\mu}_{q_0,n} = \mathrm{E}[(L\psi_{q_0,n})(\mathbf{q},\mathbf{w},t)],$$

and replace in the right-hand-side of this equation the expectations of the $\psi_{q_0,n}$ by the corresponding $\mu_{q_0,n}$. Equation (8.14) can be obtained from (8.15) by setting all the $k_j = \infty$ and consequently $\mu_{\mathit{off},n} = \mu_{\mathit{ss}_j,n} = \mu_{\mathit{d\text{-}ss}_j,n} = 0$, $\forall n, j$ since in this SHS all the modes ss_j, $d\text{-}ss_j$, and *off* are absent. Equation (8.16) can be obtained along the lines of the derivation of (8.15). Due to space limitations we omit these computations. ■

References

[1] M. Arlitt, R. Friedrich, and T. Jin. Workload characterization of a web proxy in a cable modem environment. Technical Report HPL-1999-48, Hewlett-Packard Laroratories, Palo Alto, CA, Apr. 1999.

[2] F. Baccelli and D. Hong. Flow level simulation of large IP networks. In *Proc. of the IEEE INFOCOM*, Mar. 2003.

[3] F. Baccelli and D. Hong. Interaction of TCP flows as billiards. In *Proc. of the IEEE INFOCOM*, Mar. 2003.

[4] P. Barford, A. Bestavros, A. Bradley, and M. Crovella. Changes in web client access patterns. *World Wide Web,* Special Issue on Characterization and Performance Evaluation, 2(1–2):15–28, 1999.

[5] S. Bohacek. A stochastic model of TCP and fair video transmission. In *Proc. of the IEEE INFOCOM*, Mar. 2003.

[6] S. Bohacek, J. P. Hespanha, J. Lee, and K. Obraczka. Analysis of a TCP hybrid model. In *Proc. of the 39th Annual Allerton Conf. on Comm., Contr., and Computing*, Oct. 2001.

[7] S. Bohacek, J. P. Hespanha, J. Lee, and K. Obraczka. A hybrid systems modeling framework for fast and accurate simulation of data communication networks. In *Proc. of the ACM Int. Conf. on Measurements and Modeling of Computer Systems (SIGMETRICS)*, June 2003.

[8] A. B. Downey. Evidence for long-tailed distributions in the internet. In *Proc. of ACM SIGCOMM Internet Measurement Workshop*, Nov. 2001.

[9] A. Feldmann and W. Whitt. Fitting mixtures of exponentials to long-tail distributions to analyze network performance models. In *Proc. of the IEEE INFOCOM*, Apr. 1997.

[10] S. Floyd and V. Jacobson. Random early detection gateways for congestion avoidance. *IEEE/ACM Trans. on Networking*, 1(4):397–413, Aug. 1993.

[11] J. P. Hespanha. Stochastic hybrid systems: Applications to communication networks. In R. Alur and G. J. Pappas, editors, *Hybrid Systems: Computation and Control*, number 2993 in Lect. Notes in Comput. Science, pages 387–401. Springer-Verlag, Berlin, Mar. 2004.

[12] J. P. Hespanha. A model for stochastic hybrid systems with application to communication networks. *Nonlinear Analysis,* Special Issue on Hybrid Systems, 62(8):1353–1383, Sept. 2005.

[13] J. P. Hespanha. Polynomial stochastic hybrid systems. In M. Morari and L. Thiele, editors, *Hybrid Systems: Computation and Control*, number 3414 in Lect. Notes in Comput. Science, pages 322–338. Springer-Verlag, Berlin, Mar. 2005.

[14] G. Irlam. Unix file size survey – 1993. Available at http://www.base.com /gordoni/ufs93.html, Nov. 1994.

[15] S. Kunniyur and R. Srikant. Analysis and design of an adaptive virtual queue (AVQ) algorithm for active queue management. In *Proc. of the ACM SIGCOMM*, San Diego, CA, Aug. 2001.

[16] A. Lakshmikantha, C. Beck, and R. Srikant. Robustness of real and virtual queue based active queue management schemes. In *Proc. of the 2003 Amer. Contr. Conf.*, pages 266–271, June 2003.

[17] Y. Liu, W.-B. Gong, V. Misra, and D. Towsley. On the tails of web file size distributions. In *Proc. of 39th Allerton Conference on Communication, Control, and Computing*, Oct. 2001.

[18] S. H. Low. A duality model of TCP and queue management algorithms. *IEEE/ACM Trans. on Networking*, 11(4), Aug. 2003.

[19] S. H. Low, F. Paganini, and J. C. Doyle. Internet congestion control. *IEEE Contr. Syst. Mag.*, 22(1):28–43, Feb. 2002.

[20] J. Mahdavi and S. Floyd. TCP-friendly unicast rate-based flow control. Technical note sent to the end2end-interest mailing list, Jan. 1997.

[21] M. A. Marsan, M. Garetto, P. Giaccone, E. Leonardi, E. Schiattarella, and A. Tarello. Using partial differential equations to model TCP mice and elephants in large IP networks. *Proc. of the IEEE INFOCOM*, Mar. 2004.

[22] M. Mathis, J. Semke, J. Mahdavi, and T. Ott. The macroscopic behavior of the TCP congestion avoidance algorithm. *ACM Comput. Comm. Review*, 27(3), July 1997.

[23] M. Mellia, I. Stoica, and H. Zhang. TCP model for short lived flows. *IEEE Comm. Lett.*, 6(2):85–87, Feb. 2002.

[24] V. Misra, W. Gong, and D. Towsley. Stochastic differential equation modeling and analysis of TCP-windowsize behavior. In *Proc. of PERFORMANCE '99*, Istanbul, Turkey, 1999.

[25] V. Misra, W. Gong, and D. Towsley. Fluid-based analysis of a network of AQM routers supporting TCP flows with an application to RED. In *Proc. of the ACM SIGCOMM*, Sept. 2000.

[26] T. Ott, J. H. B. Kemperman, and M. Mathis. Window size behavior in TCP/IP with constant loss probability. In *Proc. of the DIMACS Workshop on Performance of Realtime Applications on the Internet*, Nov. 1996.

[27] J. Padhye, V. Firoiu, D. Towsley, and J. Kurose. Modeling TCP throughput: a simple model and its empirical validation. In *Proc. of the ACM SIGCOMM*, Sept. 1998.

[28] J. Padhye, V. Firoiu, D. Towsley, and J. Kurose. Modeling TCP Reno performance: A simple model and its empirical validation. *IEEE/ACM Trans. on Networking*, 8(2):133–145, Apr. 2000.

[29] K. Park and W. Willinger. Self-similar network traffic: An overview. In K. Park and W. Willinger, editors, *Self-Similar Network Traffic and Performance Evaluation*. Wiley Interscience, New York, NY, 1999.

[30] S. Shakkottai and R. Srikant. How good are deterministic fluid models of Internet congestion control? In *Proc. of the IEEE INFOCOM*, June 2002.

[31] B. Sikdar, S. Kalyanaraman, and K. Vastola. Analytic models for the latency and steady-state throughput of TCP Tahoe, Reno and SACK. In *Proc. of the IEEE GLOBECOM*, pages 25–29, Nov. 2001.

[32] B. Sikdar, S. Kalyanaraman, and K. Vastola. TCP Reno with random losses: Latency, throughput and sensitivity analysis. In *Proc. of the IEEE IPCCC*, pages 188–195, Apr. 2001.

[33] W. Stallings. *High-speed networks: TCP/IP and ATM design principles*. Prentice-Hall, London, 1998.

[34] D. Zheng, G. Lazarou, and R. Hu. A stochastic model for short-lived tcp flows. In *Proc. of the IEEE Int. Conf. on Communications*, volume 1, pages 76–81, May 2003.

Chapter 9

Stochastic Hybrid Modeling of Biochemical Processes

Panagiotis Kouretas
University of Patras

Konstantinos Koutroumpas
ETH Zürich

John Lygeros
ETH Zürich

Zoi Lygerou
University of Patras

9.1 Introduction 221
9.2 Overview of PDMP 223
9.3 Subtilin Production by *B. subtilis* 228
9.4 DNA Replication in the Cell Cycle 235
9.5 Concluding Remarks 244
References 245

9.1 Introduction

One of the defining changes in molecular biology over the last decade has been the massive scaling up of its experimental techniques. The sequencing of the entire genome of organisms, the determination of the expression level of genes in a cell by means of DNA micro-arrays, and the identification of proteins and their interactions by high-throughput proteomic methods have produced enormous amounts of data on different aspects of the development and functioning of cells.

A consensus is now emerging among biologists that to exploit these data to full potential one needs to complement experimental results with formal models of biochemical networks. Mathematical models that describe gene and protein interactions in a precise and unambiguous manner can play an instrumental role in shaping the future of biology. For example, mathematical models allow computer-based simulation and analysis of biochemical networks. Such *in silico* experiments can be used for massive and rapid verification or falsification of biological hypotheses, replacing in certain cases costly and time-consuming in vitro or *in vivo* experiments. Moreover, *in silico*, *in vitro* and *in vivo* experiments can be used together in a feedback arrangement: mathematical model predictions can assist in the design of *in vitro* and *in vivo* experiments, the results of which can in turn be used to improve the fidelity of the mathematical models.

The possibility of combining new experimental methods, sophisticated mathematical techniques, and increasingly powerful computers, has given a new lease of life to an idea as appealing as it is difficult to realize: understanding how the global behavior of an organism emerges from the interactions between components at the molecular level. Although this idea of systems biology has multiple aspects [23], an ultimate challenge is the construction of a mathematical model of whole cells, that will be able to simulate *in silico* the behavior of an organism *in vivo*.

In the last few decades, a large number of approaches for modeling molecular interaction networks have been proposed. Motivated by the classification of [12], one can divide the models available in the literature in two classes:

- Models with purely continuous dynamics, for example, models that describe the evolution of concentrations of proteins, mRNAs, etc., in terms of ordinary or partial differential equations.

- Models with purely discrete dynamics, for example, graph models of the interdependencies in a regulatory network, Boolean networks and their extensions, Bayesian networks, or Markov chain models.

Common sense and experimental evidence suggest that neither of these classes alone is adequate for developing realistic models of molecular interaction networks. Timescale hierarchies cause biological processes to be more conveniently described as a mixture of continuous and discrete phenomena. For example, continuous changes in chemical concentrations or the environment of a cell often trigger discrete transitions (such as the onset of mitosis, or cell differentiation) that in turn influence the concentration dynamics. At the level of molecular interactions, the co-occurrence of discrete and continuous dynamics is exemplified by the switch-like activation or inhibition of gene expression by regulatory proteins.

The recognition that hybrid discrete-continuous dynamics can play an important role in biochemical systems has led a number of researchers to investigate how methods developed for hybrid systems in other areas (such as embedded computation and air traffic management) can be extended to biological systems [17, 1, 13, 5, 3, 15]. It is, however, fair to say that the realization of the potential of hybrid systems theory in the context of biochemical system modeling is still for the future. In addition, recently the observation that many biological processes involve considerable levels of uncertainty has been gaining momentum [27, 22]. For example, experimental observations suggest that stochastic uncertainty may play a crucial role in enhancing the robustness of biochemical processes [35], or may be behind the variability observed in the behavior of certain organisms [36, 37]. Stochasticity is even observed in fundamental processes such as the DNA replication itself [8, 26]. This has led researchers to attempt the development of stochastic hybrid models for certain biochemical processes [20, 19], that aim to couple the advantages of stochastic analysis with the generality of hybrid systems.

In this chapter we explore further stochastic, hybrid aspects in the modeling of biochemical networks. We first survey briefly a framework for modeling stochastic hybrid systems known as Piecewise Deterministic Markov Processes (PDMP; Section 9.2). We then proceed to use this framework to capture the essence of two biochemical processes: the production of subtilin by the bacterium *Bachillus subtilis* (*B. subtilis*; Section 9.3), and the process of DNA replication in eukaryotic cells (Section 9.4). The two models illustrate two different mechanisms through which stochastic features manifest themselves in biochemical processes: the uncertainty about switching genes "on" and "off" and uncertainty about the binding of protein complexes on the DNA. We also discuss how these models can be coded in simulation and present simulation results. The concluding section (Section 9.5) presents directions for further research.

9.2 Overview of PDMP

Piecewise Deterministic Markov Processes (PDMP), introduced by Mark Davis in [9, 10], are a class of continuous-time stochastic hybrid processes which covers a wide range of non-diffusion phenomena. They involve a hybrid state space, comprising continuous and discrete states. The defining feature of autonomous PDMP is that continuous motion is deterministic; between two consecutive transitions the continuous state evolves according to an ordinary differential equation (ODE). Transitions occur either when the state hits the state space boundary, or in the interior of the state space, according to a generalized Poisson process. Whenever a transition occurs, the hybrid state is reset instantaneously according to a probability distribution which depends on the hybrid state before the transition, and the process is repeated. Here we introduce formally PDMP following the notation of [6, 24]. Our treatment of PDMP is adequate for this chapter, but glosses over some of the technical subtleties introduced in [9, 10] to make the PDMP model as precise and general as possible.

9.2.1 Modeling Framework

Let Q be a countable set of discrete states, and let $d(\cdot) : Q \to \mathbb{N}$ and $X(\cdot) : Q \to \mathbb{R}^{d(\cdot)}$ be two maps assigning to each discrete state $q \in Q$ a continuous state dimension $d(q)$ and an open subset $X(q) \subseteq \mathbb{R}^{d(q)}$. We call the set

$$\mathscr{D}(Q,d,X) = \bigcup_{q \in Q} \{q\} \times X(q) = \{(q,x) \mid q \in Q,\ x \in X(q)\}$$

the hybrid state space of the PDMP and denote by $(q,x) \in \mathscr{D}(Q,d,X)$ the hybrid state. For simplicity, we use just $\mathscr{D}$ to denote the state space when the Q, d, and X

are clear from the context. We denote the complement of the hybrid state space by

$$\mathscr{D}^c = \bigcup_{q \in Q} \{q\} \times X(q)^c,$$

its closure by

$$\overline{\mathscr{D}} = \bigcup_{q \in Q} \{q\} \times \overline{X(q)},$$

and its boundary by

$$\partial \mathscr{D} = \bigcup_{q \in Q} \{q\} \times \partial X(q) = \overline{\mathscr{D}} \setminus \mathscr{D}.$$

As usual, $X(q)^c$ denotes the complement, $\overline{X(q)}$ the closure, and $\partial X(q)$ the boundary of the open set $X(q)$ in $\mathbb{R}^{d(q)}$, and $\setminus$ denotes set difference. Let $\mathscr{B}(\mathscr{D})$ denote the smallest σ-algebra on $\cup_{q \in Q}\{q\} \times \mathbb{R}^{d(q)}$ containing all sets of the form $\{q\} \times A_q$ with A_q a Borel subset of $X(q)$.

We consider a parameterized family of vector fields $f(q,\cdot) : \mathbb{R}^{d(q)} \to \mathbb{R}^{d(q)}$, $q \in Q$, assigning to each hybrid state (q,x) a direction $f(q,x) \in \mathbb{R}^{d(q)}$. As usual, we define the flow of f as the function $\Phi(q,\cdot,\cdot) : \mathbb{R}^{d(q)} \times \mathbb{R} \to \mathbb{R}^{d(q)}$ such that $\Phi(q,x,0) = x$ and for all $t \in \mathbb{R}$,

$$\frac{d}{dt}\Phi(q,x,t) = f(q,\Phi(q,x,t)). \tag{9.1}$$

Notice that we implicitly assume that the discrete state q remains constant along continuous evolution.

DEFINITION 9.1 *A PDMP is a collection $H = ((Q,d,X),f,Init,\lambda,R)$, where*

- *Q is a countable set of discrete states;*
- *$d(\cdot) : Q \to \mathbb{N}$ maps each $q \in Q$ to a continuous state space dimension;*
- *$X(\cdot) : Q \to \mathbb{R}^{d(.)}$ maps each $q \in Q$ to an open subset $X(q)$ of $\mathbb{R}^{d(q)}$;*
- *$f(q,\cdot) : \mathbb{R}^{d(q)} \to \mathbb{R}^{d(q)}$ is a family of vector fields parameterized by $q \in Q$;*
- *$\mathrm{Init}(\cdot) : \mathscr{B}(\mathscr{D}) \to [0,1]$ is an initial probability measure on $(\mathscr{D}, \mathscr{B}(\mathscr{D}))$;*
- *$\lambda(\cdot,\cdot) : \mathscr{D} \to \mathbb{R}^+$ is a transition rate function;*
- *$R(\cdot,\cdot,\cdot) : \mathscr{B}(\mathscr{D}) \times \overline{\mathscr{D}} \to [0,1]$ assigns to each $(q,x) \in \overline{\mathscr{D}}$ a measure $R(\cdot,q,x)$ on $(\mathscr{D},\mathscr{B}(\mathscr{D}))$.*

To define the PDMP executions we introduce the notions of the exit time, $t^*(\cdot,\cdot) : \mathscr{D} \to \mathbb{R}^+ \cup \{\infty\}$, defined as

$$t^*(q,x) = \inf\{t > 0 \mid \Phi(q,x,t) \notin \mathscr{D}\} \tag{9.2}$$

and of the survival function $F(\cdot,\cdot,\cdot) : \mathscr{D} \times \mathbb{R}^+ \to [0,1]$,

$$F(q,x,t) = \begin{cases} e^{-\int_0^t \lambda(q,\Phi(q,x,\tau))d\tau} & \text{if } t < t^*(q,x) \\ 0 & \text{if } t \geq t^*(q,x). \end{cases} \tag{9.3}$$

With this notation, the executions of the PDMP can be thought of as being generated by the following algorithm.

ALGORITHM 9.1 **(Generation of PDMP executions)**

set $T = 0$
extract *$\mathscr{D}$-valued random variable* $(\hat{q},\hat{x})$ *according to* $\text{Init}(\cdot)$
while $T < \infty$
  extract *$\mathbb{R}^+$-valued random variable* $\hat{T}$ *such that* $P[\hat{T} > t] = F(\hat{q},\hat{x},t)$
  set $q(t) = \hat{q}$ *and* $x(t) = \Phi(\hat{q},\hat{x},t-T)$ *for all* $t \in [T, T+\hat{T})$
  if $\hat{T} < \infty$
    extract *$\mathscr{D}$-valued random variable* (q',x')
      according to $R((\cdot,\cdot,\hat{q},\Phi(\hat{q},\hat{x},\hat{T}))$
    set $(\hat{q},\hat{x}) = (q',x')$
  end if
  set $T = T + \hat{T}$
end while

All random extractions in Algorithm 9.1 are assumed to be independent. To ensure that the algorithm produces a well-defined stochastic process a number of assumptions are introduced in [10, 9].

ASSUMPTION 9.1 The PDMP satisfies the following assumptions:

(i) *Init*$(\mathscr{D}^c) = 0$ and $R(\mathscr{D}^c, q, x) = 0$ for all $(q,x) \in \overline{\mathscr{D}}$.

(ii) For all $q \in Q$, the set $X(q) \subseteq \mathbb{R}^{d(q)}$ is open and $f(q,\cdot)$ is globally Lipschitz continuous.

(iii) $\lambda(\cdot,\cdot)$ is measurable. For all $(q,x) \in \mathscr{D}$ there exists $\varepsilon > 0$ such that the function $t \to \lambda(q,\Phi(q,x,t))$ is integrable for $t \in [0,\varepsilon)$. For all $A \in \mathscr{B}(\mathscr{D})$, $R(A,(\cdot,\cdot))$ is measurable.

(iv) The expected number of jumps in $[0,t]$ is finite for all $t < \infty$.

Most of the assumptions are technical and are needed to ensure that the transition kernels, the solutions of the differential equations, etc., are well defined. The last part of the assumption deserves some closer scrutiny. This is the stochastic variant of the non-Zeno assumption commonly imposed on hybrid systems. It states that "on the average" only a finite number of discrete transitions can take place in any finite time interval. While this assumption is generally true for real systems, it is easy to generate models that violate it due to modeling over-abstraction (see, for example, [21]). Even if a model is not Zeno, establishing this this may be difficult [25].

Under Assumption 9.1 it can be shown [10, 9] that Algorithm 9.1 defines a strong Markov process, which is continuous from the right with left limits. Based on these fundamental properties, [10, 9] proceed to completely characterize PDMP processes through their generator, and then use the generator to show how one can compute expectations, establish stability conditions and solve optimal control problems for this class of stochastic hybrid systems.

9.2.2 Simulation

The properties of PDMP are in general rather difficult to study analytically. Explicit solutions for things like expectations are impossible to derive, except in very special cases (see, for example, [11]). One therefore often resorts to numerical methods. For computing expectations, approximating distributions, etc., one of the most popular is Monte Carlo simulation.

The simulation of PDMP models presents several challenges, due to the interaction of discrete, continuous, and stochastic terms. Because the continuous dynamics are deterministic, standard algorithms used for the simulation of continuous, deterministic systems are adequate for simulating the evolution between two discrete transitions. The difficulties arise when the continuous evolution has to be interrupted so that a discrete transition can be executed.

For forced transitions (when the state attempts to leave $\mathcal{D}$) one needs to detect when the state, x, leaves an open set, $X(q)$, along continuous evolution. This is known as the event detection problem in the hybrid systems literature. Several algorithms have been developed to deal with this problem (see for example [4]) and have recently been included in standard simulation packages such as Dymola, Matlab, or the Simulink package SimEvents. Roughly speaking, the idea is to code the set $X(q)$ using a function, $g(q,x)$, of the state that changes sign at the boundary of $X(q)$. The simulation algorithm keeps track of the function $g(q,x(k))$ at each step, k, of the continuous simulation and proceeds normally as long as $g(q,x(k))$ does not change sign between one step and the next; recall that in this case q also does not change. If g changes sign (say between step k and step $k+1$) the simulation halts, a zero crossing algorithm is used to pinpoint the time at which the sign change took place, the state at this time is computed, the event is "serviced," and the simulation resumes from the new state. Zero crossing (finding the precise state just before the event) usually involves fitting a polynomial to a few values of g before and after the event (say a spline through $g(q,x(k-1))$, $g(q,x(k))$ and $g(q,x(k+1))$) and finding its roots. Servicing the event (finding the state just after the event) requires a random extraction from the transition kernel R. While it is known that for most hybrid systems initial conditions exist for which accurate event detection is problematic, it is also known that for a wide class of hybrid systems the simulation strategy outlined above works for almost all initial states [33].

For spontaneous transitions, the situation is at first sight more difficult: one needs to extract a random transition time, $\hat{T}$, such that

$$P[\hat{T} > t] = e^{-\int_0^t \lambda(\hat{q},\Phi(\hat{q},\hat{x},\tau))d\tau}.$$

It turns out, however, that this can easily be done by appending two additional continuous states, say $y \in \mathbb{R}$ and $z \in \mathbb{R}$, to the state vector. We therefore make a new PDMP, $H' = ((Q, d', X'), f', Init', \lambda', R')$ with continuous state dimension

$$d'(q) = d(q) + 2, \text{for all } q \in Q.$$

We set

$$X'(q) = X(q) \times \{(y,z) \in \mathbb{R}^2 \mid z > 0,\ y < -\ln(z)\}.$$

The continuous dynamics of these additional states are given by $\dot{y} = \lambda(q,x)$ and $\dot{z} = 0$, in other words

$$f'(q,x) = \begin{bmatrix} f(q,x) \\ \lambda(q,x) \\ 0 \end{bmatrix}.$$

We set $\lambda'(q,x,y,z) = 0$ for all $q \in Q$, $(x,y,z) \in \mathbb{R}^{d'(q)}$.

Initially, and after each discrete transition y is set to 0, whereas z is extracted uniformly in the interval $[0,1]$. For simplicity consider the first interval of continuous evolution; the same argument holds between any two transitions. Until the first discrete transition we have

$$y(t) = \int_0^t \lambda(q(0), \Phi(q(0), x(0), \tau))d\tau, \text{ and } z(t) = z(0).$$

Notice that, since λ is non-negative, the state $y(t)$ is a non-decreasing function of t.

Since λ' is identically equal to zero spontaneous transitions are not possible for the modified PDMP. Therefore the first transition will take place because either $x(t)$ leaves $X(q)$, or because $y(t) \geq -\ln(z(t))$. Assume the latter is the case, and let

$$\hat{T} = \inf\{\tau \geq 0 \mid y(\tau) \geq -\ln(z(\tau))\}.$$

Then

$$\begin{aligned} P[\hat{T} > t] &= P[y(t) < -\ln(z(t))] && (y(t) \text{ non-decreasing}) \\ &= P[\textstyle\int_0^t \lambda(q(0), \Phi(q(0), x(0), \tau))d\tau < -\ln(z(t))] \\ &= P[z(0) < e^{-\int_0^t \lambda(q(0), \Phi(q(0), x(0), \tau))d\tau}] \\ &= e^{-\int_0^t \lambda(q(0), \Phi(q(0), x(0), \tau))d\tau} && (z(0) \text{ uniform}). \end{aligned}$$

After the discrete transition the new state (q,x) is extracted according to R, y is reset to zero and z is extracted uniformly in $[0,1]$.

Therefore, for simulation purposes spontaneous transitions can be treated in very much the same way as forced transitions. In fact, the above construction is standard in the simulation of discrete event systems [7] and shows that every PDMP, H, is equivalent to another PDMP, H', that involves only forced transitions. Spontaneous transitions, however, still provide a very natural way of modeling physical phenomena and will be used extensively below.

9.3 Subtilin Production by *B. subtilis*

9.3.1 Qualitative Description

Subtilin is an antibiotic released by *B. subtilis* as a way of confronting difficult environmental conditions. The factors that govern subtilin production can be divided into internal (the physiological states of the cell) and external (local population density, nutrient levels, aeration, environmental signals, etc.). Roughly speaking, a high concentration of nutrients in the environment results in an increase in *B. subtilis* population without a remarkable change in subtilin concentration. Subtilin production starts when the amount of nutrient falls under a threshold because of excessive population growth [29]. *B. subtilis* then produces subtilin and uses it as a weapon to increase its food supply, by eliminating competing species; in addition to reducing the demand for nutrients, the decomposition of the organisms killed by subtilin releases additional nutrients in the environment.

According to the simplified model for the subtilin production process developed in [20], subtilin derives from the peptide SpaS. Responsible for the production of SpaS is the activated protein SpaRK, which in turn is produced by the SigH protein. Finally, the composition of SigH is turned on whenever the nutrient concentration falls below a certain threshold.

9.3.2 An Initial Model

An initial stochastic hybrid model for this process was proposed in [20]. The model comprises 5 continuous states: the population of *B. subtilis*, x_1, the concentration of nutrients in the environment, x_2, and the concentrations of the SigH, SpaRK and SpaS molecules (x_3, x_4, and x_5 respectively). The model also comprises $2^3 = 8$ discrete states, generated by three binary switches, which we denote by S_3, S_4 and S_5. Switch S_3 is deterministic: it goes ON when the concentration of nutrients, x_2, falls below a certain threshold (denoted by η), and OFF when it rises over this threshold. The other two switches are stochastic. In [20] this stochastic behavior is approximated by a discrete time Markov chain, with constant sampling interval Δ. Given that the switch S_4 is OFF at time $k\Delta$, the probability that it will be ON at time $(k+1)\Delta$ depends on the concentration of SigH at the time $k\Delta$ and is given by

$$a_0(x_3) = \frac{cx_3}{1+cx_3}, \tag{9.4}$$

The nonlinear form of this equation is common for chemical reactions, such as the activation of genes, that involve "binding" of proteins to the DNA. Roughly speaking, the higher x_3 is the more SigH molecules are around and the higher the probability that one of them will bind with the DNA activating the gene that produces SpaRK. The constant c is a model parameter that depends on the activation energy of the reaction (reflecting the natural "propensity" of the particular molecules to bind) and the temperature. It will be shown below that $x_3 \geq 0$ (as expected for a concentration) therefore $a_0(x_3)$ can indeed be thought of as a probability. Notice that the probability of switching ON increases to 1 as x_3 gets higher. Conversely, given that the switch S_4 is ON at time $k\Delta$, the probability that it will be OFF at time $(k+1)\Delta$ is

$$a_1(x_3) = 1 - a_0(x_3) = \frac{1}{1+cx_3}. \tag{9.5}$$

Notice that this probability increases to 1 as x_3 gets smaller. The dynamics of switch S_5 are similar, with the concentration of SpaRK, x_4, replacing x_3 and a different value, c', for the constant.

The dynamics for the *B. subtilis* population x_1 are given by the logistic equation

$$\dot{x}_1 = rx_1(1 - \frac{x_1}{D_\infty(x_2)}). \tag{9.6}$$

Under this equation, x_1 will tend to converge to

$$D_\infty(x_2) = \min\{\frac{x_2}{X_0}, D_{\max}\}, \tag{9.7}$$

the steady state population for a given nutrient amount. X_0 and $D_{\max}$ are constants of the model; the latter represents constraints on the population because of space limitations and competition within the population.

The dynamics for x_2 are governed by

$$\dot{x}_2 = -k_1x_1 + k_2x_5 \tag{9.8}$$

where k_1 denotes the rate of nutrient consumption per unit of population and k_2 the rate of nutrient production due to the action of subtilin. More precisely, the second term is proportional to the average concentration of SpaS, but for simplicity [20] assume that the average concentration is proportional to the concentration of SpaS for a single cell.

The dynamics for the remaining three states depend on the discrete state, i.e., the state of the three switches. In all three cases,

$$\dot{x}_i = \begin{cases} -l_ix_i & \text{if } S_i \text{ is OFF} \\ k_i - l_ix_i & \text{if } S_i \text{ is ON.} \end{cases} \tag{9.9}$$

It is easy to see that the concentration x_i decreases exponentially toward zero whenever the switch S_i is OFF and tends exponentially toward k_i/l_i whenever S_i is ON. Note that the model is closely related to the piecewise affine models studied by [13, 5]. The key differences are the nonlinear dynamics of x_1 and the stochastic terms used to describe the switch behavior.

9.3.3 A Formal PDMP Model

We now try to develop a PDMP, $H = ((Q,d,X), f, Init, \lambda, R)$, to capture the mechanism behind subtilin production outlined above. To do this we need to define all the quantities listed in Definition 9.1.

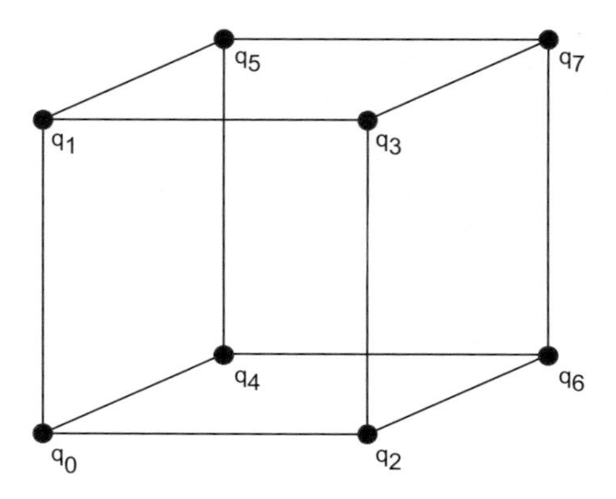

FIGURE 9.1: Visualization of discrete state space.

The presence of the three switches (S_3, S_4, and S_5) dictate that the PDMP model should have 8 discrete states (see Figure 9.1). We denote these discrete states by

$$Q = \{q_0, \dots, q_7\}, \tag{9.10}$$

so that the index (in binary) of each discrete state reflects the state of the three switches. For example, state q_0 corresponds to binary 000, i.e., all three switches being OFF. Likewise, state q_5 corresponds to binary 101, i.e., switches S_3 and S_5 being ON and switch S_4 being OFF. In the following discussion, the state names $q_0, \dots, q_7$ and the binary equivalents of their indices will be used interchangeably. A wildcard, $*$, will be used when in a statement the position of some switch is immaterial; e.g., $1**$ denotes that something holds when S_3 is ON, whatever the values of S_4 and S_5 may be.

The discussion in the previous section suggests that there are 5 continuous states and all of them are active in all discrete states. Therefore, the dimension of the continuous state space is constant

$$d(q) = 5, \text{ for all } q \in Q. \tag{9.11}$$

The definition of the survival function (9.3) suggests that the open sets $X(q) \subseteq \mathbb{R}^5$ are used to force discrete transitions to take place at certain values of state. In the subtlin production model outlined above the only forced transitions are those induced by the deterministic switch S_3: S_3 has to go ON whenever x_2 falls under the threshold η and has to go OFF whenever it rises over this threshold. These transitions can be forced by defining

$$X(0**) = \mathbb{R} \times (\eta, \infty) \times \mathbb{R}^3 \text{ and } X(1**) = \mathbb{R} \times (-\infty, \eta) \times \mathbb{R}^3. \tag{9.12}$$

The three elements defined in Equations (9.10)–(9.12) completely determine the hybrid state space, $\mathscr{D}(Q,d,X)$, of the PDMP.

The family of vector fields, $f(q,\cdot)$, is easy to infer from the above discussion. From Equations (9.6)–(9.9) we have that

$$f(q_0,x)=\begin{bmatrix} rx_1(1-\frac{x_1}{\min\{x_2/X_0,D_{\max}\}}) \\ -k_1x_1+k_2x_3 \\ -l_3x_3 \\ -l_4x_4 \\ -l_5x_5 \end{bmatrix} \quad f(q_1,x)=\begin{bmatrix} rx_1(1-\frac{x_1}{\min\{x_2/X_0,D_{\max}\}}) \\ -k_1x_1+k_2x_3 \\ -l_3x_3 \\ -l_4x_4 \\ k_5-l_5x_5 \end{bmatrix}$$

$$f(q_2,x)=\begin{bmatrix} rx_1(1-\frac{x_1}{\min\{x_2/X_0,D_{\max}\}}) \\ -k_1x_1+k_2x_3 \\ -l_3x_3 \\ k_4-l_4x_4 \\ -l_5x_5 \end{bmatrix} \quad f(q_3,x)=\begin{bmatrix} rx_1(1-\frac{x_1}{\min\{x_2/X_0,D_{\max}\}}) \\ -k_1x_1+k_2x_3 \\ -l_3x_3 \\ k_4-l_4x_4 \\ k_5-l_5x_5 \end{bmatrix}$$

$$f(q_4,x)=\begin{bmatrix} rx_1(1-\frac{x_1}{\min\{x_2/X_0,D_{\max}\}}) \\ -k_1x_1+k_2x_3 \\ k_3-l_3x_3 \\ -l_4x_4 \\ -l_5x_5 \end{bmatrix} \quad f(q_5,x)=\begin{bmatrix} rx_1(1-\frac{x_1}{\min\{x_2/X_0,D_{\max}\}}) \\ -k_1x_1+k_2x_3 \\ k_3-l_3x_3 \\ -l_4x_4 \\ k_5-l_5x_5 \end{bmatrix}$$

$$f(q_6,x)=\begin{bmatrix} rx_1(1-\frac{x_1}{\min\{x_2/X_0,D_{\max}\}}) \\ -k_1x_1+k_2x_3 \\ k_3-l_3x_3 \\ k_4-l_4x_4 \\ -l_5x_5 \end{bmatrix} \quad f(q_7,x)=\begin{bmatrix} rx_1(1-\frac{x_1}{\min\{x_2/X_0,D_{\max}\}}) \\ -k_1x_1+k_2x_3 \\ k_3-l_3x_3 \\ k_4-l_4x_4 \\ k_5-l_5x_i \end{bmatrix}.$$

Notice that most of the equations are affine (and hence globally Lipschitz) in x. The only difficulty may arise from the population equation which is nonlinear. However, the bounds on x_1 and x_2 established in Proposition 9.1 below ensure that Assumption 9.1 is met.

The probability distribution, *Init*, for the initial state of the model should respect the constraints imposed by Assumption 9.1. We therefore require that the distribution satisfies

$$Init(\{0**\}\times\{x\in\mathbb{R}^5 \mid x_2\leq\eta\})=0,\ \ Init(\{1**\}\times\{x\in\mathbb{R}^5 \mid x_2\geq\eta\})=0. \quad (9.13)$$

The initial state should also reflect any other constraints imposed by biological intuition. For example, since x_1 reflects the *B. subtilis* population, it is reasonable to assume that $x_1 \geq 0$ (at least initially). Moreover, the form of the logistic equation (9.6) suggests that another reasonable constraint is that initially $x_1 \leq D_\infty(x_2)$. Finally, since continuous states $x_2,\ldots,x_5$ reflect concentrations, it is reasonable to assume that they also start with non-negative values. These constraints can be imposed if we require that for all $q \in Q$

$$Init(\{q\}\times\{x\in\mathbb{R}^5 \mid x_1\in(0,D_\infty(x_2)) \text{ and } \min\{x_2,x_3,x_4,x_5\}>0\})=1. \quad (9.14)$$

Any probability distribution that respects constraints (9.13) and (9.14) is an acceptable initial state probability distribution for our model.

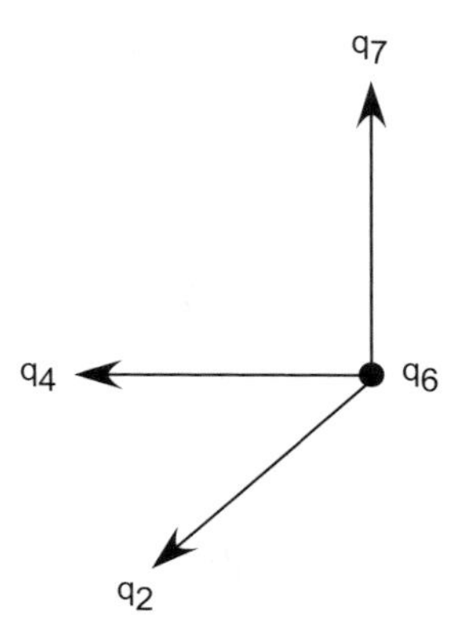

FIGURE 9.2: Transitions out of state q_6.

The main problem we confront when trying to express the subtilin production model as a PDMP is the definition of the rate function λ. Intuitively, this function indicates the "tendency" of the system to switch its discrete state. The rate function λ will govern the spontaneous transitions of the switches S_4 and S_5; recall that S_3 is deterministic and is governed by forced transitions. To present the design of an appropriate function λ we focus on discrete state q_6 (the design of λ for the other discrete states is similar). Figure 9.2 summarizes the discrete transitions out of state q_6. Notice that simultaneous switching of more than one of the switches S_3, S_4, S_5 is not allowed. This makes the PDMP model of the system more streamlined. It is also a reasonable assumption to make, since simultaneous switching of two or more switches is a null event in the underlying probability space.

Recall that q_6 corresponds to binary 110, i.e., switches S_3 and S_4 being ON and S_5 being OFF. Of the three transitions out of q_6, the one to q_2 ($S_3 \rightarrow$ OFF) is forced and does not feature in the construction of the rate function λ. For the remaining two transitions, we define two separate rate functions, $\lambda_{S_4 \rightarrow \text{ OFF}}(x)$ and $\lambda_{S_5 \rightarrow \text{ ON}}(x)$. These functions need to be linked somehow to the transition probabilities of the discrete time Markov chain with sampling period Δ used to model the probabilistic switching in [20]. This can be done via the survival function of Equation (9.3). The survival function states that the probability that the switch S_4 remains ON throughout the interval $[k\Delta, (k+1)\Delta]$ is equal to

$$\exp\left(-\int_{k\Delta}^{(k+1)\Delta} \lambda_{S_4 \rightarrow \text{ OFF}}(x(\tau))d\tau\right).$$

According to Equation (9.4) this probability should be equal to $1 - a_0(x_3(k\Delta))$. Assuming that Δ is small enough, we have that

$$1 - a_0(x_3(k\Delta)) \approx \exp\left(-\Delta\lambda_{S_4 \rightarrow \text{ OFF}}(x(k\Delta))\right).$$

Selecting

$$\lambda_{S_4 \to \text{ OFF}}(x) = \frac{\ln(1 + cx_3)}{\Delta} \tag{9.15}$$

achieves the desired effect. Likewise, we define

$$\lambda_{S_5 \to \text{ ON}}(x) = \frac{\ln(1 + c'x_4) - \ln(c'x_4)}{\Delta} \tag{9.16}$$

and set the transition rate for discrete state q_6 to

$$\lambda(q_6, x) = \lambda_{S_4 \to \text{ OFF}}(x) + \lambda_{S_5 \to \text{ ON}}(x). \tag{9.17}$$

Notice that the functions $\lambda_{S_4 \to \text{ OFF}}(x)$ and $\lambda_{S_5 \to \text{ ON}}(x)$ take non-negative values and are therefore good candidates for rate functions. $\lambda_{S_5 \to \text{ ON}}(x)$ is discontinuous at $x_3 = 0$, but the form of the vector field for x_3 ensures that there exists $\varepsilon > 0$ small enough such that if $x_3(0) > 0$, $\lambda_{S_5 \to \text{ ON}}(x(t))$ is integrable along the solutions of the differential equation over $t \in [0, \varepsilon)$. In a similar way, one can define rate functions $\lambda_{S_5 \to \text{ OFF}}(x)$ (replacing x_3 by x_4 and c by c' in Equation (9.15)) and $\lambda_{S_4 \to \text{ ON}}(x)$ (replacing x_4 by x_3 and c' by c in Equation (9.16)) and use them to define the transition rates for the remaining discrete states (in a way analogous to Equation (9.17)).

The last thing we need to define to complete the PDMP model is the probability distribution for the state after a discrete transition. The only difficulty here is removing any ambiguities that may be caused by simultaneous switches of two or more of S_3, S_4, and S_5. We do this by introducing a priority scheme: Whenever the forced transition has to take place it does, else either of the spontaneous transitions can take place. For state q_6 this leads to

$$R(q_6, x) = \delta_{(q_2, x)}(q, x) \text{ if } (q_6, x) \in \mathscr{D}^c \tag{9.18}$$

else

$$R(q_6, x) = \frac{\lambda_{S_4 \to \text{ OFF}}(x)}{\lambda(q_6, x)} \delta_{(q_4, x)}(q, x) + \frac{\lambda_{S_5 \to \text{ ON}}(x)}{\lambda(q_6, x)} \delta_{(q_7, x)}(q, x). \tag{9.19}$$

Here $\delta_{(\hat{q}, \hat{x})}(q, x)$ denotes the Dirac measure concentrated at $(\hat{q}, \hat{x})$. If desired, the two components of the measure R ((9.18) corresponding to the forced transition and (9.19) corresponding to the spontaneous transitions) can be written together using the indicator function, $I_{\mathscr{D}}(q, x)$, of the set $\mathscr{D}$.

$$\begin{aligned} R(q_6, x) = {} & (1 - I_{\mathscr{D}}(q_6, x))\, \delta_{(q_2, x)}(q, x) + \\ & I_{\mathscr{D}}(q_6, x) \left(\frac{\lambda_{S_4 \to \text{ OFF}}(x)}{\lambda(q_6, x)} \delta_{(q_4, x)}(q, x) + \frac{\lambda_{S_5 \to \text{ ON}}(x)}{\lambda(q_6, x)} \delta_{(q_7, x)}(q, x) \right). \end{aligned}$$

The measure R for the other discrete states can be defined in an analogous manner. It is easy to see that this probability measure satisfies Assumption 9.1.

The above discussion shows that the PDMP model also satisfies most of the conditions of Assumption 9.1. The only problem may be the non-Zeno condition. While this condition is likely to hold because of the structure of the vector fields and the transition rates, showing theoretically that it does is quite challenging.

9.3.4 Analysis and Simulation

To ensure that the model makes biological sense and to simplify somewhat the analysis, we impose the following restrictions on the values of the parameters.

ASSUMPTION 9.2 All model constants c, η, r, X_0, $D_{\max}$, k_i for $i = 1,\ldots,5$ and l_i for $i = 3,4,5$ are positive. Moreover, $k_1 < rX_0$.

Under these assumptions the following fact is easy to establish:

PROPOSITION 9.1 *Almost surely:*

(i) For all $t \geq 0$, $(q(t),x(t)) \in \overline{\mathscr{D}}$ *and for almost all* $t \geq 0$, $(q(t),x(t)) \in \mathscr{D}$.

(ii) For all $t \geq 0$, $x_1(t) \in [0, D_\infty(x_2(t))]$, *and* $\min\{x_2(t),x_3(t),x_4(t),x_5(t)\} > 0$.

PROOF (Outline) The first part is a general property of PDMP, and follows directly from (9.13). For the second part, the proof can be done by induction. We note first that by (9.14), the conditions hold almost surely for the initial state. The discrete transitions leave the continuous state unaffected, therefore we only have to show that the conditions remain valid along continuous evolution.

Let $x(0)$ denote the state at the beginning of an interval of continuous evolution and assume that condition 2 holds for this state. The form of the vector field is such $\dot{x}_3 \geq -l_3 x_3$. Therefore, $x_3(t) \geq e^{-l_3 t} x_3(0) > 0$ throughout the interval of continuous evolution. Similar arguments show that $x_4(t) > 0$ and $x_5(t) > 0$. Moreover, since $\dot{x}_1 = 0$ if $x_1 = 0$ or $x_1 = D_\infty(x_2)$, $x_1(t)$ remains in the interval $[0, D_\infty(x_2)]$ if it is initially in this interval.

It remains to show that $x_2(t) > 0$. Consider the function $V(x) = \frac{x_1}{x_2}$. Differentiating and assuming $x_2 \leq D_{\max}/X_0$ we get

$$
\begin{aligned}
\dot{V}(x) &= \tfrac{rx_1}{x_2}\left(1 - \tfrac{X_0 x_1}{x_2}\right) - \tfrac{x_1}{x_2^2}(-k_1 x_1 + k_2 x_3) \\
&= r\tfrac{x_1}{x_2} - (rX_0 - k_1)\left(\tfrac{x_1}{x_2}\right)^2 - k_2 \tfrac{x_1 x_3}{x_2^2} \\
&\leq r\tfrac{x_1}{x_2} - (rX_0 - k_1)\left(\tfrac{x_1}{x_2}\right)^2
\end{aligned}
$$

If we let $\alpha = x_1/x_2$ the last inequality reads $\dot{\alpha} \leq r\alpha - (rX_0 - k_1)\alpha^2$. If $k_1 < rX_0$, then for α large enough (equivalently, x_2 small enough since x_1 is bounded) the quadratic term dominates and keeps α bounded. Therefore, $x_2(t) > 0$ along continuous evolution. ■

Even though some additional facts about this model can be established analytically, the most productive way to analyze the model (especially its stochastic behavior) is by simulation. The model can easily be coded in simulation, using the methods

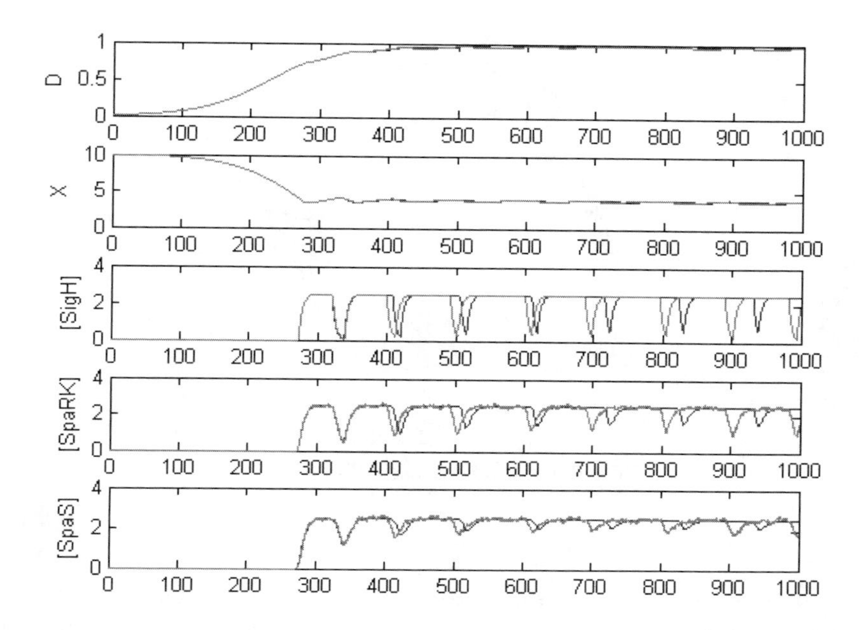

FIGURE 9.3: Sample solution for PDMP model of subtilin production.

outlined in Section 9.2.2. In this case the only forced transitions are those governing the switch S_3. An obvious choice for a function to code these forced transitions as zero crossings is $g(q,x) = x_2 - \eta$; the same function can be used for switching S_3 ON (crossing zero from above) and OFF (crossing zero from below). Servicing the event simply involves switching the state of S_3.

The model was coded in Matlab using ode45 with events enabled. Typical trajectories of the system is shown in Figure 9.3.

9.4 DNA Replication in the Cell Cycle

9.4.1 Qualitative Description

DNA replication, the process of duplication of the cells genetic material, is central to the life of every living cell, and is always carried out prior to cell division to ensure that the cells genetic information is maintained. Replication takes place during a specific stage in the life cycle of a cell (the cell cycle). The cell cycle (as shown in

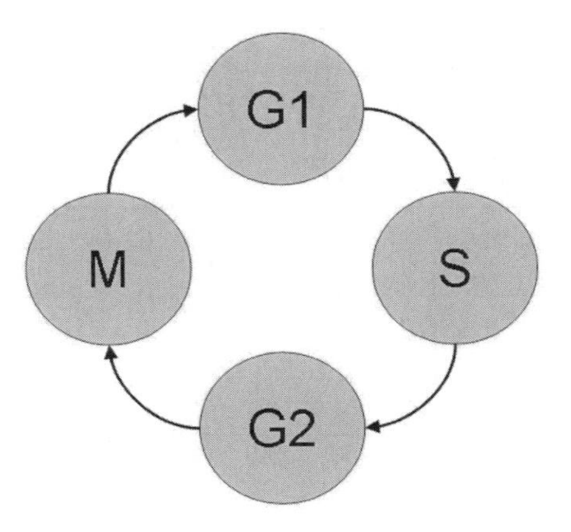

FIGURE 9.4: The phases of cell cycle.

Figure 9.4) can be subdivided in four phases: G1, a growth phase in which the cell increases its mass; S (synthesis), when DNA replication takes place; G2, a second growth phase; and finally M phase (mitosis), during which the cell divides and gives rise to two daughter cells.

Cell cycle events are regulated by the periodic fluctuations in the activity of protein complexes called Cyclin Dependent Kinases (CDK). CDK are the master regulators of the cell cycle [30]. In Figure 9.5, the so-called quantitative model of cell cycle regulation is illustrated. There are two identified thresholds in CDK activity, threshold 1 associated with entry into S phase and threshold 2 associated with entry into mitosis. Complex models have already been developed for the biochemical network regulating the fluctuation of CDK activity during the cell cycle [34].

Because daughter cells must have the same genetic information as their progenitor, during S-phase, every base of the genome must be replicated once and only once. Genomes of eukaryotic cells are large in size and the speed of replication is limited. To accelerate the process, DNA replication initiates from multiple points along the chromosomes, called origins of replication. Following initiation from a given origin, replication continues bi-directionally along the genome, giving rise to two replication forks moving in opposite directions.

To be able to ensure that each region of the genome is replicated once and only once, a cell must be able to distinguish a replicated from an unreplicated region. Before replication, and while CDK activity is low, origins are present in the pre-replicative state and can initiate DNA replication when CDK activity passes threshold 1. When an origin fires (or when it is passively replicated by a passing replication fork from a nearby origin) it automatically switches to the post-replicating state and can no longer support initiation of replication. CDK activities over threshold 1 inhibit conversion of the post-replicating state to the pre-replicative state. To re-acquire the

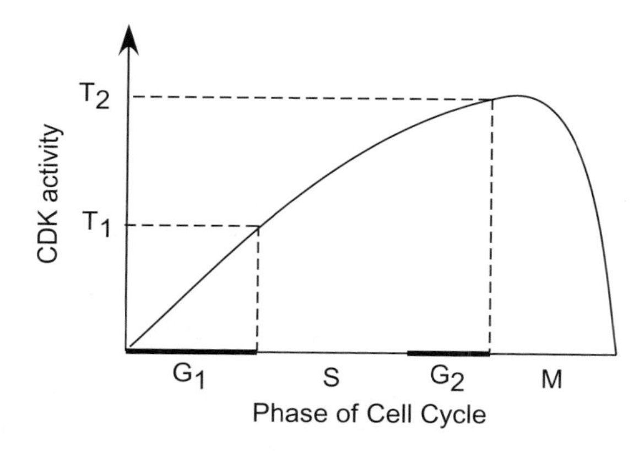

FIGURE 9.5: Quantitative model of cell cycle regulation.

pre-replicative state origins must wait for the end of the M phase, when CDK activity resets to zero. With this simple mechanism, re-replication is prevented.

Initial models, influenced by the replication of bacterial genomes, postulated that defined regions in the genome would act as origins of replication in every cell cycle [16]. Indeed, initial work from the budding yeast (*Saccharomyces cerevisiae*) identified specific sequences which acted as origins of replication with high efficiency [31]. This simple deterministic model of origin selection however is reappraised following more recent findings which show that, especially in higher eukaryotes, a large number of potential origins exist, and active origins are stochastically selected during each S phase [8, 26]. For example, recent work on the fission yeast *Schizosaccharomyces pombe* [26] clearly showed that origins fire stochastically during the S phase. The fission yeast genome has many hundreds of potential origins, but only a few of them fire in any given cell cycle. Multicellular eukaryotes are also believed to exhibit similar behavior.

9.4.2 Stochastic Hybrid Features

There are two main sources of uncertainty in the DNA replication process. The first has to do with which origins of replication fire in a particular cell cycle and the second with the times at which they fire.

Not all the origins participate in every S phase [8, 26]. Origins are classified as *strong* and *weak*, according to the frequency with which they fire. Given a population of cells undergoing an S-phase, strong origins are observed to fire in many cells, whereas weak ones fire in only a few. This firing probability is typically encoded as a number between 0 and 1 for each origin, reflecting the percentage of cells in which the particular origin is observed to fire.

Even if an origin does fire, the time during the S phase when it will do so is still uncertain. Some origins have been observed to fire earlier in the S phase, while

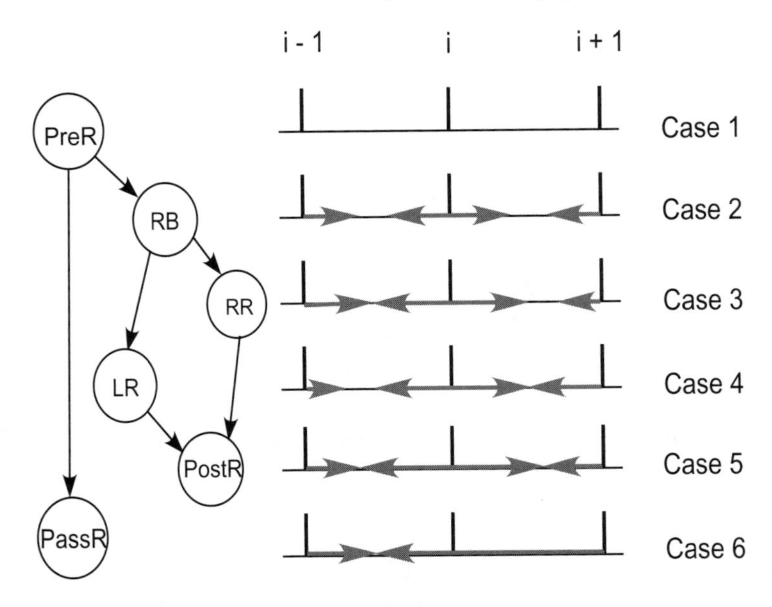

FIGURE 9.6: Possible states of an origin.

others tend to fire later [14]. This timing aspect is usually encoded by a number in minutes, reflecting the average firing time of the origin in a population of cells.

The two uncertainty parameters, efficiency and firing time, are clearly correlated. Late firing origins will also tend to have smaller efficiencies. This is because origins that tend to fire later during S phase give the chance to nearby early firing origins to replicate the part of the genome around them [18]. It is an on going debate among the cell cycle community as to whether these two manifestations of uncertainty are in fact one and the same, or whether there are separate biological mechanisms that determine if an origin is weak versus strong and early firing versus late firing. The hope is that mathematical models, like the one presented below, will assist biologists in answering such questions.

During the S phase an origin of replication may find itself in a number of "states". These are summarized in Figure 9.6, where we concentrate on an origin i and its neighbors, denoted by $i-1$ (left) and $i+1$ (right). We distinguish a number of cases.

Case 1: Pre-replicative. Every origin is in this mode until the time it becomes active (firing time). In this case it does not replicate in any direction.

Case 2: Replicating on both sides. When the origin firing time is reached, the origin gets activated and begins to replicate the DNA to its left and right. The points of replication ("forks") move away from the origin with a certain speed ("fork velocity").

Case 3: Right replicating. When the section of DNA that i has replicated to its left reaches the section of DNA that $i-1$ has replicated to its right, then the whole section between the two origins has been replicated. Origin i does not need to do any more replication to its left and so it continues only to the right.

Case 4: Left replicating. This is symmetric to Case 3. Origin i stops replication to the right and continues to the left.

Case 5: Post replicating. The replication has finished in both directions and the origin has completed its job.

Case 6: Passively replicated. The section replicated by an active origin ($i+1$ in the figure) reaches origin i before it has had a chance to fire. Replication of $i+1$ continues, overtaking and destroying origin i.

The above discussion suggests that DNA replication is a complex process that involves different types of dynamics: discrete dynamics due to the firing of the origins, continuous dynamics from the evolution of the replication forks, and stochastic terms needed to capture origin efficiencies and uncertainty about their firing times. In the next section we present a stochastic hybrid model to deal with these diverse dynamics.

9.4.3 A PDMP Model

The model splits the genome in pieces whose replication is assumed to be independent of one another. Examples of pieces are chromosomes. Chromosomes may be further divided into smaller pieces, to exclude, for example, rDNA repeats in the middle of a chromosome which are usually excluded in sequencing databases and micro-array data. The model for each piece of the genome requires the following data:

- The length, L, of the piece of the genome, in bases. We will assume that L is large enough so that, if we normalize by L, we can approximate the position along the genome with a continuous quantity, $l \in [0,1]$. This is a reasonable approximation even for the simplest organisms.

- The normalized positions, $O_i \in (0,1)$, $i = 1,2,\ldots,N$, of the origins of replication along the genome. For notational convenience, we append two dummy origins to the list of true origins, situated at the ends of each genome piece, $O_0 = 0$ and $O_{N+1} = 1$.

- The firing rate of the origins, $\lambda_i \in \mathbb{R}_+$, $i = 1,2,\ldots,N$, in $minutes^{-1}$. We set $\lambda_0 = \lambda_{N+1} = 0$.

- The fork velocity, $v(l) \in \mathbb{R}_+$ as a function of the location, $l \in [0,1]$, in the genome.

Using micro-array techniques, values for all these parameters are now becoming available for a number of organisms.

The above discussion reveals that during the S phase, each origin of replication can find itself in one of six discrete states: pre-replicative, *PreR*, replicating on both sides, *RB*, replicating to the right only, *RR*, replicating to the left only, *LR*, post

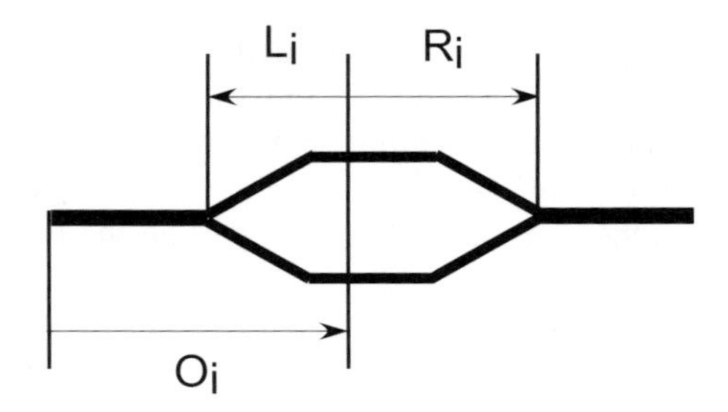

FIGURE 9.7: Definition of continuous states of an origin.

replicating, *PostR*, and passively replicated, *PassR*. The discrete state space of our model will therefore be

$$Q = \{PreR, RB, RR, LR, PostR, PassR\}^N.$$

The discrete state, $q \in Q$ will be denoted as a N-tuple, $q = (q_1, q_2, \ldots, q_N)$ with $q_i \in \{PreR, RB, RR, LR, PostR, PassR\}$. The dummy origins introduced at the beginning and the end of the section of DNA are not reflected in the discrete state, we simply set $q_0 = q_{N+1} = PreR$. Note that the number of discrete states, 6^N, grows exponentially with the number of origins. Even simple organisms have several hundreds of origins and even though only a small fraction of the possible states get visited in any one S phase, the total number of discrete states reached can be enormous.

The number of continuous states depends on the discrete state and will change during the evolution of the system. Since the continuous state reflects the progress of the replication forks, we introduce one continuous state for each origin replicating only to the left, or only to the right and two continuous states for each origin replicating in both directions. Therefore the dimension of the continuous state space for a given discrete state $q \in Q$ will be

$$d(q) = |\{i \mid q_i \in \{RR, LR\}\}| + 2|\{i \mid q_i = RB\}|,$$

where, as usual, $|\cdot|$ denotes the cardinality of a set. For an origin with $q_i \in \{RR, RB\}$ we will use R_i to denote the length of DNA it has replicated to its right. Likewise for an origin with $q_i \in \{LR, RB\}$ we will use L_i to denote the length of DNA it has replicated to its left (see Figure 9.7). For a discrete state, $q \in Q$, the continuous state $x \in \mathbb{R}^{d(q)}$ will be a $d(q)$-tuple consisting of the R_i and L_i listed in the order of increasing i; if $q_i = RB$ we assume that the R_i is listed before L_i. Notice that initially all origins will be in the pre-replicative mode and after the completion of the S phase all origins will be in either post replicating or passively replicated. Therefore both at the beginning and at the end of the S phase we will have $d(q) = 0$ and the continuous state space will be trivial.

The open sets $X(q)$ are used to force discrete transitions to take place. Figure 9.6 also summarizes the discrete transitions that can take place for each origin of replication. All transitions except the one from *PreR* to *RB* are forced and have to do with the relation between the replication forks of origin i and those of other replicating origins to its left and to its right. For a discrete state $q \in Q$ and an origin $i = 1,\ldots,N$ we denote these replicating neighbors to the left and right of origin i by

$$LN_i(q) = \max\{j < i \mid q_j \in \{RR, RB\}\}$$
$$RN_i(q) = \min\{j > i \mid q_j \in \{LR, RB\}\}.$$

Whenever the sets are empty we set $LN_i(q) = 0$ and $RN_i(q) = N+1$.

We build the set $X(q)$ out of sets, $X_i(q)$, one of each active origin. Forced transitions occur when replication forks meet. For example, if origin i is only replicating to its right, $q_i = RR$, and its right replication fork, R_i, meets the left replication fork, $L_{RN_i(q)}$, of its right neighbor, $RN_i(q)$, then origin i must stop replicating and switch to $q_i = PostR$. Therefore

$$q_i = RR \Rightarrow X_i(q) = \{O_{RN_i(q)} - L_{RN_i(q)} > O_i + R_i\} \subseteq \mathbb{R}^{d(q)}.$$

Notice that the set is well defined: because $q_i = RR$ and, by definition, $q_{RN_i(q)} \in \{LR, RB\}$ both R_i and $L_{RN_i(q)}$ are included among the continuous states. Likewise, we define

$$q_i = LR \Rightarrow X_i(q) = \{O_{LN_i(q)} + R_{RN_i(q)} < O_i - L_i\}$$
$$q_i = RB \Rightarrow X_i(q) = \{O_{LN_i(q)} + R_{RN_i(q)} < O_i - L_i\} \cap \{O_{RN_i(q)} - L_{RN_i(q)} > O_i + R_i\}$$
$$q_i = PreR \Rightarrow X_i(q) = \{O_{LN_i(q)} + R_{RN_i(q)} < O_i\} \cap \{O_{RN_i(q)} - L_{RN_i(q)} > O_i\}$$
$$q_i \in \{PostR, PassR\} \Rightarrow X_i(q) = \mathbb{R}^{d(q)}.$$

We define the overall set by

$$X(q) = \bigcap_{i=1}^{N} X_i(q).$$

$X(q)$ is clearly an open set.

The vector field, f, reflects the continuous progress of the replication forks along the genome. It is again defined one origin at a time. We set

$$f_i(q,x) = \begin{cases} v(O_i + R_i) & \in \mathbb{R} \quad \text{if } q_i = RR \\ \begin{bmatrix} v(O_i + R_i) \\ v(O_i - L_i) \end{bmatrix} & \in \mathbb{R}^2 \quad \text{if } q_i = RB \\ v(O_i - L_i) & \in \mathbb{R} \quad \text{if } q_i = LR. \end{cases}$$

Recall that all other discrete states do not give rise to any continuous states. The overall vector field $f(q,x) \in \mathbb{R}^{d(q)}$ is obtained by stacking the $f_i(q,x)$ for the individual replicating origins one on top of the other. Under mild assumptions on the fork velocity it is easy to see that $f(q,\cdot)$ satisfies Assumption 9.1.

The initial state measure is trivial. Biological intuition suggests that at the beginning of the S phase all origins are pre-replicative and no replication forks are active. The initial probability measure is therefore just the Dirac measure

$$Init(q,x) = \delta_{PreR^N \times \{0\}}(q,x).$$

Recall that when $q = PreR^N$ the continuous state is trivial $x \in \mathbb{R}^0 = \{0\}$.

The only spontaneous transition in our model is the one from state *PreR* to the state *RB*; all other transitions are forced. The transition rate, λ, governing spontaneous transitions reflects the randomness in the firing times of the origins. Therefore λ is only important for origins in state *PreR*. We define λ one origin at a time, setting

$$\lambda_i(q,x) = \begin{cases} \lambda_i & \text{if } q_i = PreR \\ 0 & \text{otherwise.} \end{cases}$$

This implies that the firing time, T_i, of origin i have a survival function of the form

$$P[T_i \geq t] = e^{-\lambda_i t}. \tag{9.20}$$

Notice that here T_i refers to the time origin i would fire in the absence of interference from other origins, not the observed firing times. In practice, origin i will sometimes get passively replicated by adjacent origins before it gets a chance to fire. Therefore the observed firing times will show a bias toward smaller values that the $1/\lambda_i$ anticipated by (9.20). We set the overall rate to

$$\lambda(q,x) = \sum_{i=1}^{N} \lambda_i(q,x).$$

Finally, for the transition measure R we distinguish two cases: either no transition is forced (i.e., state before the transition is in $\mathscr{D}$), or a transition is forced (i.e., state before the transition in $\partial\mathscr{D}$). In the former case, for $q \in Q$ let

$$d_i(q) = |\{j < i \mid q_j \in \{RR, LR\}\}| + 2|\{j < i \mid q_j = RB\}|.$$

For $(\hat{q},\hat{x}) \in \mathscr{D}$ with $\hat{q}_i = PreR$ define the measure

$$\delta_{q_i \to RB}(q,x)$$

as the Dirac measure concentrated on $(q,x) \in \mathscr{D}$ with $q_i = RB$, $q_j = \hat{q}_j$ for $j \neq i$, $x_j = \hat{x}_j$ for $j < d_i(\hat{q})$, $x_{d_i(\hat{q})} = x_{d_i(\hat{q})+1} = 0$, and $x_{j+2} = \hat{x}_j$ for $j \geq d_i(\hat{q})$. In words, if origin i fires spontaneously, its discrete state changes to *RB* and two new continuous states are introduced to store the progress of its replication forks. Since a spontaneous transition takes place whenever one of the origins in state *PreR* can fire, the overall reset measure from state $(\hat{q},\hat{x}) \in \mathscr{D}$ can be written as

$$R((q,x),(\hat{q},\hat{x})) = \frac{\sum_{\{i \mid \hat{q}_i = PreR\}} \lambda_i \delta_{q_i \to RB}(q,x)}{\lambda(q,x)}. \tag{9.21}$$

Finally, if $(\hat{q},\hat{x}) \in \partial\mathscr{D}$, i.e., a transition is forced for at least one origin, define the "guard" conditions

$$\begin{aligned}
G_{q_i\to PassR}(\hat{q},\hat{x}) &= (\hat{q}_i = PreR)\wedge \\
&\quad [(O_{LN_i(q)} + R_{RN_i(q)} \geq O_i) \vee (O_{RN_i(q)} - L_{RN_i(q)} \leq O_i)] \\
G_{q_i\to RR}(\hat{q},\hat{x}) &= (\hat{q}_i = RB)\wedge \\
&\quad (O_{LN_i(q)} + R_{RN_i(q)} \geq O_i - L_i) \wedge (O_{RN_i(q)} - L_{RN_i(q)} > O_i + R_i) \\
G_{q_i\to LR}(\hat{q},\hat{x}) &= (\hat{q}_i = RB)\wedge \\
&\quad (O_{RN_i(q)} - L_{RN_i(q)} \leq O_i + R_i) \wedge (O_{LN_i(q)} + R_{RN_i(q)} < O_i - L_i) \\
G_{q_i\to PostR}(\hat{q},\hat{x}) &= [(\hat{q}_i = RB)\wedge \\
&\quad (O_{RN_i(q)} - L_{RN_i(q)} \leq O_i + R_i) \wedge (O_{LN_i(q)} + R_{RN_i(q)} \geq O_i - L_i)] \\
&\quad \vee[(\hat{q}_i = RR) \wedge (O_{RN_i(q)} - L_{RN_i(q)} \leq O_i + R_i)] \\
&\quad \vee[(\hat{q}_i = LR) \wedge (O_{LN_i(q)} + R_{RN_i(q)} \geq O_i - L_i)].
\end{aligned}$$

We can then define $R((\cdot,\cdot),(\hat{q},\hat{x}))$ as a Dirac measure concentrated on (q,x) with

$$q_i = \begin{cases} PassR & \text{if } G_{q_i\to PassR}(\hat{q},\hat{x}) \text{ is true} \\ RR & \text{if } G_{q_i\to RR}(\hat{q},\hat{x}) \text{ is true} \\ LR & \text{if } G_{q_i\to LR}(\hat{q},\hat{x}) \text{ is true} \\ PostR & \text{if } G_{q_i\to PostR}(\hat{q},\hat{x}) \text{ is true} \end{cases}$$

and x same as $\hat{x}$, with the elements corresponding to i with $q_i \neq \hat{q}_i$ dropped. Notice that, as in the case of *B. subtilis*, if forced transitions are available they are taken, preempting any spontaneous transitions.

9.4.4 Implementation in Simulation and Results

The model of DNA replication is very complex, with a potentially enormous number of discrete (6^N) and continuous ($2N$) states. The model has the advantage that it is naturally decomposed to fairly independent components (the models for the individual origins) which interact via their continuous states (the progress of the replication forks). Current research concentrates on exploiting compositional frameworks for stochastic hybrid systems ([2, 28, 32], see also Chapter 3 of this volume) to model and analyze the behavior of the DNA replication mechanism.

In the meantime, the best way to analyze the behavior of this system is through simulation. A simulator of the DNA replication process was developed which simulates the DNA replication process genome wide, given a specific genome size, specific origin positions and efficiencies and specific fork velocities. Event detection was accomplished by computing the zero crossings of functions of the form

$$g(q,x) = O_{RN_i(q)} - L_{RN_i(q)} - O_i - R_i$$

(for the discrete transition from *RR* to *PostR*, and similar functions for the remaining transitions). Servicing the events involved switching the discrete state, but also changing the continuous state dimension, by dropping or adding states.

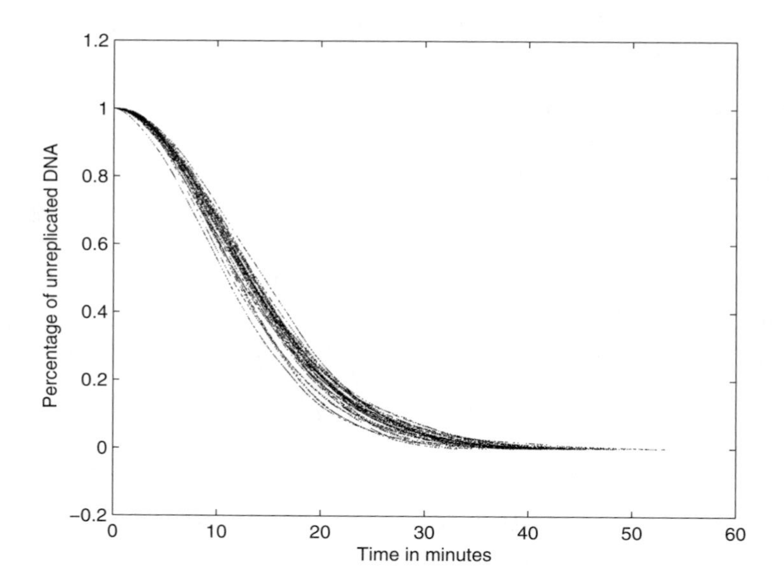

FIGURE 9.8: Evolution of unreplicated DNA.

Simulation results for a number of runs of the model are shown in Figure 9.8. Here the genome size used was 12 million bases, with 900 origins introduced at random locations and with random efficiencies. The fork velocity was constant at 5500 bases per minute. The figure clearly shows the randomness in the DNA replication process predicted by the model.

9.5 Concluding Remarks

We have presented an overview of stochastic hybrid modeling issues that arise in biochemical processes. We have argued that stochastic hybrid dynamics play a crucial role in this context and illustrated this point by developing PDMP models for two biochemical processes, subtilin production by the organism *B. subtilis* and DNA replication in eukaryotes. We also discussed how the models can be analyzed by Monte Carlo simulation. Current research focuses on tuning the parameters of the models based on experimental data and exploiting the analysis and simulation results obtained with the models (in particular the DNA replication model) to gain biological insight. Already the results of the DNA replication model have led biologists to re-think long held conventional opinions about the duration of the S phase and the role of different mechanisms that play a role in cell cycle regulation.

Acknowledgments. The authors are grateful to S. Dimopoulos, G. Ferrari-Trecate, C. Heichinger, H. de Jong, I. Legouras, and P. Nurse for extensive discussions providing insight into the topic. Research supported by the European Commission under the HYGEIA project, FP6-NEST-4995.

References

[1] R. Alur, C. Belta, F. Ivancic, V. Kumar, M. Mintz, G. Pappas, H. Rubin, J. Schug, and G.J. Pappas. Hybrid modeling and simulation of biological systems. In M. Di Benedetto and A. Sangiovanni-Vincentelli, editors, *Hybrid Systems: Computation and Control*, number 2034 in LNCS, pages 19–32. Springer-Verlag, Berlin, 2001.

[2] R. Alur, R. Grosu, I. Lee, and O. Sokolsky. Compositional refinement for hierarchical hybrid systems. In M. Di Benedetto and A. Sangiovanni-Vincentelli, editors, *Hybrid Systems: Computation and Control*, number 2034 in LNCS, pages 33–48. Springer-Verlag, Berlin, 2001.

[3] K. Amonlirdviman, N. A. Khare, D. R. P. Tree, W.-S. Chen, J. D. Axelrod, and C. J. Tomlin. Mathematical modeling of planar cell polarity to understand domineering nonautonomy. *Science*, 307(5708):423–426, Jan. 2005.

[4] M. Andersson. *Object-Oriented Modeling and Simulation of Hybrid Systems*. PhD thesis, Lund Institute of Technology, Lund, Sweden, December 1994.

[5] G. Batt, D. Ropers, H. de Jong, J. Geiselmann, M. Page, and D. Schneider. Qualitative analysis and verification of hybrid models of genetic regulatory networks: Nutritional stress response in *Escherichia coli*. In L. Thiele and M. Morari, editors, *Hybrid Systems: Computation and Control*, number 3414 in LNCS, pages 134–150. Springer-Verlag, Berlin, 2005.

[6] M.L. Bujorianu and J. Lygeros. Reachability questions in piecewise deterministic Markov processes. In O. Maler and A. Pnueli, editors, *Hybrid Systems: Computation and Control*, number 2623 in LNCS, pages 126–140. Springer-Verlag, Berlin, 2003.

[7] C.G. Cassandras and S. Lafortune. *Introduction to Discrete Event Systems*. Kluwer Academic Publishers, Norwell, MA, 1999.

[8] J. Dai, R.Y. Chuang, and T.J. Kelly. DNA replication origins in the *schizosaccharomyces pombe* genome. *PNAS*, 102(102):337–342, January 2005.

[9] M.H.A. Davis. Piecewise-deterministic Markov processes: A general class of

non-diffusion stochastic models. *Journal of the Royal Statistical Society, B*, 46(3):353–388, 1984.

[10] M.H.A. Davis. *Markov Processes and Optimization*. Chapman & Hall, London, 1993.

[11] M.H.A. Davis and M.H. Vellekoop. Permanent health insurance: a case study in piecewise-deterministic Markov modelling. *Mitteilungen der Schweiz. Vereinigung der Versicherungsmathematiker*, 2:177–212, 1995.

[12] H. de Jong. Modeling and simulation of genetic regulatory systems: A literature review. *Journal of Computational Biology*, 79:726–739, 2002.

[13] H. de Jong, J.-L. Gouze, C. Hernandez, M. Page, T. Sari, and J. Geiselmann. Hybrid modeling and simulation of genetic regulatory networks: A qualitative approach. In O. Maler and A. Pnueli, editors, *Hybrid Systems: Computation and Control*, number 2623 in LNCS, pages 267–282. Springer Verlag, Berlin, 2003.

[14] J.F.X. Diffley and K. Labib. The chromosome replication cycle. *Journal of Cell Science* 115:869–872, 2002.

[15] S. Drulhe, G. Ferrari-Trecate, H. de Jong, and A. Viari. Reconstruction of switching thresholds in piecewise-affne models of genetic regulatory networks. In J. Hespanha and A. Tiwari, editors, *Hybrid Systems: Computation and Control*, volume 3927 of *Lecture Notes in Computer Science*, pages 184–199. Springer Verlag, Berlin, 2006.

[16] E. Fanning M.I. Aladjem. The replicon revisited: an old model learns new tricks in metazoan chromosomes. *EMBO Rep.*, 5(7):686–691, July 2004.

[17] R. Ghosh and C. Tomlin. Lateral inhibition through delta-notch signalling: A piecewise affine hybrid models. In M. Di Benedetto and A. Sangiovanni-Vincentelli, editors, *Hybrid Systems: Computation and Control*, number 2034 in LNCS, pages 232–246. Springer-Verlag, Berlin, 2001.

[18] D.M. Gilbert. Making sense of eukaryotic DNA replication origins. *Science*, 294:96–100, October 2001.

[19] J. Hespanha and A. Singh. Stochastic models for chemically reacting systems using polynomial stochastic hybrid systems. *International Journal on Robust Control*, 15(15):669–689, September 2005.

[20] J. Hu, W.C. Wu, and S.S. Sastry. Modeling subtilin production in *bacillus subtilis* using stochastic hybrid systems. In R. Alur and G.J. Pappas, editors, *Hybrid Systems: Computation and Control*, number 2993 in LNCS, pages 417–431. Springer-Verlag, Berlin, 2004.

[21] K.H. Johansson, M. Egerstedt, J. Lygeros, and S.S. Sastry. On the regularization of Zeno hybrid automata. *Systems and Control Letters*, 38(3):141–150, 1999.

[22] M. Kaern, T.C. Elston, W.J. Blake, and J.J. Collins. Stochasticity in gene expression: From theories to phenotypes. *Nature Reviews Genetics*, 6(6):451–464, 2005.

[23] H. Kitano. Looking beyond the details: a rise in system-oriented approaches in genetics and molecular biology. *Curr. Genet.*, 41:1–10, 2002.

[24] J. Lygeros, K.H. Johansson, S.N. Simić, J. Zhang, and S.S. Sastry. Dynamical properties of hybrid automata. *IEEE Transactions on Automatic Control*, 48(1):2–17, January 2003.

[25] J.S. Miller. Decidability and complexity results for timed automata and semi-linear hybrid automata. In Nancy Lynch and Bruce H. Krogh, editors, *Hybrid Systems: Computation and Control*, number 1790 in LNCS, pages 296–309. Springer-Verlag, Berlin, 2000.

[26] P.K. Patel, B. Arcangioli, SP. Baker, A. Bensimon, and N. Rhind. DNA replication origins fire stochastically in the fission yeast. *Molecular Biology of the Cell*, 17(1):308–316, January 2006.

[27] C.V. Rao, D.M. Wolf, and A.P. Arkin. Control, exploitation and tolerance of intracellular noise. *Nature*, 420(6912):231–237, 2002.

[28] R. Segala. *Modelling and Verification of Randomized Distributed Real-Time Systems*. PhD thesis, Laboratory for Computer Science, Massachusetts Institute of Techology, 1998.

[29] T. Stein, S. Borchert, P. Kiesau, S. Heinzmann, S. Kloss, M. Helfrich, and K.D. Entian. Dual control of subtilin biosynthesis and immunity in *bacillus subtilis*. *Molecular Microbiology*, 44(2):403–416, 2002.

[30] B. Stern and P. Nurse. A quantitative model for the cdc2 control of S phase and mitosis in fission yeast. *Trends in Genetics*, 12(9), September 1996.

[31] B. Stillman. Origin recognition and the chromosome cycle. *FEBS Lett.*, 579:877–884, February 2005.

[32] S.N. Strubbe. *Compositional Modelling of Stochastic Hybrid Systems*. PhD thesis, Twente University, 2005.

[33] L. Tavernini. Differential Automata and their Simulators. *Nonlinear Analysis, Theory, Methods and Applications*, 11(6):665–683, 1997.

[34] J.J. Tyson, A. Csikasz-Nagy, and B. Novak. The dynamics of cell cycle regulation. *BioEssays*, 24:1095–1109, 2002.

[35] JM. Vilar, HY. Kueh, N. Barkai, and S. Leibler. Mechanisms of noise-resistance in genetic oscillators. *PNAS*, 99:5988–5992, 2002.

[36] D. Wolf, V.V. Vazirani, and A.P. Arkin. Diversity in times of adversity: probabilistic strategies in microbial survival games. *Journal of Theoretcial Biology*, 234(2):227–253, 2005.

[37] L. Wu, J.C. Burnett, J.E. Toettcher, A.P. Arkin, and D.V. Schaffer. Stochastic gene expression in a lentiviral positive-feedback loop: HIV-1 tat fluctuations drive phenotypic diversity. *Cell*, 122(2):169–182, 2005.

Chapter 10

Free Flight Collision Risk Estimation by Sequential MC Simulation

Henk A.P. Blom
National Aerospace Laboratory NLR

Jaroslav Krystul
University of Twente

G.J. (Bert) Bakker, Margriet B. Klompstra
National Aerospace Laboratory NLR

Bart Klein Obbink
National Aerospace Laboratory NLR

10.1 Introduction 249
10.2 Sequential MC Estimation of Collision Risk 253
10.3 Development of a Petri Net Model of Free Flight 259
10.4 Simulated Scenarios and Collision Risk Estimates 271
10.5 Concluding Remarks 276
References 277

10.1 Introduction

10.1.1 Safety Verification of Free Flight Air Traffic

Technology allows aircraft to broadcast information about its own-ship position and velocity to surrounding aircraft, and to receive similar information from surrounding aircraft. This development has stimulated the rethinking of the overall concept for future Air Traffic Management (ATM), e.g., to transfer responsibility for conflict prevention from ground to air. As the aircrews thus obtain the freedom to select their trajectory, this conceptual idea is called free flight [57]. It changes ATM in such a fundamental way, that one could speak of a paradigm shift: centralised control becomes distributed, responsibilities transfer from ground to air, fixed air traffic routes are removed, and appropriate new technologies are brought in. Each aircrew has the responsibility to timely detect and solve conflicts, thereby assisted by navigation means, surveillance processing, and conflict resolution systems. Due to the potentially many aircraft involved, the system is highly distributed. This free flight concept idea has motivated the study of multiple operational concepts and implementation choices [33], [37], [41], [44], [54]. One of the key outstanding issues is the safety verification of free flight design, and in particular when air traffic demand is high.

For en-route traffic, the International Civil Aviation Organisation (ICAO) has established thresholds on the acceptable probability of a mid-air collision. Hence, the en-route free flight safety verification problem consists of estimating the collision probability of free flight operations, and subsequently comparing this estimated level with the ICAO established thresholds [34]. The civil aviation community also has established some approximate models to estimate (an upper-bound of) the risk of collision between aircraft flying within a given parallel route structure [32], [38], [40]. Additional methods have been exploited to develop some valuable extensions of this approach, e.g., using fault trees see [22] and using stochastic analysis and Monte Carlo (MC) simulation [3], [4], [29]. Andrews et al. [1] have shown how statistical data in combination with a fault tree of the functionalities of the advanced operation can serve to predict how reliability of free flight supported systems impact contributions to collision risk of an advanced operation [24], neglecting other contributions to collision risk. The challenge is to analyse the risk of collision between aircraft in free flight without the limitation of a fixed route structure. We aim to improve this situation by developing a novel approach toward collision risk assessment for advanced air traffic designs. An initial shorter paper on this development is [7].

10.1.2 Probabilistic Reachability Analysis

In air traffic, a mid-air collision event happens at the moment in time that the physical shapes of two airborne aircraft hit each other. Such event can be represented as a moment in time that the joint state of aircraft involved hit a certain subset of their joint state space. With this, the problem to estimate the probability of collision between two aircraft within a finite time period is to analyse the probability that this collision subset is reached by their joint aircraft state within that time period. In systems theory, the estimation of the probability of reaching a given subset of the state space within a given time period is known as a problem of probabilistic reachability analysis, e.g., see [49].

Hu et al. [39], Prandini and Hu [56], and Chapter 5 of this volume apply probabilistic reachability analysis for the development of a grid based computation to evaluate the probability that two aircraft come closer to each other than some established minimum separation criteria. The numerical challenge of this problem, however, differs from free flight collision risk estimation on the following aspects:

- The collision subset is more than three orders smaller in volume than the conflict subset is.
- A safety directed model of an air traffic operation includes per aircraft also the states of the technical systems and the pilot models, which increases the size

of the state space by many orders in magnitude.

- There are multiple aircraft, not just two, inducing a non-zero probability of a chain collision.

If we would follow the numerical approach of Chapter 5 of this volume to estimate collision risk in free flight operations, then these aspects would imply a blow up of the number of grid points to a practically unmanageably large number.

In most safety critical industries, e.g., nuclear, chemical, etc., reachability analysis is addressed by methods that are known as dynamical approaches towards Probabilistic Risk Analysis (PRA). For an overview of these dynamical methods in PRA, see [50]. These dynamical PRA methods make explicitly use of the fact that between two discrete events the dynamical evolution satisfies an ordinary differential equation. Essentially this means that these dynamical PRA methods apply to the class of stochastic hybrid system models that do not involve Brownian motion. In the hybrid systems control community these are known as piecewise deterministic Markov process [9], [17].

For proper safety modelling of air traffic operations, however, it is needed to incorporate Brownian motion in the piecewise deterministic Markov process models, e.g., to represent the effect of random wind disturbances on aircraft trajectories [55]. The class of systems which incorporates Brownian motion within piecewise deterministic Markov processes, has been defined as a Generalised Stochastic Hybrid System (GSHS) [10]. GSHS is the class of non-linear stochastic continuous-time hybrid dynamical systems, having a hybrid state consisting of two components: a continuous valued state component and a discrete valued state component. The continuous state evolves according to an SDE whose vector field and drift factor depend on both hybrid state components. Switching from one discrete state to another discrete state is governed by a probability law or occurs when the continuous state hits a pre-specified boundary. Whenever a switching occurs, the hybrid state is reset instantly to a new state according to a probability measure which depends itself on the past hybrid state. GSHS contain, as a subclass, the switching diffusion process, the probabilistic reachability of which is studied in Chapter 5 of this volume. Important complementary dynamics is induced by the interaction between the hybrid state components.

10.1.3 Sequential Monte Carlo Simulation

Shah et al. [58] explain very well that the advantage of using MC simulation in evaluating advanced operations is its capability to identify and evaluate emergent behaviour, i.e., novel behaviour which is exhibited by complex safety critical systems and emerges from the combined dynamical actions and reactions by individual systems and humans within the system. This emergent behaviour typically cannot be foreseen and evaluated by examining the individuals behaviour alone. Shah et al. [58] explain that agent based MC simulation is able to predict the impact of revolutionary changes in air transportation; it integrates cognitive models of technology behaviour and description of their operating environment. Simulation of these individual models acting together can predict the results of completely new transformations in procedures and technology. Their MC simulations reach up to the level of novel emerging hazardous events. For safety risk assessment however, it is required to go further with the MC simulations up to the level of emerging catastrophic events. In en-route air traffic these catastrophic events are mid-air collisions.

A seemingly simple approach toward the estimation of mid-air collision probability is to run many MC simulations with a free flight stochastic hybrid model and count the fraction of runs for which a collision occurs. The advantage of a MC simulation approach is that this does not require specific assumptions or limitations regarding the behaviour of the system under consideration. A key problem is that in order to obtain accurate estimates of rare event probabilities, say about 10^{-9} per flying hour, it is required to simulate 10^{11} flying hours or more. Taking into account that an appropriate free flight model is large, this would require an impractically huge simulation time.

Del Moral and co-workers [13], [14], [18] developed a sequential MC simulation approach for estimating small reachability probabilities, including a characterisation of convergence behaviour. The idea behind this approach is to express the small probability to be estimated as the product of a certain number of larger probabilities, which can be efficiently estimated by the MC approach. This can be achieved by introducing sets of intermediate states that are visited one set after the other, in an ordered sequence, before reaching the final set of states of interest. The reachability probability of interest is then given by the product of the conditional probabilities of reaching a set of intermediate states given that the previous set of intermediate states has been reached. Each conditional probability is estimated by simulating in parallel several copies of the system, i.e., each copy is considered as a particle following the trajectory generated through the system dynamics. To ensure unbiased estimation, the simulated process must have the strong Markov property. Hence, we extend the approach of [13]–[14] for application to free flight, and illustrate its application to free flight scenarios.

10.1.4 Development of MC Simulation Model

For the modelling of accident risk of safety-critical operations in nuclear and chemical industries, the most advanced approaches use Petri nets as model specification formalism, and stochastic analysis and Monte Carlo simulation to evaluate the specified model, e.g., see [50]. Since their introduction as a systematic way to specify large discrete event systems that one meets in computer science, Petri nets have shown their usefulness for many practical applications in different industries, e.g., see [16]. Various Petri net extensions and generalisations and numerous supporting computer tools have been developed, which further increased their modelling opportunities. Nevertheless, literature on Petri nets appeared to fall short for modelling the class of GSHS [10] that was needed to model air traffic safety aspects well [55].

Cassandras and Lafortune [12] provide a control systems introduction to Petri nets and a comparison with other discrete event modelling formalisms like automata. Both Petri nets and automata have their specific advantages. Petri net is more powerful in the development of a model of a complex system, whereas automata are more powerful in supporting analysis. In order to combine the advantages offered by both approaches, there is need for a systematic way of transforming a Petri net model into an automata model. Such a transformation would allow using Petri nets for the specification and automata for the analysis. For a timed or stochastic Petri net with a bounded number of tokens and deterministic or Poisson process firing, such a transformation exists [12]. In order to make the Petri net formalism useful in modelling air traffic operations, we need an extension of the Petri net formalism including a one-to-one transformation to and from GSHS. Everdij and Blom [26]–[28] have developed such extension in the form of (Stochastically and) Dynamically Coloured Petri Net, or for short (S)DCPN.

Jensen [42] introduced the idea of attaching to each token in a basic Petri net (i.e., with logic transitions only), a colour which assumes values from a finite set. Tokens and the attached colours determine which transitions are enabled. Upon firing by a transition, new tokens and attached colours are produced as a function of the removed tokens and colours. Haas [36] extended this colour idea to (stochastically) timed Petri nets where the time period between enabling and firing depends of the input tokens and their attached colours. In [36], [42] a colour does not change as long as the token to which it is attached remains at its place. Everdij and Blom [26], [27] defined a Dynamically Coloured Petri Net (DCPN) by incorporating the following extensions: (1) a colour assumes values from a Euclidean state space, its value evolves as solution of a differential equation and influences the time period between enabling and firing; (2) the new tokens and attached colours are produced as random functions of the removed tokens and colours. An SDCPN extends an DCPN in the sense that colours evolve as solutions of a stochastic differential equation [28].

This chapter is organised as follows. Section 10.2 develops the sequential MC simulation approach toward probabilistic reachability analysis of a GSHS model of free flight air traffic. Section 10.3 explains how an initial GSHS model has been developed for a specific free flight air traffic concept of operation. Section 10.4 applies the sequential MC simulation approach of Section 10.2 to the GSHS model of Section 10.3. Section 10.5 draws conclusions.

10.2 Sequential MC Estimation of Collision Risk

10.2.1 Stochastic Hybrid Process Considered

Throughout this and the following sections, all stochastic processes are defined on a complete stochastic basis $(\Omega, \mathcal{F}, \mathbb{F}, \mathsf{P}, \mathsf{T})$ with $(\Omega, \mathcal{F}, \mathsf{P})$ a complete probability space, and $\mathbb{F}$ an increasing sequence of sub-σ-algebra's on the positive time line $\mathsf{T} = \mathbb{R}_+$, i.e., $\mathbb{F} \triangleq \{\mathcal{J}, (\mathcal{F}_t, t \in \mathsf{T}), \mathcal{F}\}$, $\mathcal{J}$ containing all P-null sets of $\mathcal{F}$ and $\mathcal{J} \subset \mathcal{F}_s \subset \mathcal{F}_t \subset \mathcal{F}$ for every $s < t$.

We assume that air traffic operations are represented by a stochastic hybrid process $\{x_t, \theta_t\}$ which satisfies the strong Markov property. In [10], [11], [46] and in Chapter 2 of this volume, this property has been shown to hold true for the processes generated as execution of a GSHS. For an N-aircraft free flight traffic scenario the stochastic hybrid process $\{x_t, \theta_t\}$ consists of Euclidean valued components $x_t \triangleq \text{Col}\{x_t^0, x_t^1, \ldots, x_t^N\}$ and discrete valued components $\theta_t \triangleq \text{Col}\{\theta_t^0, \theta_t^1, \ldots, \theta_t^N\}$, where x_t^i assumes values from $\mathbb{R}^{n_i}$, and θ_t^i assumes values from a finite set (M^i). Physically, $\{x_t^i, \theta_t^i\}$, $i = 1, \ldots, N$, is the hybrid state process related to the i-th aircraft, and $\{x_t^0, \theta_t^0\}$ is a hybrid state process of all non-aircraft components. The process $\{x_t, \theta_t\}$ is $\mathbb{R}^n \times M$-valued with $n = \sum_{i=0}^{N} n_i$ and $M = \bigotimes_{i=0}^{N} M^i$. In order to model collisions between aircraft, we introduce mappings from the Euclidean valued process $\{x_t\}$ into the relative position and velocity between a pair of two aircraft (i, j). The relative horizontal position is obtained through the mapping $y^{ij}(x_t)$, the relative horizontal velocity is obtained through the mapping $v^{ij}(x_t)$. The relative vertical position is obtained through the mapping $z^{ij}(x_t)$, and relative vertical rate of climb is obtained through the mapping $r^{ij}(x_t)$. The relation between the position and velocity mappings satisfies:

$$dy^{ij}(x_t) = v^{ij}(x_t)\,dt \tag{10.1}$$

$$dz^{ij}(x_t) = r^{ij}(x_t)\,dt. \tag{10.2}$$

A collision between aircraft (i, j) means that the process $\{y^{ij}(x_t), z^{ij}(x_t)\}$ hits the boundary of an area where the distance between aircraft i and j is smaller than their physical size. Under the assumption that the length of an aircraft equals the width of an aircraft, and that the volume of an aircraft is represented by a cylinder the orientation of which does not change in time, then aircraft (i, j) have zero separation if $x_t \in D^{ij}$ with:

$$D^{ij} = \left\{x \in \mathbb{R}^n; |y^{ij}(x)| \leq (l_i + l_j)/2 \text{ AND } |z^{ij}(x)| \leq (s_i + s_j)/2\right\}, \quad i \neq j \tag{10.3}$$

where l_j and s_j are length and height of aircraft j. For simplicity we assume that all aircraft have the same size, by which (10.3) becomes:

$$D^{ij} = \left\{x \in \mathbb{R}^n; |y^{ij}(x)| \leq l \text{ AND } |z^{ij}(x)| \leq s\right\}, \quad i \neq j \tag{10.4}$$

Although all aircraft have the same size, notice that in (10.4), D^{ij} still depends of (i, j). If x_t hits D^{ij} at time τ^{ij}, then we say a collision event between aircraft (i, j) occurs at τ^{ij}, i.e.,

$$\tau^{ij} = \inf\{t > 0\,;\, x_t \in D^{ij}\}, \quad i \neq j \tag{10.5}$$

Next we define the first moment τ^i of collision with any other aircraft, i.e.,

$$\tau^i = \inf_{j \neq i}\{\tau^{ij}\} = \inf_{j \neq i}\{t > 0\,;\, x_t \in D^{ij}\} = \inf\{t > 0\,;\, x_t \in D^i\}, \tag{10.6}$$

with $D^i \triangleq \cup_{j \neq i} D^{ij}$. From this moment τ^i on, we assume that the differential equations for $\{x_t^i, \theta_t^i\}$ stop evolving.

An unbiased estimation procedure of the risk would be to simulate many times aircraft i amidst other aircraft over a period of length T and count all cases in which the realization of the moment τ^i is smaller than T. An estimator for the collision risk of aircraft i per unit T of time then is the fraction of simulations for which $\tau^i < T$.

10.2.2 Risk Factorisation Using Multiple Conflict Levels

Cérou et al. [13]–[14] developed a novel way of speeding up Monte Carlo simulation to estimate the probability that an $\mathbb{R}^n$-valued strong Markov process x_t hits a given "small" subset $D \in \mathbb{R}^n$ within a given time period $(0,T)$. This method essentially consists of taking advantage of an appropriately nested sequence of closed subsets of $\mathbb{R}^n$: $D = D_m \subset D_{m-1} \subset \ldots \subset D_1$, and then start simulation from outside D_1, and subsequently simulate from D_1 to D_2, from D_2 to D_3, ..., and finally from D_{m-1} to D_m. Krystul and Blom [45], [47] extended this Interacting Particle System (IPS) approach to switching diffusions. For probabilistic reachability analysis of an air traffic design, this IPS approach is now further extended to the class of SHS the execution of which satisfies the strong Markov property as adressed in Chapter 2 of this volume.

Prior to a collision of aircraft i with aircraft j, a sequence of conflicts ranging from long term to short term always occurs. In order to incorporate this explicitly in the MC simulation, we formalise this sequence of conflict levels through a sequence of closed subsets of $\mathbb{R}^n$: $D^{ij} = D^{ij}_m \subset D^{ij}_{m-1} \subset \ldots \subset D^{ij}_1$ with for $k = 1,\ldots,m$:

$$\begin{aligned} D^{ij}_k = \{x \in \mathbb{R}^n; |y^{ij}(x) + \Delta v^{ij}(x)| \le d_k \text{ AND} \\ |z^{ij}(x) + \Delta r^{ij}(x)| \le h_k, \text{ for some } \Delta \in [0,\Delta_k]\}, \end{aligned} \tag{10.7}$$

for $i \neq j$, with d_k, h_k and Δ_k the parameters of the conflict definition at level k, and with $d_m = l$, $h_m = s$ and $\Delta_m = 0$, and with $d_{k+1} \le d_k$, $h_{k+1} \le h_k$ and $\Delta_{k+1} \le \Delta_k$. If x_t hits D^{ij}_k at time τ^{ij}_k, then we say the first level k conflict event between aircraft (i,j) occurs at moment τ^{ij}_k, i.e.,

$$\tau^{ij}_k = \inf\{t > 0\,;\, x_t \in D^{ij}_k\}. \tag{10.8}$$

Similarly as we did for reaching the collision level by aircraft i, we consider the first moment τ^i_k that aircraft i reaches conflict level k with any of the other aircraft, i.e.,

$$\tau^i_k = \inf_{j \neq i}\{\tau^{ij}_k\} = \inf_{j \neq i}\{t > 0\,;\, x_t \in D^{ij}_k\} = \inf\{t > 0\,;\, x_t \in D^i_k\}, \tag{10.9}$$

with $D^i_k \triangleq \cup_{j \neq i} D^{ij}_k$.

Next, we define $\{0,1\}$-valued random variables $\{\chi^i_k, k = 0,1,\ldots,m\}$ as follows:

$$\begin{aligned} \chi^i_k &= 1, \text{ if } \tau^i_k < T \text{ or } k = 0 \\ &= 0, \text{ else.} \end{aligned}$$

By using this χ^i_k definition we can write the probability of collision of aircraft i with any of the other aircraft as a product of conditional probabilities of reaching the next conflict level given the current conflict level has been reached:

$$\begin{aligned} \mathbb{P}\left(\tau^i_m < T\right) &= \mathbb{E}\left[\chi^i_m\right] = \mathbb{E}\left[\prod_{k=1}^m \chi^i_k\right] = \prod_{k=1}^m \mathbb{E}\left[\chi^i_k \,|\, \chi^i_{k-1} = 1\right] \\ &= \prod_{k=1}^m \mathbb{P}\left(\tau^i_k < T \,|\, \tau^i_{k-1} < T\right) = \prod_{k=1}^m \gamma^i_k, \end{aligned} \tag{10.10}$$

with $\gamma_k^i \triangleq \mathbb{P}\left(\tau_k^i < T \mid \tau_{k-1}^i < T\right)$.

With this, the problem can be seen as one to estimate the conditional probabilities γ_k^i in such a way that the product of these estimators is unbiased. Because of the multiplication of the various individual γ_k^i estimators, which depend on each other, in general such a product may be heavily biased. The key novelty in [13] was to show that such a product may be evaluated in an unbiased way when $\{x_t\}$ makes part of a larger stochastic process that satisfies the strong Markov property. This approach is explained next.

10.2.3 Characterisation of the Risk Factors

Let us denote $E' = \mathbb{R}^{n+1} \times M$, and let $\mathcal{E}'$ be the Borel σ-algebra of E'. For any $B \in \mathcal{E}'$, $\pi_k^i(B)$ denotes the conditional probability of $\xi_k \triangleq (\tau_k, x_{\tau_k}, \theta_{\tau_k}) \in B$ given $\chi_l^i = 1$ for $1 \leq l \leq k$.

Define $Q_k^i = (0,T) \times D_k^i \times M, k = 1, \ldots, m$. Then the estimation of the probability for ξ_k to arrive at the k-th nested Borel set Q_k^i is characterised through the following recursive set of transformations

$$
\begin{array}{ccccc}
 & \text{prediction} & & \text{conditioning} & \\
\pi_{k-1}^i(\cdot) & \longrightarrow & p_k^i(\cdot) & \longrightarrow & \pi_k^i(\cdot) \\
 & & \downarrow & & \\
 & & \gamma_k^i & &
\end{array}
$$

where $p_k^i(B)$ is the conditional probability of $\xi_k \in B$ given $\chi_l^i = 1$ for $0 \leq l \leq k-1$.

Because $\{x_t, \theta_t\}$ is a strong Markov process, $\{\xi_k\}$ is a Markov sequence. Hence the one step prediction of ξ_k satisfies a Chapman-Kolmogorov equation:

$$
p_k^i(B) = \int_{E'} p_{\xi_k \mid \xi_{k-1}}(B|\xi)\, \pi_{k-1}^i(d\xi) \text{ for all } B \in \mathcal{E}'. \tag{10.11}
$$

Next we characterise the conditional probability of reaching the next level

$$
\begin{aligned}
\gamma_k^i &= \mathbb{P}\left(\tau_k^i < T \mid \tau_{k-1}^i < T\right) \\
&= \mathbb{E}\left[\chi_k^i \mid \chi_{k-1}^i = 1\right] \\
&= \int_{E'} 1_{\{\xi \in Q_k^i\}}\, p_k^i(d\xi),
\end{aligned} \tag{10.12}
$$

and the conditioning satisfies:

$$
\pi_k^i(B) = \frac{\int_B 1_{\{\xi \in Q_k^i\}}\, p_k^i(d\xi)}{\int_{E'} 1_{\{\xi' \in Q_k^i\}}\, p_k^i(d\xi')} \text{ for all } B \in \mathcal{E}'. \tag{10.13}
$$

With this, each of the m terms γ_k^i in (10.10) is characterised as a solution of a sequence of "filtering" kind of equations (10.11)–(10.13). However, an important difference with "filtering" equations is that (10.11)–(10.13) are ordinary integral equations, i.e., they have no stochastic term entering them.

10.2.4 Interacting Particle System Based Risk Estimation

Based on this theory, an Interacting Particle System (IPS) simulation algorithm is explained next for an arbitrary hybrid state strong Markov process model of air traffic. The transformations (10.11)–(10.13) lead to the IPS algorithm of [13] to estimate $\mathbb{P}(\tau_m^i < T)$, where $\bar{\gamma}_k^i$, $\bar{p}_k^i$ and $\bar{\pi}_k^i$ denote the numerical approximations of γ_k^i, p_k^i and π_k^i respectively. When simulating from D_{k-1}^i to D_k^i, a fraction $\bar{\gamma}_k^i$ of the Monte Carlo simulated trajectories only will reach D_k^i within the time period $(0,T)$.

IPS Step 0. Initial sampling for $k = 0$.

- For $l = 1,\ldots,N_p$ generate initial state value outside Q_1^i by independent drawings (x_0^l, θ_0^l) from $p_{x_0,\theta_0}(\cdot)$ and set $\xi_0^l = (0, x_0^l, \theta_0^l)$.
- For $l = 1,\ldots,N_p$, set the initial weights: $\omega_0^l = 1/N_p$.
- Then $\bar{\pi}_0^i = \sum_{l=1}^{N_p} \omega_0^l \, \delta_{\{\xi_0^l\}}$.

IPS Iteration cycle: For $k = 1,\ldots,m$ perform step 1 (prediction), step 2 (assess fraction), step 3 (conditioning), and step 4 (resampling).

IPS Step 1. Prediction of $\pi_{k-1}^i \longrightarrow p_k^i$, based on (10.11);

- For $l = 1,\ldots,N_p$ simulate a new path of the hybrid state Markov process, starting at ξ_{k-1}^l until the k-th set Q_k^i is hit or $t = T$ (the first component of ξ_k^l counts time).
- This yields new particles $\{\hat{\xi}_k^l, \omega_{k-1}^l\}_{l=1}^{N_p}$.
- $\bar{p}_k^i$ is the empirical distribution associated with the new cloud of particles: $\bar{p}_k^i = \sum_{l=1}^{N_p} \omega_{k-1}^l \delta_{\xi_k^l}$.

IPS Step 2. Assess fraction γ_k^i, based on (10.12);

- The particles that do not reach the set Q_k^i are killed, i.e., we set $\hat{\omega}_k^l = 0$ if $\hat{\xi}_k^l \notin Q_k^i$ and $\hat{\omega}_k^l = 0$ if $\hat{\omega}_k^l = w_{k-1}^l$ if $\hat{\xi}_k^l \in Q_k^i$.
- Approximation: $\gamma_k^i \approx \bar{\gamma}_k^i = \sum_{l=1}^{N_p} \hat{\omega}_k^l$. If all particles are killed, i.e., $\bar{\gamma}_k^i = 0$, then the algorithm stops without $\mathbb{P}(\tau^i < T)$ estimate.

IPS Step 3. Conditioning of $p_k^i \longrightarrow \pi_k^i$, based on (10.13);

The non-killed particles form a set S_k^i, i.e., iff $\hat{\xi}_k^l \in Q_k^i$, then particle $\{\hat{\xi}_k^l, \hat{\omega}_k^l\}$ is stored in S_k^i.

Renumbering the particles in S_k^i yields a set of particles $\{\tilde{\xi}_k^l, \tilde{\omega}_k^l\}_{l=1}^{N_{S_k}}$ with N_{S_k} the number of particles in S_k^i. Hence, we also have $\bar{\gamma}_k^i = \sum_{l=1}^{N_{S_k}} \tilde{\omega}_k^l$.

IPS Step 4. Resampling of π_k^i

Draw N_p particles ξ_k^l independently from the empirical measure $\bar{\pi}_k^i = \sum_{l=1}^{N_{S_k}} \tilde{\omega}_k^l \delta_{\{\tilde{\xi}_k^l\}}$ each of which gets weight $\omega_k^l = \frac{1}{N_p}$.

After step 4, the new set of particles is $\{\xi_k^l, \omega_k^l\}_{l=1}^{N_p}$. If $k < m$ then repeat steps 1, 2, 3, 4 for $k := k+1$. Otherwise, stop with $\mathbb{P}(\tau^i < T) \approx \prod_{k=1}^{m} \bar{\gamma}_k^i$.

REMARK 10.1 Cérou et al. [13]–[14] have proven, under certain conditions, how to manage the simulations from D_{k-1}^i to D_k^i, such that the product of these fractions $\bar{\gamma}_k^i$ forms an unbiased estimate of the probability of x_t to hit the set D^i within the time period $(0,T)$, i.e.,

$$\mathbb{E}\left[\prod_{k=1}^{m} \bar{\gamma}_k^i\right] = \prod_{k=1}^{m} \gamma_k^i = \mathbb{P}(\tau^i < T),$$

and there also is some bound on the expected estimation error, i.e.,

$$\left(\mathbb{E}(\prod_{k=1}^{m} \bar{\gamma}_k^i - \prod_{k=1}^{m} \gamma_k^i)^q\right)^{\frac{1}{q}} \leq \frac{a_q b_q}{\sqrt{N_p}},$$

for some finite constants a_q and b_q, which depend on the simulated scenario and the sequence of intermediate levels adopted. ■

10.2.5 Modification of IPS Resampling Step 4

A well known problem with particle systems is the possibility of particle depletion or impoverishment. In order to reduce the sensitivity of the above algorithm on these points, we modify step 4 in two ways: (1) we reduce the chance of impoverishment by not throwing away any particle; and (2) we make copies of particles, but avoid that these copies take away too much weight from the original particles.

Modified IPS Step 4

Resample N_p particles from S_k^i as follows:

If $\frac{1}{2}N_p \leq N_{S_k} \leq N_p$, then copy the N_{S_k} particles, i.e., $\xi_k^l = \tilde{\xi}_k^l$ and set $\omega_k^l = \tilde{\omega}_k^l \dfrac{N_{S_k}}{\bar{\gamma}_k^i N_p}$ for $l = 1,\ldots,N_{S_k}$; the total weight of these particles is $\frac{N_{S_k}}{N_p}$. Subsequently, draw $N_p - N_{S_k}$ particles ξ_k^l independently from the empirical measure $\bar{\pi}_k^i = \sum_{l=1}^{N_{S_k}} \tilde{\omega}_k^l \delta_{\{\tilde{\xi}_k^l\}}$ and set $\omega_k^l = \dfrac{\sum_{l=1}^{N_{S_k}} \tilde{\omega}_k^l}{\bar{\gamma}_k^i N_p} = \dfrac{1}{N_p}$; the total weight of this is $1 - \frac{N_{S_k}}{N_p}$.

Else, i.e., if $N_{S_k} < \frac{1}{2}N_p$, then copy the N_{S_k} particles, i.e., $\xi_k^l = \tilde{\xi}_k^l$, and set $\omega_k^l = \frac{1}{2}\tilde{\omega}_k^l / \bar{\gamma}_k^i$ for $l = 1, \ldots, N_{S_k}$; the total weight of these particles is $\frac{1}{2}$. For the remaining weight of $\frac{1}{2}$, we independently draw $N_p - N_{S_k}$ particles ξ_k^l from the empirical measure $\bar{\pi}_k^i = \sum_{l=1}^{N_{S_k}} \tilde{\omega}_k^l \delta_{\{\xi_k^l\}}$ and set $\omega_k^l = \frac{\frac{1}{2}\sum_{l=1}^{N_{S_k}} \tilde{\omega}_k^l}{\bar{\gamma}_k^i (N_p - N_{S_k})} = \frac{\frac{1}{2}}{N_p - N_{S_k}}$.

10.3 Development of a Petri Net Model of Free Flight

In order to apply the IPS algorithm toward the assessment of collision risk of free flight, we need to develop a MC simulator of these operations such that the simulated trajectories constitute realizations of a hybrid state strong Markov process. Everdij and Blom [26]–[28] have developed a Stochastically and Dynamically Coloured Petri Net (SDCPN) formalism that ensures the specification of a free flight MC simulation model which is of the appropriate class. This section explains how the SDCPN formalism has been used to develop a MC simulation model of a particular free flight design.

The specific free flight design for which we wish to estimate the collision risk by sequential MC simulation is the Autonomous Mediterranean Free Flight (AMFF) operation [43]. AMFF has been developed to study the introduction of autonomous free flight operation in Mediterranean airspace. In parallel to the current study, the safety of the AMFF operation has been addressed in [53], following a fault tree analysis approach. These results show that application of AMFF seems feasible to accommodate low en-route traffic conditions over the Mediterranean. However, this study also concludes that the fault tree approach has limited analysis capabilities in showing whether AMFF can safely accommodate a higher traffic density. For this, a need was identified to use a more advanced safety risk assessment approach that considers complex situations involving dynamic interactions between multiple human and systems. The current study addresses this for AMFF under relatively high traffic densities.

For the development of a Petri net model of the AMFF operation, two key challenges have to be addressed: a syntactical challenge of developing a model that is consistent, complete, and unambiguous; and a semantics challenge of representing the AMFF operation sufficiently well. This section shows how the (S)DCPN formalism has been used to address the syntactical challenge. Addressing the semantics challenge falls outside the scope of this study.

10.3.1 Specification of Petri Net Model

In using the (S)DCPN formalism [26], [27], [28] in modelling more and more complex multi-agent hybrid systems, it was found that the compositional specification power of Petri nets reaches its limitations. More specifically, the following problems were identified:

1. For the modelling of a complete Petri net for complex systems, a hierarchical approach is necessary in order to be able to separate local modelling issues from global or interaction modelling issues.

2. Often the addition of an interconnection between two low-level Petri nets leads to a duplication of transitions and arcs in the receiving Petri net.

3. The number of interconnections between the different low level Petri nets tends to grow quadratically with the size of the Petri net.

Everdij et al. [30] explained which Petri net model specification approaches from literature solve problem 1, and developed novel approaches to solve problems 2 and 3. Together, these approaches are integrated into a compositional specification approach for SDCPN, which is explained below.

In order to avoid problem 1, the compositional specification of an SDCPN for a complex process or operation starts with developing a Local Petri Net (LPN) for each agent that exists in the process or operation (e.g., air traffic controller, pilot, navigation and surveillance equipment). Essential is that these LPNs are allowed to be connected with other Petri net parts in such a way that the number of tokens residing in an LPN is not influenced by these interconnections. We use two types of interconnections between nodes and arcs in different LPNs:

- Enabling arc (or inhibitor arc) from one place in one LPN to one transition in another LPN. These types of arcs have been used widely in Petri net literature.

- Interaction Petri Net (IPN) from one (or more) transition(s) in one LPN to one (or more) transition(s) in another LPN.

In order to avoid problems 2 and 3, high level interconnection arcs have been introduced that allow, with well-defined meanings, arcs to initiate and/or to end on the edge of the box surrounding an LPN [30]. The meaning of these interconnections from or to an edge of a box allows several arcs or transitions to be represented by only one arc or transition.

10.3.2 High Level Interconnection Arcs

As an illustration of how high level interconnection arcs avoid duplication of arcs and transitions within an LPN and duplication of arcs between LPNs, we give three examples of these high level interconnection arcs. See [30] for a complete overview of these high level interconnection arcs.

In the first example, Figure 10.1, an enabling arc starts on the edge of an LPN box and ends on a transition in another LPN box, means that enabling arcs initiate from all places in the first LPN and end on duplications of this transition in the second LPN. The duplicated transitions should have the same guard or delay function and the same firing function and their input places should have the same colour type. This

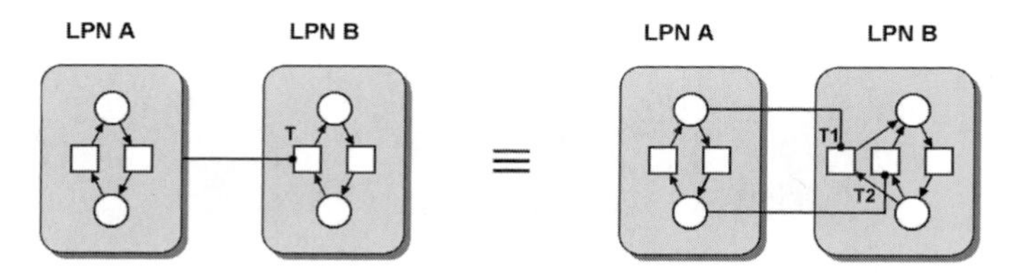

FIGURE 10.1: High level enabling arc starts at the edge of an LPN box.

high level interconnection arc is not defined for inhibitor or ordinary arcs instead of enabling arcs.

In the second example, Figure 10.2, an enabling arc ends on the edge of an LPN box.[1] This means that for each transition in the receiving LPN a copy of this enabling arc should be in place. Figure 10.2 shows an example of this high level interconnection arc. This type of high level arc can also be used with inhibitor arcs instead of enabling arcs. It cannot be used with ordinary arcs, due to the restriction that the number of tokens in an LPN should remain the same.

In the third example, Figure 10.3, an ordinary arc starts on the edge of an LPN box and ends on a transition inside the same box. This means that ordinary arcs start from all places in the LPN box to duplications of this transition. The duplicated transitions should have the same guard or delay function and the same firing function and their set of input places should have the same set of colour types. Figure 10.3 illustrates how this avoids both the duplication of transitions and arcs within an LPN, and the duplication of arcs between LPNs.

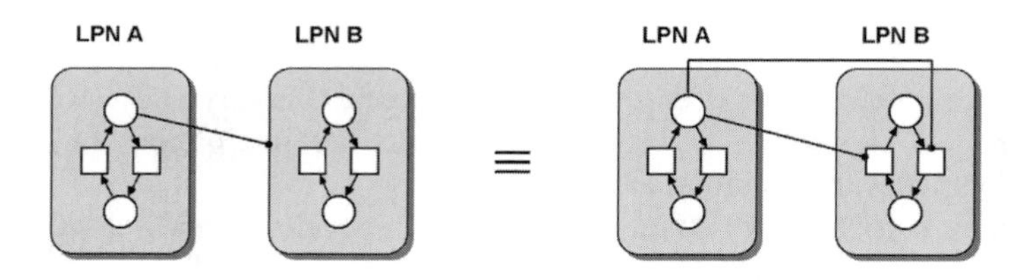

FIGURE 10.2: High level enabling arc ends at the edge of an LPN box.

10.3.3 Agents and LPNs to Represent AMFF Operations

In the Petri net modelling of AMFF operations for the purpose of an initial collision risk assessment, the following agents are taken into account:

[1]Figures 10.2, 10.3, 10.4, and 10.5 are from [8], with can kind permission of Springer Science and Business Media.

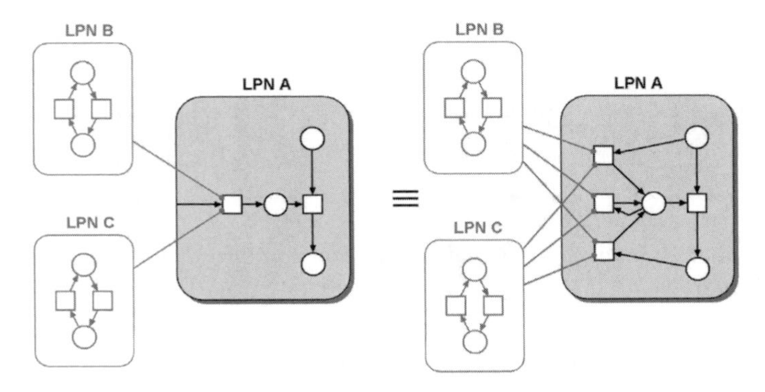

FIGURE 10.3: High level ordinary arc starts on the edge of an LPN box and ends on a transition inside the same LPN box.

- Aircraft
- Pilot-Flying (PF)
- Pilot-Not-Flying (PNF)
- Airborne Guidance, Navigation and Control (AGNC)
- Airborne Separation Assistance System (ASAS)
- Communication, Navigation and Surveillance (CNS)

It should be noticed that our initial model representing AMFF operations, does not yet incorporate other relevant agents such as Airborne Collision Avoidance System (ACAS), Airline Operations Centre (AOC), Air Traffic Control (ATC), or an environmental model. This should be taken into account when interpreting the simulation results obtained with this initial model.

Per agent, particular LPNs and IPNs have been developed and subsequently the interactions between these LPNs and IPNs have been specified. A listing of LPNs per agent reads as follows:

- Aircraft LPNs:
  - Type
  - Evolution mode
  - Systems mode
  - Emergency mode
- Pilot-Flying (PF) LPNs:
  - State situation awareness

- Intent situation awareness
- Goal memory
- Current goal
- Task performance
- Cognitive mode

- Pilot-Not-Flying (PNF) LPNs:
  - Current goal
  - Task performance
- Airborne Separation Assistance System (ASAS) LPNs:
  - Processing
  - Alerting
  - Audio alerting
  - Surveillance
  - System mode
  - Priority switch mode
  - Anti-priority switch mode
  - Predictive alerting (of other aircraft)
- Airborne Guidance, Navigation and Control (AGNC) LPNs:
  - Indicators failure mode for PF
  - Engine failure mode for PF
  - Navigation failure indicator for PF
  - ASAS failure indicator for PF
  - Automatic Dependent Surveillance-Broadcast (ADS-B) receiver failure indicator for PF
  - ADS-B transmitter failure indicator for PF
  - Indicator failure mode for PNF
  - Guidance mode
  - Horizontal guidance configuration mode
  - Vertical guidance configuration mode
  - Flight Management System (FMS) flight plan
  - Airborne Global Positioning System (GPS) receiver
  - Airborne Inertial Reference System (IRS)
  - Altimeter

- Horizontal position processing
- Vertical position processing
- ADS-B transmission
- ADS-B receiver

- Communication, Navigation and Surveillance (CNS) LPNs:
  - Global GPS / satellites
  - Global ADS-B ether frequency
  - Secondary Surveillance Radar (SSR) mode-S frequency

The actual number of LPNs in the whole model then equals $38N + 3$, where N is the number of aircraft. In addition there are 35 IPNs per aircraft; hence the number of IPNs equals $35N$. Brownian motion enters in each of the aircraft evolution mode LNPs. In this initial model these Brownian motions are assumed to be independent.

10.3.4 Interconnected LPNs of ASAS

The approach taken in developing the AMFF concept of operation [43] is to avoid much information exchange between aircraft and to avoid dedicated decision-making by artificial intelligent machines. Although the conflict detection and resolution approach developed for AMFF has its roots in the modified potential field approach [37], it has some significant deviations from this. The main deviation is that conflict resolution in AMFF is intentionally designed not to take the potential field of all aircraft into account. The resulting AMFF design can be summarised as follows:

- All aircraft are supposed to be equipped with Automatic Dependent Surveillance-Broadcast (ADS-B), which is a system that periodically broadcasts own aircraft state information, and continuously receives the state information messages broadcasted by aircraft that fly within broadcasting range ($\sim$ 100 Nm).

- To comply with pilot preferences, conflict resolution algorithms are designed to solve multiple conflicts one by one rather than according to a full concurrent way, e.g., see [37].

- Conflict detection and resolution are state-based, i.e., intent information, such as information at which point surrounding aircraft will change course or height, is supposed to be unknown.

- The vertical separation minimum is 1000 ft and the horizontal separation minimum is 5 Nm. A conflict is detected if these separation minima will be violated within 6 minutes.

- The conflict resolution process consists of two phases. During the first phase, one of the aircraft crews should make a resolution maneuver. If this does not work, then during the second phase, both crews should make a resolution maneuver.

- Prior to the first phase, the crew is warned when an ASAS alert is expected to occur if no preventive action would be timely implemented; this prediction is done by a system referred to as P-ASAS (Predictive ASAS).
- Conflict co-ordination does not take place explicitly, i.e., there is no communication on when and how a resolution maneuver will be executed.
- All aircraft are supposed to use the same resolution algorithm, and all crew are assumed to use ASAS and to collaborate in line with the procedures.
- Two conflict resolution maneuver options are presented: one in vertical and one in horizontal direction. The pilot decides which option to execute.
- ASAS related information is presented to the crew through a Cockpit Display of Traffic Information (CDTI).

ASAS is modelled through the SDCPN depicted in Figure 10.4. The ADS-B information received from other aircraft is processed by the *LPN ASAS surveillance*. Together with the information about its own aircraft state information (from AGNC), the *LPN ASAS processing* uses this information to perform conflict detection and resolution functionalities. Subsequently, the *LPN ASAS alerting* and the *LPN P-ASAS processing* informs the PF and PNF through *ASAS audio alerting* about any aircraft that is in potential ASAS conflict with its own aircraft, and suggests resolution options including a prioritization. Three complementary LPNs represent non-nominal behaviour modes, each combination of which has a specific influence on the ASAS alerting LPN:

- *ASAS system mode* may be working, failed, or corrupted (failed or corrupted mode also influences the ASAS processing LPN).
- *ASAS priority switching mode*; under emergency, the PF switches this from "off" to "on."
- *ASAS anti-priority switch*; this is switched from "off" to "on" when own ADS-B is not working.

10.3.5 Interconnected LPNs of "Pilot Flying"

This subsection illustrates the specific Petri net model developed for the Pilot Flying. For the semantical basis behind this type of model, see [2], [5], [6], [15], [59]. A graphical representation of all LPNs the Pilot-Flying consists of, is given in Figure 10.5.

The Human-Machine-Interface where sound or visual clues might indicate that attention should be paid to a particular issue, is represented by a LPN that does not belong to the Pilot-Flying as agent and is therefore not depicted in the figure. Similarly, the arcs to or from any other agent are not shown in Figure 10.5. Because of the

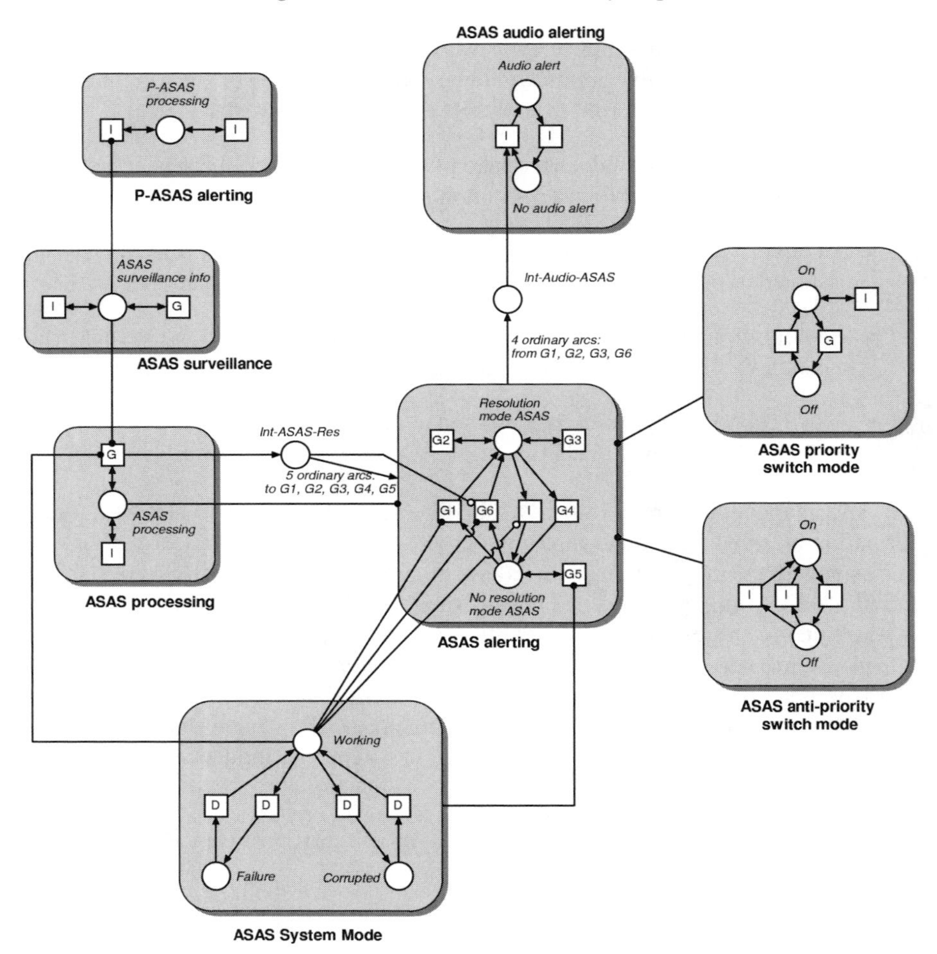

FIGURE 10.4: The agent ASAS in AMFF is modelled by eight LPNs, a number of ordinary and enabling arcs, and two IPNs (with one place each).

very nature of Petri nets, these arcs can easily be added during the follow-up specification cycle. To get an understanding of the different LPNs, a good starting point might be the LPN "Current Goal" (at the bottom of the figure) as it represents the objective the Pilot-Flying is currently working on. Examples of such goals are "Collision Avoidance," "Conflict Resolution," and "Horizontal Navigation." For each of these goals, the pilot executes a number of tasks in a prescribed or conditional order, represented in the LPN "Task Performance." Examples of such tasks are "Monitoring and Decision," "Execution," and "Execution Monitoring." If all relevant tasks for the current goal are considered executed, the pilot chooses another goal, thereby using his memory (where goals deserving attention might be stored, represented by

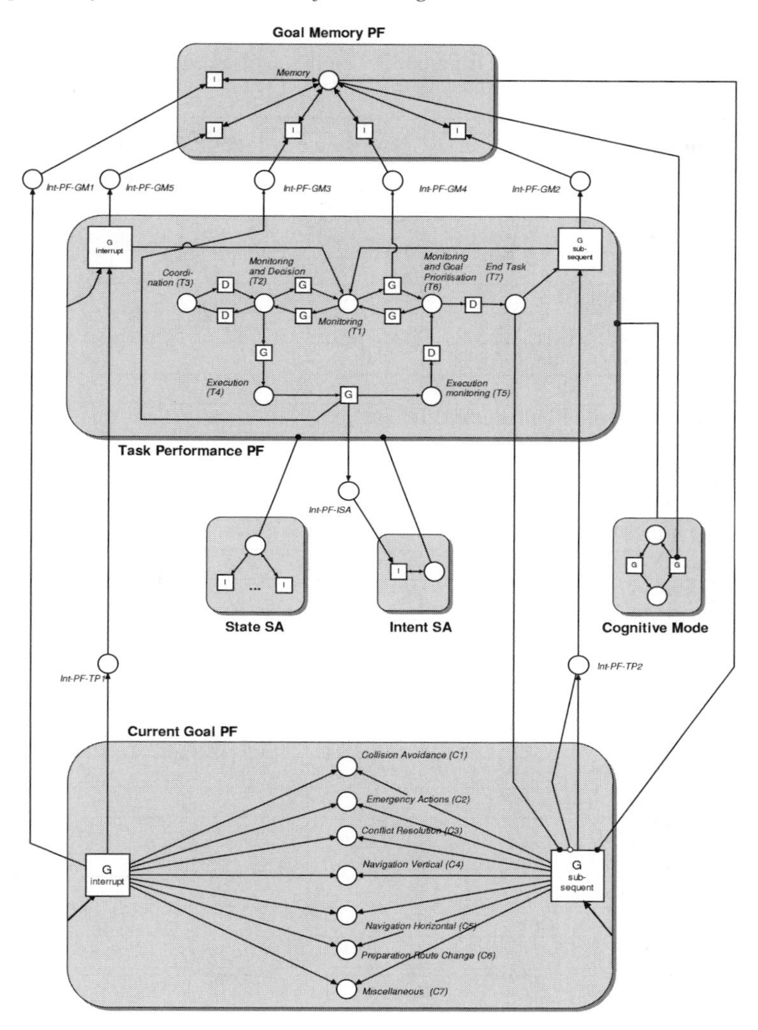

FIGURE 10.5: The agent Pilot-Flying in AMFF is modelled by six LPNs, and a number of ordinary and enabling arcs and some IPNs, consisting of one place and input and output arcs. The interconnections with other agents are not shown.

the LPN "Goal Memory") and the Human-Machine-Interface. His memory where goals deserving attention might be stored is represented as the LPN "Goal Memory" in Figure 10.5.

So, the LPNs "Current Goal," "Task Performance," and "Goal Memory" are important in the modelling of which task the Pilot-Flying is executing. The other three LPNs are important in the modelling on how the Pilot-Flying is executing the tasks. The LPN "State SA", where SA stands for Situation Awareness, represents the rel-

Table 10.1: Dimensional analysis of agent PF.

Pilot-Flying (PF) LPNs and IPNs	**Number of places**	**Maximum colour state space**
Pilot Flying (PF) LPNs:		
State Situation Awareness	1	$\mathbb{R}^3$
Intent Situation Awareness	1	$\mathbb{R}^5$
Goal memory	1	$\mathbb{R}$
Current goal	7	$\mathbb{R}$
Task performance	7	$\mathbb{R}^2$
Cognitive mode	2	$\mathbb{R}$
Pilot Flying (PF) internal IPNs:		
Int-PF-GM1	1	$\mathbb{R}$
Int-PF-GM2	1	$\mathbb{R}$
Int-PF-GM3	1	$\mathbb{R}$
Int-PF-GM4	1	$\mathbb{R}$
Int-PF-GM5	1	$\mathbb{R}$
Int-PF-TP1	1	$\mathbb{R}^2$
Int-PF-TP2	1	$\mathbb{R}$
Int-PF-ISA	1	$\mathbb{R}$
Pilot Flying (PF) external IPNs:		
Ext-PF-Audio-PF	5	
Ext-PF-PNF	1	$\mathbb{R}$
Ext-PF-PASAS	1	$\emptyset$
Ext-PF-SSA-1	1	$\emptyset$
Ext-PF-SSA-2	1	$\mathbb{R}$
Ext-PF-SSA-3	1	$\mathbb{R}$
Ext-PF-SSA-4	1	$\mathbb{R}$
Ext-PF-SSA-5	1	$\mathbb{R}$
PRODUCT	490	$\mathbb{R}^{28}$

evant perception of the pilot about the states of elements in his environment, e.g., whether he is aware of an engine failure. The LPN "Intent SA" represents the intent, e.g., whether he intends to leave the free flight airspace. The LPN "Cognitive mode" represents whether the pilot is in an opportunistic mode, leading to a high but error-prone throughput, or in a tactical mode, leading to a moderate throughput with a low error probability.

10.3.6 Model Verification, Parameterisation, and Validation

The compositionally specified SDCPN model enables a systematic implementation, verification and validation of the resulting Monte Carlo simulator. This is done through the following systematic steps:

- Software code testing. This is done through conducting the following sequence of testing: random number generation, statistical distributions, common functions, each LPN implementation, each agent implementation, interactions between all agents, full MC simulation.
- Numerical approximation testing. This is needed to learn choosing an appropriate numerical integration step size and an appropriate number of particular MC simulations.
- Graphical user interface testing. This is to verify that the input and output of data works well.
- Parameterisation. This is done through a search for literature and statistical sources, and complemented by expert interviews. The fusion of these different pieces of information is accomplished following a Bayesian approach.
- Initial model validation through studying MC simulator behaviour and sensitivities to parameter changes under dedicated scenarios.
- Overall validation, which is directed to the evaluation of differences between model and reality and what effect these differences have at the assessed risk level.

The last validation step typically is done at a later stage in the risk assessment process, with the help of active participation of operational experts [25], [31].

10.3.7 Dimensions of MC Simulation Model

Now, we analyse the dimensions of the joint state space of the resulting MC simulation model. In Table 10.1 and Table 10.2, this is done for the agents PF and ASAS respectively, including all LPNs and all IPNs that end on one of these LPNs. The second column gives the number of places in the LPN or IPN. The third column gives the maximum state space of the colour used within an LPN or IPN. We also perform this analysis to the LPNs and IPNs of the other agents. The resulting number of product places and product state spaces is given in Table 10.3. This table brings into account that of each agent, except global CNS, there is one per aircraft.

The product places of the global CNS agent form the θ_t^0 state space M^0. The corresponding state space is empty, which means that there is no x_t^0. The product places of the other agents form the state space $\bigotimes_{i=1}^{N} M^i$ of the process components θ_t^i, $i = 1,\ldots,N$. Per aircraft, the number of product places is $|M^i| \approx 0.777 \times 10^{12}$. The colours attached to the places in the other agents form the process components x_t^i, $i = 1,\ldots,N$, each of which assumes values in $\mathbb{R}^{126+21N}$.

Each of the scenarios considered in the next subsection has eight aircraft, so $N = 8$. This means that the number of product places equals $\approx 16 \times (0.777 \times 10^{12})^8 \approx 2.13 \times 10^{96}$, and that the product of the colour state spaces equals $\mathbb{R}^{2352}$.

Table 10.2: Dimensional analysis of agent ASAS.

ASAS LPNs and IPNs	**Number of places**	**Maximum colour state space**
ASAS LPNs:		
Processing	1	$\mathbb{R}^{13+12N}$
Alerting	2	$\mathbb{R}^7$
Audio alerting	2	$\emptyset$
Surveillance	1	$\mathbb{R}^{11+9N}$
System mode	3	$\emptyset$
Priority switch mode	2	$\emptyset$
Anti-priority switch mode	2	$\emptyset$
Predictive alerting	1	$\mathbb{R}^3$
ASAS internal IPNs:		
Int-ASAS-Resolution	1	$\emptyset$
Int-ASAS-Audio	1	$\emptyset$
ASAS external IPNs:		
Ext-ASAS-PF	1	$\mathbb{R}^3$
Ext-PASAS-PNF	1	$\emptyset$
Ext-ASASProc-PNF	1	$\emptyset$
Ext-ASASurv-ADSB-Global	1	$\mathbb{R}$
Ext-ASASprio-PNF	1	$\emptyset$
PRODUCT	48	$\mathbb{R}^{38+21N}$

Table 10.3: Dimensional analysis of complete SDCPN.

Agent	**Number of product places**	**Maximum colour product state space**
Aircraft	24^N	$\mathbb{R}^{13N}$
Pilot Flying (PF)	490^N	$\mathbb{R}^{28N}$
Pilot-not-Flying (PNF)	7^N	$\mathbb{R}^{3N}$
AGNC	$(15 \times 2^{16})^N$	$\mathbb{R}^{45N}$
ASAS	48^N	$\mathbb{R}^{37N+21N^2}$
Global CNS	16	$\emptyset$
PRODUCT	$\approx 16 \times (0.777 \times 10^{12})^N$	$\mathbb{R}^{126N+21N^2}$

10.4 Simulated Scenarios and Collision Risk Estimates

The IPS algorithm of Section 10.2 is now applied to three hypothetical AMFF air traffic scenarios. The first scenario has eight aircraft that fly at the same flight level and their flight plans cause them to fly through the same point in airspace at the same moment in time. The second scenario has one aircraft flying through a virtual infinite airspace of randomly distributed aircraft, with a density 3 times as high as in a current high capacity en route area. The third scenario is the same as the second, except that the aircraft density is four times lower. Prior to describing these scenarios and simulation results, we explain the parameterisation of the IPS algorithm used.

10.4.1 Parameterisation of the IPS Simulations

The main safety critical parameter settings of the free flight enabling technical systems (GPS, ADS-B and ASAS) are given in Table 10.4; in the table, global ADS-B down refers to frequency congestion/overload of the data transfer technology used for ADS-B. The IPS conflict levels k are defined by parameter values for lateral conflict distance d_k, conflict height h_k, and time to conflict Δ_k. These values have been determined through two steps. The first was to let an operational expert make a best guess of proper parameter values. Next, during initial simulations with the IPS some fine tuning of the number of levels and of parameter values per level has been done. The resulting values are given in Table 10.5.

10.4.2 Eight Aircraft on Collision Course

In this simulation eight aircraft start at the same flight level, some 135 Nm (250 km) out of each other, and fly in eight 45 degrees differing directions with a ground speed of 466 knots (= 240 m/s), all aiming to pass through the same point in airspace. By running ten times the IPS algorithm the collision risk is estimated ten times. The number of particles per IPS simulation run is 12,000. The total simulation time took about 20 hours on two machines, and the load of computer memory per machine was about 2.0 gigabyte. For the first eight IPS runs, the estimated fractions $\bar{\gamma}_k^i$ are given in Table 10.6 for each of the conflict levels, $k = 1, \ldots, 8$, and aircraft $i = 1$. It can be seen that the first and sixth IPS runs have zero particles that reach the last (8th level). Hence the first and sixth IPS runs yield $\bar{\gamma}_8^i = 0$. This is a clear example of particle depletion.

The IPS estimated mean probability for one aircraft to collide with any of the other seven aircraft equals 2.2×10^{-5}. The minimum and maximum values now are respectively a factor 250 lower and a factor 4 higher than the mean value. We also verified that this risk value was not sensitive at all to the failure rates of the ASAS related technical systems.

In [37] a similar eight aircraft encounter scenario has been simulated some hundred times, for varying initial aircraft positions, without noticing any collision event.

Table 10.4: Parameter values of free flight enabling technical systems.

Model Parameter	Probability
Global GPS down	1.0×10^{-5}
Global ADS-B down	1.0×10^{-6}
Aircraft ADS-B receiver down	5.0×10^{-5}
Aircraft ADS-B transmitter down	5.0×10^{-5}
Aircraft ASAS system mode corrupted	5.0×10^{-5}
Aircraft ASAS system mode failure	5.0×10^{-5}

Table 10.5: IPS conflict level parameter values.

k	1	2	3	4	5	6	7	8
d_k (Nm)	4.5	4.5	4.5	4.5	2.5	1.25	0.5	0.054
h_k (ft)	900	900	900	900	900	500	250	131
Δ_k (min)	8	2.5	1.5	0	0	0	0	0

Table 10.6: Fractions counted during eight IPS runs of scenario 1.

Level	1st IPS	2nd IPS	3rd IPS	4th IPS	5th IPS	6th IPS	7th IPS	8th IPS
1	1.000	1.000	1.000	1.000	1.000	1.000	1.000	1.000
2	0.528	0.529	0.539	0.533	0.537	0.538	0.536	0.539
3	0.426	0.429	0.424	0.431	0.421	0.428	0.426	0.418
4	0.033	0.036	0.035	0.037	0.039	0.031	0.044	0.039
5	0.175	0.180	0.183	0.181	0.142	0.157	0.181	0.147
6	0.267	0.158	0.177	0.144	0.255	0.138	0.295	0.146
7	0.150	0.268	0.281	0.427	0.645	0.208	0.253	0.295
8	0.000	0.009	0.233	0.043	0.455	0.000	0.006	0.815
Product of fractions	0.0	5.58×10^{-7}	1.67×10^{-5}	4.01×10^{-6}	9.33×10^{-5}	0.0	8.00×10^{-7}	4.48×10^{-5}

At a collision probability value of 2.2×10^{-5}, the chance to count at least one collision would be less than 1%. As such the current results agree quite well with the fact that in these earlier simulations for an eight aircraft scenario no collision has been observed. We also verified that the novel simulation results for an eight aircraft scenario agreed quite well with the expectation of the designers of the AMFF operational concept.

Table 10.7: Fractions counted during eight IPS runs of scenario 2.

Level	1st IPS	2nd IPS	3rd IPS	4th IPS	5th IPS	6th IPS	7th IPS	8th IPS
1	0.922	0.917	0.929	0.926	0.925	0.925	0.925	0.921
2	0.567	0.551	0.560	0.559	0.554	0.551	0.561	0.556
3	0.665	0.666	0.674	0.676	0.672	0.673	0.664	0.670
4	0.319	0.331	0.323	0.321	0.328	0.321	0.334	0.331
5	0.370	0.367	0.371	0.379	0.363	0.345	0.366	0.343
6	0.181	0.158	0.162	0.171	0.164	0.181	0.148	0.191
7	0.130	0.209	0.174	0.145	0.162	0.170	0.214	0.215
8	0.067	0.005	0.094	0.066	0.002	0.150	0.015	0.019
Product of fractions	6.42×10^{-5}	6.76×10^{-6}	1.11×10^{-4}	6.99×10^{-5}	2.57×10^{-6}	1.75×10^{-4}	1.99×10^{-5}	2.98×10^{-5}

10.4.3 Free Flight Through an Artificially Constructed Airspace

In this simulation the complete airspace is divided into packed containers. Within each container a fixed number of seven aircraft ($i = 2,\ldots,8$) fly at arbitrary position and in arbitrary direction at a ground speed of 466 Nm/hr. One additional aircraft ($i = 1$) aims to fly straight through a sequence of connected containers, at the same speed, and the aim is to estimate its probability of collision with any of the other aircraft per unit time of flying.

Per container, the aircraft within it behave the same. This means that we have to simulate each aircraft in one container only, as long as we apply the ASAS conflict prediction and resolution also to aircraft copies in the neighbouring containers. In principle this can mean that an aircraft experiences a conflict with its own copy in a neighbouring container. This also means that the size of a container should not go below a certain minimum size.

By changing container size we can vary traffic density. To choose the appropriate traffic density, our reference point is the highest number (17) of aircraft counted at 23rd July 1999 in an en-route area near Frankfurt of size 1 degree $\times$ 1 degree $\times$ FL290-FL420. This comes down to 0.0032 a/c per Nm^3. For our simulation we assume a 3 times higher traffic density, i.e., 0.01 a/c per Nm^3. This resulted in choosing containers having a length of 40 Nm, a width of 40 Nm, and a height of 3000 feet and with 8 aircraft flying in such a container.

By running the IPS algorithm ten times (+ one extra later on) over 25 minutes (5 minutes to allow convergence and 20 minutes to estimate collision probability) the collision probability per unit time of flying has been estimated. The number of particles per IPS simulation run is 10,000. The total simulation time took about 300 hours on two machines, and the load of computer memory per machine was about 2.0 gigabyte. For the first eight IPS runs, the estimated fractions $\bar{\gamma}_k^i$ are given in Table 10.7 for each of the conflict levels, $k = 1,\ldots,8$, for aircraft $i = 1$.

The estimated mean probability of collisions per 20 minutes of aircraft flight equals 5.22×10^{-5}, which is equal to a probability of collisions per aircraft flight hour of 1.6×10^{-4}, with minimum and maximum values respectively a factor four lower and higher. We also verified that this risk value was not sensitive at all to the failure rates of the ASAS related technical systems.

One should be aware that this value has been estimated for the simulation model of the intended AMFF operation. Hence the question is what this means for the intended AMFF operation itself? By definition a simulation model of AMFF differs from the intended AMFF operation. If it can be shown that the combined effect of these differences on the risk level is small, then the results obtained for the simulation model may be considered as a good representation of the accident risk of the intended operation. In order to assess the combined effect of these differences there is need to perform a bias and uncertainty assessment [25].

In order to better learn understanding of what causes the collision risk of the simulation model to be relatively high, we performed an extra IPS run, and memorised in static memory for each particle the ancestor history at each of the eight levels. This allowed us to trace back what happened for the particles that hit the last level set (i.e., collision). There appeared to be five different collision events. Evaluation of these five collision events showed that all five happened under nominal safety critical conditions. Four of the five collisions were due to a growing number of multiple conflicts that could not be solved in time under the operational concept adopted. The fifth collision was of another type: at quite a late moment finally a conflict between two aircraft was solved with a maneuver by one of the two aircraft. However because of this maneuver there was a sudden collision with a third nearby aircraft.

These detailed evaluations of the five collision events of the 11th IPS run also showed that a significant increase of collision risk is caused by the relatively small height (4000 ft) of a container. Because of this small height it happened that an aircraft in one container came in conflict with a copy of its own in a neighbouring container, and in such a situation there was an undesired limitation in conflict resolution options, and thus an undesired artificial increase in collision risk.

The results in this section seem to indicate that the key factor in the increased risk of collision for encounters with homogeneous traffic in the background — as opposed to the eight encountering aircraft only scenario — are the multiple conflicts. Under the far higher traffic densities than what the AMFF operational concept was designed for, it is not always possible to timely solve a sufficiently high fraction of those multiple conflicts. On the basis of this finding one would expect that the collision risk would decrease faster than linear with a decrease in traffic density. The validity of this expectation is verified by the next scenario.

10.4.4 Reduction of the Aircraft Density by a Factor Four

Now we enlarge the length and width of each container by a factor two. This means that the traffic density has gone down by a factor four. Hence the density is now $\frac{3}{4}$ of the density counted on 23rd July 1999 in the en-route area near Frankfurt. This still is a factor 2.5 higher than current average density above Europe. At the

Table 10.8: Fractions counted during four IPS runs of scenario 3.

Level	1st IPS	2nd IPS	3rd IPS	4th IPS
1	0.755	0.750	0.752	0.749
2	0.295	0.292	0.286	0.285
3	0.476	0.475	0.497	0.487
4	0.263	0.258	0.266	0.267
5	0.321	0.315	0.300	0.328
6	0.068	0.088	0.082	0.096
7	0.156	0.367	0.290	0.254
8	0.011	0.059	0.021	0.005
Product of fractions	1.07×10^{-6}	1.61×10^{-5}	4.31×10^{-6}	1.07×10^{-6}

same time simulated flying time has been increased to 60 minutes (with 10 minutes prior flying to guarantee convergence).

By running four times the IPS algorithm the collision risk is estimated four times. The number of particles per IPS simulation run is 10,000. The total simulation time took about 280 hours on two machines, and the load of computer memory per machine was about 2.0 gigabyte. For these IPS runs, the estimated fractions $\bar{\gamma}_k^i$ are given in Table 10.8 for each of the conflict levels, $k = 1, \ldots, 8$, for aircraft $i = 1$. The estimated mean probability of collision per aircraft flight hour equals 5.64×10^{-6}, with minimum and maximum values respectively a factor five lower and higher. This is about a factor 30 lower than the previous scenario with a four times higher aircraft density. Thus, for the model there is a steep decrease of collision probability with decrease of traffic density, and this agrees well with the expectation at the end of the previous section.

10.4.5 Discussion of IPS Simulation Results

Because of the IPS simulation approach we were able to estimate collision risk for complex multiple aircraft scenarios. The large increase in handling complexity of multiple aircraft encounter situations is a major improvement over what was feasible before for two aircraft flying in a parallel route structure [4], [29]. Inherent to the IPS way of simulation, the dynamic memory of the computers used appeared to pose the main limitation on the full exploitation of the novel sequential MC simulation approach. This also prevented performing a bias and uncertainty assessment for the differences between the simulation model and the AMFF operation. As long as such a bias and uncertainty assessment has not been performed, any conclusion drawn from the simulation apply to the simulation model only, and need not apply to the intended AMFF operation.

The simulations performed for a model of AMFF allow free flight operational concept developers to learn characteristics of the simulation model. Because of the sequential MC simulation based speed up, these simulations can show events that have not been observed before in MC simulations of an AMFF model. Under far higher traffic densities than what the AMFF operational concept has been designed for, the simulations of the model shows it is not always possible to timely solve multiple conflicts. As a result of this, at high traffic levels there is a significant chance that multiple conflicts are clogging together, and this eventually may cause a non-negligible chance of collision between aircraft in the simulation model. It has also been shown that by lowering traffic density, the chance of collision for the model rapidly goes down.

10.5 Concluding Remarks

We studied collision risk estimation of a free flight operation through a sequential Monte Carlo simulation. First a Monte Carlo simulation model of this free flight operational concept has been specified in a compositional way using the Stochastically and Dynamically Coloured Petri Net (SDCPN). Subsequently a novel sequential MC simulation method [13], [14] has been extended for application to collision risk estimation in air traffic, and has subsequently been applied to an SDCPN model of free flight.

The results obtained show that the novel simulation model specification and collision risk estimation method allow to speed up the Monte Carlo simulations for much more complex air traffic encounter situations than what was possible before, e.g., [4], [29]. Moreover, for the simulation model of the free flight operational concept considered, behaviour has been made visible that was expected by free flight concept designers, but could not be observed in straightforward Monte Carlo simulations of free flight concepts (e.g., Hoekstra [37]): the rare chance of clogging multiple conflicts at far higher traffic density levels than where the particular concept has been designed for. Hence, further attention has to be drawn toward the development and incorporation in the particular operational concept design of advanced methods in handling multiple conflicts. For example, Hoekstra [37] studied a conflict resolution approach that performs better than the one adopted in the AMFF concept. In addition, there are some complementary developments that aim to develop complex conflict resolution solvers with some guaranteed level of performance [20], [51] under nominal conditions, and ways to incorporate situation awareness views by human operators (pilots and/or controllers) in these combinatorial conflict resolution problems [21].

The initial collision risk estimation results obtained with our sequential MC simulation of free flight provides valuable feedback to the design team and allows them to learn from Monte Carlo simulation results they have never seen before. This allows them to significantly improve their understanding of when and why multiple conflicts are not solved in time anymore in the simulation model. Subsequently the operational concept designers can use their better understanding for adapting the free flight design such that it can better bring into account future high traffic levels.

In its current form the sequential MC simulation approach works well, but at the same time poses very high requirements on the availability of dynamic computer memory and simulation time. The good message is that in literature on sequential MC simulation, e.g., see [18], [19], [23], [35], [48], [52], complementary directions have been developed which remain to be explored for application to free flight collision risk estimation. These potential improvements of sequential MC simulation will be studied in follow-up research.

Acknowledgement. The authors thank Mariken Everdij (NLR) for valuable discussions and a thorough review of a draft version of this chapter.

References

[1] J.W. Andrews, J.D. Welch, H. Erzberger. Safety analysis for advanced separation concepts. In *Proceedings of USA/Europe ATM R&D Seminar*, Baltimore, MD, 27–30 June 2005.

[2] H.A.P. Blom, J. Daams and H.B. Nijhuis. Human cognition modeling in Air Traffic Management safety assessment. In *Air Transportation Systems Engineering*, edited by G.L. Donohue and A.G. Zellweger, Vol. 193 in Progress in Astronautics and Aeronautics, Paul Zarchan, Editor-in-Chief, Chapter 29, pages 481–511, 2001.

[3] H.A.P. Blom, G.J. Bakker. Conflict probability and incrossing probability in air traffic management. In *Proc. IEEE Conf. on Decision and Control*, Las Vegas, December 2002.

[4] H.A.P. Blom, G.J. Bakker, M.H.C. Everdij, M.N.J. van der Park. Collision risk modeling of air traffic. In *Proceedings of European Control Conference*, Cambridge, UK, 2003.

[5] H.A.P. Blom, S.H. Stroeve, M.H.C. Everdij, M.N.J. Van der Park. Human cognition performance model to evaluate safe spacing in air traffic. *Human Factors and Aerospace Safety*, Vol. 2, pages 59–82, 2003.

[6] H.A.P. Blom, K.M. Corker, S.H. Stroeve. On the integration of human performance and collision risk simulation models of runway operation. In *Proceedings of the 6th USA/Europe Air Traffic Management R&D Seminar*, Baltimore, MD, 27–30 June 2005.

[7] H.A.P. Blom, G.J. Bakker, B. Klein Obbink, M.B. Klompstra. Free flight safety risk modelling and simulation. In *Proceedings of International Conference on Research in Air Transportation (ICRAT)*, Belgrade, 26–28 June 2006.

[8] H.A.P. Blom, J. Lygeros. *Stochastic Hybrid Systems: Theory and Safety Critical Applications*, LNCIS series, Springer, Berlin, July 2006.

[9] M.L. Bujorianu, J. Lygeros. Reachability questions in piecewise deterministic Markov processes. In *Proceedings of Hybrid Systems Computation and Control*, Eds. O. Mahler, A. Pnuelli, LNCIS number 2623, Springer, Berlin, pages 126–140, 2003.

[10] M.L. Bujorianu. Extended stochastic hybrid systems. In *Proceedings of Hybrid Systems Computation and Control*, Eds. O. Mahler, A. Pnuelli, LNCIS number 2993, Springer, Berlin, pages 234–249, 2004.

[11] M.L. Bujorianu, J. Lygeros. Towards a general theory of stochastic hybrid systems. In [8], pages 3–30, 2006.

[12] C.G. Cassandras, S. Lafortune. *Introduction to Discrete Event Systems*, Kluwer Academic Publishers, Boston, 1999.

[13] F. Cérou, P. Del Moral, F. Le Gland and P. Lezaud. Genetic genealogical models in rare event analysis, Publications du Laboratoire de Statistiques et Probabilites, Toulouse III, 2002.

[14] F. Cérou, P. Del Moral, F. Le Gland, P. Lezaud. Limit theorems for the multilevel splitting algorithms in the simulation of rare events. In *Proceedings of Winter Simulation Conference*, Orlando, FL, 2005.

[15] K. Corker. Cognitive Models & Control: Human & System Dynamics in Advanced Airspace Operations, Eds: N. Sarter and R. Amalberti, *Cognitive Engineering in the Aviation Domain*, Lawrence Earlbaum Associates, Hillsdale, New Jersey, 2000.

[16] R. David, H. Alla. Petri Nets for the modeling of dynamic systems - A survey, *Automatica*, Vol. 30, No. 2, pages 175–202, 1994.

[17] M.H.A. Davis. *Markov Models and Optimization*, Chapman & Hall, London, 1993.

[18] P. Del Moral. *Feynman-Kac Formulae. Genealogical and Interacting Particle Systems with Applications*, Springer Verlag, New York, 2004.

[19] P. Del Moral, P. Lezaud. Branching and interacting particle interpretations of rare event probabilities. In [8], pages 277–324, 2006.

[20] D.V. Dimarogonas, S.G. Loizou, K.J. Kyriapoulos. Multirobot navigation functions II: towards decentralization. In [8], pages 209–256, 2006.

[21] E. De Santis, M.D. Di Benedetto, S. Di Gennaro, A.D.'Innocenzo, G. Pola. Critical observability of a class of hybrid systems and application to air traffic management. In [8], pages 141–170, 2006.

[22] DNV. Safety assessment of P-RNAV route spacing and aircraft separation, Final report TRS 052/01, Eurocontrol, 2003.

[23] A. Doucet, N. de Freitas and N. Gordon. *Sequential Monte Carlo Methods in Practice*, Springer-Verlag, New York, NY, 2001.

[24] H. Erzberger. Transforming the NAS: The next generation air traffic control system. In *Proceedings of the 24th Int. Congress of the Aeronautical Sciences (ICAS)*, Yokohoma, Japan, 2004.

[25] M.H.C. Everdij, H.A.P. Blom. Bias and uncertainty in accident risk assessment, NLR report CR-2002-137, National Aerospace Laboratory NLR, 2002.

[26] M.H.C. Everdij, H.A.P. Blom. Petri nets and hybrid state Markov processes in a power-hierarchy of dependability models. In *Proceedings of IFAC Conference on Analysis and Design of Hybrid Systems*, Saint-Malo Brittany, France, pages 355–360, June 2003.

[27] M.H.C. Everdij, H.A.P. Blom. Piecewise deterministic Markov processes represented by Dynamically Coloured Petri Nets, *Stochastics*, Vol. 77, pages 1–29, 2005.

[28] M.H.C. Everdij, H.A.P. Blom. Hybrid Petri nets with diffusion that have into-mappings with generalised stochastic hybrid processes. In [8], pages 31–64, 2006.

[29] M.H.C. Everdij, H.A.P. Blom, G.J. (Bert) Bakker. Modeling lateral spacing and separation for airborne separation assurance using Petri nets. Forthcoming in: *Simulation, Transactions of the Society for Modeling and Simulation International*, Vol. 82, 2006.

[30] M.H.C. Everdij, M.B. Klompstra, H.A.P. Blom, B. Klein Obbink. Compositional specification of a multi-agent system by stochastically and Dynamically Coloured Petri Nets. In [8], pages 325–350, 2006.

[31] M.H.C. Everdij, H.A.P. Blom, S.H. Stroeve. Structured assessment of bias and uncertainty in Monte Carlo simulated accident risk. In *Proceedings of the 8th Int. Conf. on Probabilistic Safety Assessment and Management (PSAM8)*, New Orleans, LA, May 2006.

[32] FAA/Eurocontrol. A concept paper for separation safety modeling, Cooperative R&D Action Plan 3 report. Available at http://www.faa.gov/asd/ia-or/pdf/cpcomplete.pdf, 20 May 1998.

[33] FAA/Eurocontrol. Principles of Operations for the Use of ASAS, Cooperative R&D Action Plan 1 report, Version 7.1, 2001.

[34] FAA/Eurocontrol. Safety and ASAS applications, Co-operative R&D Action Plan 1 report, version 4.1, 2004.

[35] P. Glasserman. *Monte Carlo Methods in Financial Engineering*, Stochastic Modeling and Applied Probability, Vol. 53, Springer, New York, NY, 2003.

[36] P.J. Haas. *Stochastic Petri Nets, Modeling, Stability, Simulation*, Springer-Verlag, New York, 2002.

[37] J. Hoekstra. *Designing for Safety, the Free Flight Air Traffic Management concept*, PhD Thesis, Delft University of Technology, November 2001.

[38] D.A. Hsu. The evaluation of aircraft collision probabilities at intersecting air routes. *Journal of Navigation*, Vol. 34, pages 78–102, 1981.

[39] J. Hu, M. Prandini, S. Sastry. Probabilistic safety analysis in three dimensional aircraft flight. In *Proceedings of the 42nd IEEE Conference on Decision and Control (CDC)*, Maui, 2003.

[40] ICAO. Manual on airspace planning methodology for the determination of separation minima, ICAO Doc. 9689-AN/953, 1998.

[41] ICAO. Airborne separation assistance system (ASAS) circular, Draft, version 3, SCRSP, WGW/1 WP/5.0, International Civil Aviation Organization, May 2003.

[42] K. Jensen. *Coloured Petri Nets: Basic Concepts, Analysis Methods and Practical Use*, Vol. 1, Springer, London, UK, 1992.

[43] B. Klein Obbink. MFF airborne self separation assurance OSED, Report MFF R733D. Available at http://www.medff.it/public/index.asp, April 2005.

[44] J. Krozel. Free flight research issues and literature search. Under NASA contract NAS2-98005, 2000.

[45] J. Krystul, H.A.P. Blom. Monte Carlo simulation of rare events in hybrid systems, Hybridge Report D8.3. Available at http://www.nlr.nl/public/hosted-sites/hybridge, 2004.

[46] J. Krystul, H.A.P. Blom. Generalised stochastic hybrid processes as strong solutions of stochastic differential equations, Hybridge report D2.3. Available at http://www.nlr.nl/public/hostedsites/hybridge/, 2005.

[47] J. Krystul, H.A.P. Blom. Sequential Monte Carlo simulation of rare event probability in stochastic hybrid systems. In *Proceedings of the 16th IFAC World Congress*, Prague, Czech Republic, June 4-8, 2005.

[48] J. Krystul, H.A.P. Blom. Sequential Monte Carlo simulation for the estimation of small reachability probabilities for stochastic hybrid systems. In *Proceedings of IEEE-EURASIP Int. Symposium on Control, Communications and Signal Processing*, Marrakech, Morocco, March 13-15, 2006.

[49] A.B. Kurzhanski, P. Varaiya. On reachability under uncertainty. *SIAM Journal on Control and Optimization*, Vol. 41, pages 181–216, 2002.

[50] P.E. Labeau, C. Smidts and S. Swaminathan. Dynamic reliability: towards an integrated platform for probabilistic risk assessment. *Reliability Engineering and System Safety*, Vol. 68, pages 219–254, 2000.

[51] A. Lecchini, W. Glover, J. Lygeros, J. Maciejowski. Monte Carlo optimisation for conflcit resolution in air traffic control. In [8], pages 257–276, 2006.

[52] F. Le Gland, N. Oudjane. A sequential particle algorithm that keeps the particle system alive. In [8], pages 351–400, 2006.

[53] MFF. MFF Final safety case, Report MFF D734, ed. 1.0. Available at http://www.medff.it/public/index.asp, November 2005.

[54] NASA. Concept definition for distributed air-/ground traffic management (DAG-TM), Version 1.0, Advanced Air Transportation Technologies project, Aviation System Capacity Program, National Aeronautics and Space Administration, NASA, 1999.

[55] G. Pola, M.L. Bujorianu, J. Lygeros, M.D. Di Benedetto. Stochastic hybrid models: an overview with applications to air traffic management. In *Proceedings of IFAC Conf. Analysis and Design of Hybrid Systems (ADHS)*, Saint-Malo, Brittany, France, 2003.

[56] M. Prandini, J. Hu. A stochastic approxmation method for reachability computations. In [8], pages 107–139, 2006.

[57] RTCA. Free Flight Implementation, Task Force 3 Final Technical Report, Washington DC, 1995.

[58] A.P. Shah, A.R. Pritchett, K.M. Feigh, S.A. Kalarev, A. Jadhav, K.M. Corker, D.M. Holl, R.C. Bea. Analyzing air traffic management systems using agent-based modeling and simulation. In *Proceedings of the 6th USA/Europe Seminar on Air Traffic Management Research and Development*, Baltimore, MD, 27-30 June 2005.

[59] S.H. Stroeve, H.A.P. Blom and M.N. van der Park. Multi-agent situation awareness error evolution in accident risk modeling. In *Proceedings of the 5th USA/Europe Seminar on Air Traffic Management Research and Development*, Budapest, Hungary, 23-27 June 2003.

Index

$\mathscr{F}_t$-measurable, 18, 25, 26

active transition, 56
agent, 251, 260, 262, 265, 269
Air Traffic
  Control (ATC), 262
  Management (ATM), 3, 249
air traffic operations, 251
Airborne
  Collision Avoidance System (ACAS), 262
  Guidance, Navigation and Control (AGNC), 262, 263
  Separation Assistance System (ASAS), 262, 263
Airline Operations Centre (AOC), 262
arc
  enabling, 260
  inhibitor, 260
  interconnection, 260
Automatic Dependent Surveillance-Broadcast (ADS-B), 263
Autonomous Mediterranean Free Flight (AMFF), 259

Bachillus subtilis, 228
  PDMP model, 229
biochemical networks, 11, 221
bisimulation, 95, 96, 98

cell cycle, 235
Chapman-Kolmogorov equation, 256
characteristics, 18, 20, 22
Cockpit Display of Traffic Information (CDTI), 265
collision, 254, 255
  risk, 250, 251, 253, 254, 259, 270, 271, 274–276
  risk assessment, 250, 261
Communicating PDP (CPDP), 55
  flow map, 60
  its CFSJS/NTS, 60
  output semantics, 61
  state/output space, 59
  value passing, 69
Communication, Navigation and Surveillance (CNS), 262
composition
  commutativity/associativity, 64
  for CPDPs, 61
  for value passing CPDPs, 70
  for CFSJSs, 53
conditioning, 257
congestion control, 191
Continuous Flow Spontaneous Jump System (CFSJS), 50
continuous state, 213
continuous-time Markov chain, 87
  bisimulation, 97
  inhomogenity, 92
  rewards, 89
  simulation preorder, 98
  transient probability, 88
  uniformization, 88
continuous-time Markov reward model, 89
  time inhomogenity, 92
controlled switching diffusion, 16
counting process, 16
CSP, 68

Dirac measure, 233, 242
Discrete Event System (DES), 1
discrete state, 213
discrete-time Markov chain, 82

bisimulation, 95
multiple rewards, 85
rewards, 84
simulation preorder, 98
transient probability, 84
weak bisimulation, 96
discrete-time Markov reward model, 84
transient reward probability, 86
DNA replication, 235
PDMP model, 239
Dynamically Coloured Petri Net (DCPN), 253

empty location, 59
enabled state, 53
enabling arc, 260
event
detection, 226
exogenous and endogenous, 143
event-driven system, 2
execution path of TMS, 49
exit time, 224
exogenous and endogenous events, 143
extended generator, 214

Flight Management System (FMS), 263
forced transition, 53, 226, 233, 240
Forced Transition Structure (FTS), 53
free flight, 249, 250, 252, 253, 259, 270, 275, 276

Generalised Stochastic Hybrid System (GSHS), 251
Global Positioning System (GPS), 263
gradient estimation, 141
guard, 56

hazard rate, 201
hybrid
automaton, 2, 5, 143
jump, 17, 29, 31
state space, 16, 17, 27, 33, 43, 223
system, 259

Inertial Reference System (IRS), 263
Infinitesimal Perturbation Analysis (IPA), 10, 141, 150, 169
inhibitor arc, 260
Interaction Petri Net (IPN), 260
interconnection arc, 260, 261
International Civil Aviation Organisation (ICAO), 250
Internet, 11, 140

jump, 17, 19, 21, 28, 32, 34, 38
at a boundary, 39
diffusion, 20, 27
instantaneous hybrid, 31
process, 213
rate, 50, 57

Local Petri Net (LPN), 260
long-lived flows, 192, 193, 201, 203, 205, 209
LOTOS, 68

Markov
chain, 228
decision drift process, 16
process, 15, 16, 34, 40
property, 15, 18, 40
mid-air collision, 250
model checking, 10, 79
moment
closure, 192, 205
closure, log-normal, 205, 206
dynamics, 192, 198, 203, 204, 209
Monte Carlo (MC) simulation, 9, 11, 226, 250, 252, 255, 276
sequential, 252, 253, 259

network packet drops, 191, 193, 197, 200, 208
non-determinism, 54
non-deterministic transition, 54
Non-deterministic Transition System (NTS), 54

on-off flows, 192, 197, 199, 201, 210
open system, 54

output variable, 56

passive transition, 56
Petri net, 2, 10, 252, 253, 259–261
Piecewise Deterministic Markov Process (PDMP), 7, 10, 11, 16, 223, 251
    executions, 225
Pilot-Flying (PF), 262
Pilot-Not-Flying (PNF), 262, 263
Poisson random measure, 17, 21, 22, 26, 27, 31, 33, 36, 43, 111, 112
prediction, 257
Predictive ASAS (P-ASAS), 265
Probabilistic
    Risk Analysis (PRA), 251
probabilistic reachability analysis, 83, 87, 91, 250, 253

Random Early Detection (RED), 194, 196, 197, 201
reachability analysis, 10, 108, 129, 130, 133
    Markov chain approximation, 108, 110, 115–117, 120, 122, 123, 129, 130, 134, 135
reachability computations, 108, 135
    finite horizon case, 83, 87, 91, 126
    infinite horizon case, 126
    iterative algorithm, 116, 125, 129
reachability problem, 108, 114, 128, 134, 135
    probabilistic safety, 99, 108, 128, 130, 135
    regulation, 108, 129, 131, 135
reset map, 56, 212
round-trip time (RTT), 192, 193, 197, 199

Secondary Surveillance Radar (SSR), 264
semimartingale, 16–19
sensitivity analysis, 142
sequential Monte Carlo simulation, 252, 253, 259
simultaneous jump, 28, 33, 35, 38
Situation Awareness (SA), 267
spontaneous transition, 226, 233, 242
    of CFSJS, 50
    of CPDP, 57
state
    continuous, 213
    discrete, 213
    enabled, 53
    variable, 56
stochastic
    differential equation (SDE), 16, 18, 23, 25, 26, 29, 35, 38
    flow systems, 139
    fluid systems, 139
    hybrid model, 18, 33, 34, 36, 38
    hybrid process, 16, 17, 27, 43, 254
    hybrid system, 16, 17
Stochastic Fluid Model (SFM), 7, 10, 140, 170
Stochastic Hybrid System (SHS), 4, 7
Stochastically and Dynamically Coloured Petri Net (SDCPN), 253
stopping time, 31
strong
    existence, 22, 23
    Markov property, 18, 40, 42, 252, 254, 256
    solution, 22, 26, 29, 35, 38
    uniqueness, 23, 26
survival function, 50, 225, 230, 242
switch, 28, 30, 43
switching diffusion, 10, 16, 17, 110, 112, 115, 117, 123, 128, 129, 131, 134

temporal logic, 80
    CSL, 87
    CSRL, 90
    PCTL, 82
    PRCTL, 84
time-driven system, 2

transfer-size distribution, 192, 197, 198, 201, 206, 210
  exponential, 201
  mixture of exponentials, 202, 203, 206, 208
  Pareto, 202, 204
transition
  active, 56
  counter, 213
  forced, 53, 226, 233, 240
  intensity, 212
  measure, 50
  mechanism, 49
  non-deterministic, 54
  passive, 56
  spontaneous, 50, 226, 233, 242
Transition Mechanism Structure (TMS), 49
  of CFSJS/FTS, 53
Transmission Control Protocol (TCP), 11, 191
  ACK packet, 193, 199, 201
  congestion-avoidance mode, 191, 194, 197, 200, 208
  fast-recovery mode, 195, 197, 201
  slow-start mode, 191, 195, 197, 199, 202, 208
  TCP-friendly formula, 192, 193, 206, 212

unbiasedness, 153
uniformization, 88

valuation space, 59
value passing, 68
  element, 69

weak
  existence, 22, 23
  uniqueness, 23
Wiener process, 21–23, 26, 27, 36

Zeno, 225